Reinforced Concrete Design

This book explains behavioral aspects of Reinforced Concrete (RCC) structures along with brief research to understand the codal recommendations in different countries. Related issues of RCC design are also discussed and these are supplemented by numerical/design examples and fundamental review questions in each chapter. The subject matter in this book also critically discusses the considerations for non-elastic behavior in design procedures to accommodate design objectives. A comparison of design methodology and output as per IS, BS, EURO and ACI Codes is also included.

Features:

- Covers the basic behavioral aspects of reinforced concrete structures.
- Includes design examples to understand the theoretical concepts of different modules.
- Discusses considerations for non-elastic behavior for making simple design procedures to accommodate design objectives within codal provisions.
- Provides the basic insights necessary for effective development of a design.
- Includes a number of design examples along with working drawings.

This book is aimed at researchers; professionals; graduate students in RCC structures, civil and infrastructure engineering; and those involved in project execution and consultancy firms.

Reinforced Concrete Design

Somnath Ghosh and Kushal Ghosh

CRC Press is an imprint of the
Taylor & Francis Group, an **informa** business

Designed cover image: www.pexels.com

First edition published 2025
by CRC Press
2385 NW Executive Center Drive, Suite 320, Boca Raton FL 33431

and by CRC Press
4 Park Square, Milton Park, Abingdon, Oxon, OX14 4RN

CRC Press is an imprint of Taylor & Francis Group, LLC

Library of Congress Cataloging-in-Publication Data
Names: Ghosh, Somnath, author. | Ghosh, Kushal, author.
Title: Reinforced concrete design / Somnath Ghosh and Kushal Ghosh.
Description: First edition. | Boca Raton : CRC Press, 2025. |
Includes bibliographical references and index. | Identifiers: LCCN 2024032307 (print) |
LCCN 2024032308 (ebook) | ISBN 9781032076669 (hbk) |
ISBN 9781032076676 (pbk) | ISBN 9781003208204 (ebk)
Subjects: LCSH: Reinforced concrete construction.
Classification: LCC TA683 .G56 2025 (print) | LCC TA683 (ebook) |
DDC 624.1/8341—dc23/eng/20241104
LC record available at https://lccn.loc.gov/2024032307
LC ebook record available at https://lccn.loc.gov/2024032308

ISBN: 9781032076669 (hbk)
ISBN: 9781032076676 (pbk)
ISBN: 9781003208204 (ebk)

DOI: 10.1201/9781003208204

Typeset in Times
by codeMantra

Dedicated to our parents, wives, children and students

Contents

Preface

Reinforced concrete as a construction material provides enormous architectural freedom. The basic concepts for the design of reinforced concrete structural elements are, fundamentally, more or less the same, but students and practicing engineers have access to a number of computer software packages, hundreds of textbooks, articles and research papers and large amounts of online information. It is difficult to recommend the best and most appropriate design resources. The present engineering education system focuses more on computer-based mathematical models without understanding their shortcomings. However, such analyses would be more powerful and useful if they had been developed in the light of realistic engineering concepts. Intuitive skill and experience, along with computer-aided mathematical analysis, need to be given due attention. The design concept needs to be cost-effective and better than its alternative if it is to be accepted. This book explains the issues of reinforced concrete design in brief and provides design examples based on the recommendations of the latest Indian Standard codes, which are essential and mandatory documents for industrial design and in order to achieve an acceptable common platform of understanding. It also provides detailed working drawings; several such typical design examples of buildings are given. This book provides the basic insights necessary for the effective development of a design. The most intricate issues of reinforced concrete design are discussed, supplemented by a number of real-life examples for deeper understanding of the subject. The target readers of this book are practicing structural engineers and architects, students and teachers of civil engineering and architecture, striving to understand reinforced concrete design. Detail design calculations and reinforcement detailing as per recommendation of different relevant codes of practices have been furnished. Reinforcement detailing in the form of working drawings is also included to reduce the gap of understanding between different groups of professionals in structural engineering. It is a user-friendly complete package for all who deal with reinforced concrete design and provides numerical design examples, along with working drawings based on the recommendations of the latest IS codes.

Acknowledgments

We would like to take this opportunity to remember the sacrifice and encouragement received from our parents, wives, friends and relatives.

We are thankful to a number of persons who helped us in preparing the manuscript.

We are extremely thankful to Dr. Gagandeep Singh of CRC Press, Taylor & Francis, for guidance and help in writing this book.

About the Authors

Somnath Ghosh was Dean of the Engineering Faculty and Head of the Civil Engineering Department at Jadavpur University, India. He is also an Emeritus Professor of Civil Engineering at Haldia Institute of Technology, India. Dr. Ghosh obtained his B.E. in Civil Engineering from Jadavpur University and his M.Tech and Ph.D from the Indian Institute of Technology Kharagpur, India. He is a member and chartered engineer of the Institute of Structural Engineers in the United Kingdom. He previously published a book called *Design of Wind and Earthquake Resistant Reinforced Concrete Buildings* with CRC Press.

He has carried out research in the United States and Australia and delivered invited lectures in the United Kingdom, Australia, Singapore, Malaysia, Thailand and the United States. He has also delivered a huge number of lectures as a resource person in different Indian Institutes of Technology, National Institutes of Technology and universities. He has served as an expert member on several occasions for many institutes and universities. He has served as a member of several high-powered committees in the All India Council for Technical Education, the University Grants Commission, the Council of Scientific and Industrial Research, the Union Public Service Commission, etc. and at Jadavpur University level. He has contributed significantly in the areas of structural engineering and materials. He has also been a structural consultant to a number of key projects at the national level. Based on his research works, Dr. Ghosh has published a number of papers in peer-reviewed national and international journals and published six monographs at international level. Apart from his research activities, Dr. Ghosh has demonstrated his technical skills by providing advice on industrial projects, and these have been implemented successfully. The repair and restoration techniques adopted for the earthquake-damaged structures of Kandla Special Economic Zone through his expertise deserve special mention. His other noteworthy contribution is the restoration of the earthquake-damaged assembly building in Sikkim. In addition, his skill in computer-aided structural analysis has been demonstrated through the design of a 52-m tall Buddha statue at the top of a hill at Rabangla, Namchi, Sikkim and a cricket stadium at Guwahati, Assam. His selection as country head for a division of a multinational company in Nigeria, speaks volumes about his administrative abilities as well as his academic skill and expertise.

Kushal Ghosh has received Bachelor's degree in Civil Engineering with distinction from Pune University, India and Doctor of Philosophy from Jadavpur University, India. At present, he is involved in research activities at Waikato University, Hamilton, New Zealand. He has served as Assistant Professor in the National Institute of Technology (NIT), Sikkim, India. He has worked as a

Structural Design Engineer in Skematic Consultant, India. His major research area is "Sustainable High Performance Green construction material". He previously published a book called *Alkali Activated Fly Ash: Blast Furnace Slag Composites* with CRC Press. He has guided a number of students for their thesis at ME level. He has also published a number of papers in peer-reviewed journals and conferences.

1 Behavior of Concrete

PREAMBLE

Reinforced concrete is extensively used in the construction of different types of structures, such as buildings, stadiums, auditoriums, bridges, berthing structures, retaining walls, waterways, water tanks, swimming pools, cooling towers, bunkers and silos, pavements, chimneys, pipes, tunnels, etc. The performance of concrete under aggressive environment is reasonably good to provide a life of around 100 years. The life of the structure can be increased by periodical maintenance. However, "Prevention is better than cure". Therefore, the highest level of workmanship and good quality of ingredients need to be used. It is highly durable. Generally, the concrete is made out of ingredients like sand as fine aggregate, stone chips/gravel as coarse aggregate, cement as binder, admixtures/additives/plasticizers as special property provider, and water for hydration to take place. It is possible to produce a wide range of concrete of high strength, high flow/workability, higher durability, etc. having the bulk density of concrete ranging between 1200 and 2400 kg/m^3 and compressive strength ranging between 25 and 100 MPa (N/mm^2) for primary structural elements. The steel bars are embedded in concrete for taking care of tensile stresses. The bond between steel and the surrounding concrete is considered perfect. It is assumed that the strain at any point in the steel is equal to that in the surrounding concrete. This is generally called as reinforced concrete. A minimum amount of reinforcing steel provides ductility to brittle concrete, but a higher amount of steel may reduce ductility. It is generally said that consumption of concrete is next to water. Longitudinal reinforcing steel is also carrying compressive stresses. The longitudinal bars need to be confined by transverse steel ties/stirrups, in order to maintain their positions and to prevent their lateral buckling. The lateral ties/stirrups also provide confinement to concrete, which enhances shear capacity and provides more ductility.

Concrete is very weak in tension. The direct tensile strength is only about 7%–15% of its compressive strength. However, tensile stresses develop in reinforced concrete members as a result of flexure, shrinkage, and temperature changes. Principal tensile stresses may also result from multi-axial states of stress. Pure shear in concrete causes tension on diagonal planes; thus, the value of direct tensile strength of concrete is useful in estimating the shear strength of beams, etc. Also, the flexural tensile strength of concrete is required for the estimation of the moment at first crack required for the computation of deflections and crack widths in flexural members. It is possible to predict the elastic and inelastic performance of reinforced concrete elements. However, more research is required to come closer to an exact one considering uncertainties for several reasons. Variation in the strength of in-situ concrete is more compared to steel,

DOI: 10.1201/9781003208204-1

which is not manufactured under good quality control. Properties of concrete depend on the properties of aggregates, mixing, placing, compacting, and curing. However, the uncertainty is taken care by providing an appropriate factor of safety in the mix design methodology. The development of reliable design and construction techniques has enabled the construction of a wide variety of reinforced concrete structures. The Reinforced Cement Concrete (RCC) structure should be durable, corrosion-resistant, etc. Economy and aesthetics are important, but stability, strength, and serviceability should get priority.

Cube specimens are generally cast and cured from the mix prepared and used. Then the cube specimen is tested under direct compression. Variation in strength basically depends on the degree of quality control during manufacturing. Standard deviation may be obtained and checked against tolerable limits. The probability distribution of concrete strength approximately follows normal (Gaussian). The coefficient of variation generally ranges from 0.01 to 0.02.

Characteristic strength is defined as the strength of concrete below which not more than 5% of the test results are expected to fall. Generally, 28-day compressive strength has to be significantly greater than the 5%–10% characteristic strength f_{ck} considered while designing. Generally, compressive strength, tensile strength, shear strength, bond strength, density, permeability, durability, etc. are verified before recommendation. Many properties of concrete can be obtained from the compressive strength, using correlations. The quality or grade of concrete is designated in terms of characteristic 28-day compressive strength of 150 mm cubes in MPa (N/mm^2). For example, M 25-grade concrete means the characteristic strength of concrete is 25 MPa. Minimum grade of concrete nowadays used is M25 for structural elements like slabs, beams, columns, etc. However, for piles, it may be M30 depending on the chemical composition of the soil. Basic requirements of design are as follows

- **Stability** against overturning, sliding or buckling of the structure, or parts of it.
- **Strength** to resist the stresses induced and good ductile performance.
- **Serviceability** to ensure satisfactory performance under service load. Deflections, crack widths, and vibration amplitude and frequencies should be within acceptable limits.

Structural elements may be load-carrying, like slabs, beams, columns, etc. and nonstructural elements like false ceilings, partitions, doors, etc. The entire structural system has to carry dead load, live load, wind load, seismic load, etc. and finally to be transferred to soil through foundation. The structural system of a reinforced concrete overhead water tank structure comprises the tank, the staging, and the foundation. Generally, the **INTZ type** water tank has a top dome roof, a cylindrical wall with ring beams at top and bottom, a bottom dome, and a main ring beam with ring beam, which takes support on columns. The columns are tied up with beams (bracing members) at different levels. The foundation may be a raft or pile cap supported by piles, etc.

Building is the most common RCC structure. Structural system and its load transmission mechanism are extremely important especially due to wind and earthquake. Structural system is primarily decided from relative cost. However, it should effectively resist lateral loads and gravity loads. The building and columns may be rectangular, square, circular, L-shaped, etc. Generally, the column is designed to resist axial compression/tension combined with biaxial bending moments. If the column length is on the higher side, the columns should be properly designed as slender columns. The floor slabs, generally 100–200 mm thick with spans ranging from 3 to 8 m, may be supported on load-bearing brick walls. Generally, the slab panels are rectangular.

When the slab panel is supported on all four sides and the longer span/shorter span is less than 2, it is called a **Two-way slab.** If the longer span/shorter span, is greater than 2, it is called a **One-way slab.** Sometimes, the slab is supported directly by columns generally called **Flat slab**. Drop panels and/or column capitals are sometimes provided. Generally, slab thickness may be 200–300 mm for spans of 5–8 m. **Grid floor,** i.e., slab with ribbed beams in both directions, closely spaced (less than 1–1.5 m apart). The width of rib beams may be 200, and slab depth may be 100 mm. Column-to-column distance is generally 10–15 m. This type of two-way ribbed slab is sometimes called **Waffle slab.**

The horizontal and vertical structural elements jointly resist gravity loads and lateral loads. In case of high rise buildings, effect of wind and earthquake forces are governing the selection of the structural system. Shear walls are solid RCC walls, generally extend up to the top of the building, commonly located at the lift/staircase regions. Shear walls are generally provided along the shorter direction of a building. The walls are resisting mostly lateral loads due to wind, earthquake, etc., and building frame bends like vertical cantilevers, fixed at the base level. Sometimes closely spaced columns are provided along the periphery of a building and generally termed as **framed tube** with a high flexural rigidity against lateral loads. The outer framed tube may be combined with an inner tube generally called as **Tube-in-Tube**, which is adopted for tall buildings.

The subsystems or components of the tall building structural systems are essentially the following. Floor systems, vertical load resisting systems, lateral load resisting systems, connections, energy dissipation systems and damping. These are broadly defined as follows: moment resisting frames, shear wall-frame systems, shear truss-outrigger braced systems, framed-tubes, tube-in-tube systems with interior columns, bundled tubes, truss tubes without interior columns, and modular tubes. The structural system should be able to carry different types of loads, such as gravity, lateral, temperature, blast, and impact loads. The drift of the tower should be kept within limits, such as $H/500$.

Most popular vertical structural systems to resist horizontal seismic forces are shear walls, moment resisting frames, and braced frames. These systems are capable of resisting gravity forces and providing horizontal resistance. These basically act as a vertical cantilever more or less fixed at foundation level and primarily resist horizontal forces due to earthquake/wind, etc.

Building structures are generally a three-dimensional framework of structural elements, which act integrally to resist loads. Vertical elements, diaphragms, etc., are commonly the resisting systems adopted. **Moment resisting frames** are generally selected as one of the seismic resisting systems to allow flexibility of architectural planning. Moment resisting frames are commonly used as lateral resisting systems when sufficient ductility and deformability demand need to be fulfilled. Specially moment resisting frame (SMRF) must be specifically detailed to provide ductile behavior and comply with the provisions of the guideline of the codes. SMRF is a moment resisting frame that fulfills special detailing as per IS 13920 and IS 4326. Ordinary moment resisting frame (OMRF) is a moment resisting frame that does not fulfill the special detailing requirements for ductile behavior of the frame. A highly ductile building frame of high degree of redundancy can allow freedom in architectural planning of internal spaces, etc. Its flexibility and associated long period may serve to detune the structure from the forcing motions on stiff soil or rock sites. However, poorly designed, moment resisting frames have been observed to fail catastrophically during earthquakes, due to formation of weak stories, failure of beam-column joints, etc. Beam-column joints are zones of high stress concentration and therefore needs special attention and care during designing it.

Proportioning and detailing requirements of specially moment frames need to done carefully so that it becomes capable of multiple cycles of inelastic stresses without major loss of strength. Moment resisting frames may be considered as the structures with a satisfactory behavior under severe earthquakes. These frames can provide a large number of dissipative zones, where plastic hinges form with potentially high dissipation capacity. In order to maximize the energy dissipation capacity, it ensures failure as a whole. Moment resisting frames can provide different level of strength and ductility, if recommendations of code and expert opinions are strictly followed.

Reinforced concrete, especially moment frames, are made to improve the performance of beams, columns, and joints, in order to achieve more overall ductility of the structure. It is required to achieve strong column-weak beam structural system that ensures ductile response in the yielding regions. Nonstructural elements, infill need to be detailed in a way, so that the target behavior could be achieved. Failure of a column is of greater impact than failure of a beam. Building codes are recommended so that columns shall be stronger than the beams. This strong column-weak beam concept is the fundamental approach to achieve safety during strong earthquake ground shaking. Many codes adopt the strong column-weak beam concept and provide guidelines by keeping the sum of column moment strengths exceed the sum of beam moment strengths at each joint. Non-ductile failures can be avoided using a capacity design approach. Shear failure, especially in columns, is relatively brittle and needs to be avoided. A capacity design approach that requires the design shear strength should be at least equal to the shear that occurs when yielded sections reach near the ultimate moment capacity. The strength of lap/spliced longitudinal reinforcement loses cover during earthquake shaking. Therefore, lap/splices should be located away from sections of maximum bending moment, i.e., away from beam-column junction, and closed ties/stirrups need to be provided to confine the lap/splice zone. Different codes restrict diameter of reinforcement as well. Bars anchored in exterior joints must have hooks extended as recommended by different codes of practices. Different requirements apply to interior and exterior joints. The cracked stiffness of the beams, columns, and joints must be appropriately considered while analyzing special moment frame, which provides reasonable building periods, and thus, base shear, story drifts etc. can be assessed correctly. Some code recommends the column dimension parallel to the longitudinal reinforcement of the beam should be at least 20 times the diameter of longitudinal bars for normal concrete. A few more general guidelines are followed for a beam element of a specially moment resisting frame, lateral ties or stirrups and recommended to provide closely in the locations where flexural yielding is expected to occur, particularly at the end of the beams or the locations at lap/spliced bars. The transverse reinforcements simultaneously act as confinement reinforcement to achieve more ductility of beams, columns and their joints. Transverse reinforcements at joints are provided to confine the joint core and to improve anchorage of the longitudinal reinforcement of both beams and columns. The maximum spacing of lateral ties/stirrups is recommended not to exceed 75 mm as per IS 1893-Part I-2016.

RCC braced frame is not very common vertical seismic resisting system but offers architects an alternative to walls and moment resisting frames, often concealed within the walls of building core braced frames to resist the seismic forces transferred from diaphragms. Braced frames resist lateral loads by the transfer of axial forces (tension or compression) through diagonal bracing members. These brace members transfer forces from roof to the foundation. Concentric frames have braces connecting at the ends of elements. The use of RCC bracing member has potential advantage over other bracing like higher stiffness and stability. The bracing system is generally provided on the periphery between the columns. It may be X-braced frames, V-braced frames, etc. The most important parameters are base shear and storey displacement. It is observed that X-braced frames are more efficient and safe during earthquake shaking when compared with moment resisting frames with V-brace members. Moment resisting frame undergoes higher storey displacement compared to braced frame buildings, as both strength and stiffness of the braced frame are much greater compared to un-braced building frames. Base shear of braced frame buildings increases compared to building without brace member. Storey displacement of the building is reduced by to a great extent. By providing braces in the frame, the horizontal load at node is distributed among brace members along with beams and columns. Due to provision of the bracing system in the building, bending moment comparatively reduced.

Shear wall is a structural member in a reinforced concrete framed structure to resist lateral forces such as wind and earthquake forces. Shear walls provide large strength and stiffness to buildings in the direction of their orientation, which significantly reduces lateral sway of the building and as a result damage reduces. Moment resisting frames carry lateral loads primarily by flexure. Joints are designed in such a way that they are completely rigid and therefore any lateral deflection of the frame occurs due to bending of the three-dimensional frame. Generally, shear walls are provided in high rise/ slender buildings. Moment resisting frames are more flexible than shear wall structures or braced frames, but the horizontal deflection or drift is greater. Adjacent buildings cannot be located too close to each other. But, in general, the energy dissipating capacity of RCC shear walls is not that good, and it is found that using the bracing system may provide a better solution. It is observed from experimental results that braced and infill bare frames have shown a higher lateral strength compared to bare frames depending on the type of bracing and infill. The energy dissipation for the braced and infill frames is always higher than that for the bare frame up to failure.

Potentially high ductile system with a higher degree of redundancy allows freedom in the architectural planning of internal spaces and external cladding. Its flexibility and associated long period may serve to detune the structure from the forcing motions on stiff soil or rock sites. However, poorly designed, moment resisting frames have been observed to fail catastrophically in earthquakes, due to formation of weak stories and failure of beam-column joints. Beam-column joints are zones of high stress concentration. Although most of the shear walls are rectangular in plan. It may be a gentle curve or C, L, and I shapes, which are usually structurally feasible. However, shear wall is effective in the direction of its length only. A structurally adequate wall possesses sufficient strength to resist both shear forces and bending moments. Cutouts for windows and doors may reduce the strength of a shear wall significantly. Therefore, such cutouts may be avoided at highly stressed zones, particularly near its base.

Any structure is made up of structural elements (load-carrying, such as beams and columns) and nonstructural elements (such as partitions, false ceilings, and doors). The structural elements, put together, constitute the 'structural system'. Its function is to effectively resist the action of gravitational and environmental loads and to transmit the resulting forces to the supporting ground, without significantly disturbing the geometry, integrity, and serviceability of the structure.

This book is confined to reinforced concrete design. It covers the basic principles of designing structural members for flexure, shear, torsion, and axial compression — with applications to beams, slabs, staircases, columns, footings, and retaining walls. Applications to special structures, such as bridges, chimneys, etc. are not covered here, although the basic principles of design remain the same.

BEHAVIOR OF CONCRETE UNDER UNIAXIAL COMPRESSION

Uniaxial compressive strength is obtained by testing standard cubes (150 mm) tested to destruction in a compression testing machine following IS 516: 1959. Generally, test specimens are tested on 28th day after casting and continuous curing with water. Strain-controlled machine is generally used, and load is applied at a uniform strain rate of 0.001 mm/mm per minute. Standard test cylinders (150 mm diameter and 300 mm high) are also used instead of cubes. The cylinder strength is lower than the cube strength. Compressive strength can be correlated to many engineering properties such as tensile strength, shear strength, modulus of elasticity, etc. The cylinder test specimen is used to get stress-strain curve of concrete, using extensometer.

Stress-strain curve is linear in the very initial stage of loading but becomes prominently nonlinear when the stress level is near to one-third of the maximum stress level. The maximum stress is reached at a particular stain depending on the type and grade of concrete, which is generally considered 0.0035 in bending compression and 0.002 under direct compression as per IS456, for design purposes. High-grade concrete shows a bit rigid behavior compared to low-grade concrete.

MODULUS OF ELASTICITY AND POISSON'S RATIO

The concrete tends to expand laterally, and longitudinal cracks become visible when the lateral strain due to the **Poisson effect** exceeds the limiting tensile strain of concrete (0.0001–0.0002). Cracks occur at the interface of the aggregates. The fall in stress with increasing strain with extensive micro-cracking is extremely difficult to understand. This stage is called softening of concrete. The stress-strain behavior of the compression zone of a reinforced concrete flexural member is very close to stress-strain behavior under uniaxial compression and different codes adopted simplified stress-strain curves as observed in the case of uniaxial compression, with a little bit of modifications.

There is a distinct effect on the compressive strength due to height/width ratio and the cross-sectional dimensions of the test specimen in the uniaxial compression test. Cylinder strength is lower than cube strength, due to different fracture patterns.

Concrete is not really an elastic material. Modulus of elasticity and Poisson's ratio are not applicable. Still with this limitation, these parameters are used in linear elastic analysis. In the case of concrete under uniaxial compression, when the load intensity is low, and of very short duration, initial tangent modulus concept exists. If the loading is sustained, elastic creep effects need to be considered even at a relatively low-stress level. The initial tangent modulus is sometimes considered a measure of the dynamic modulus of elasticity of concrete, which is important in case of cyclic loading due to wind, earthquake, etc. Induced stress conditions. However, particularly if high-intensity cyclic loads are involved a pronounced hysteresis effect is observed.

Based on linear static analysis, the static modulus of elasticity is used. The long-term effects of creep reduce the effective modulus of elasticity significantly. It is not yet possible to separate the long-term strains induced by creep and shrinkage from the short-term elastic strains. The total deflection is assumed to be a sum of an instantaneous elastic deflection termed as short-term deflection and the long-term deflections due to creep and shrinkage. The short-term static modulus of elasticity (E_c) is generally considered as initial tangent modulus.

IS 456 (clause no. 6.2.3.1) recommends an empirical expression for the static modulus of elasticity E_c in terms of the characteristic cube strength f_{ck}, as follows:

$$E_c = 5000 \sqrt{f_{ck}}, \text{ both } E_c \text{ and } f_{ck} \text{ are in N/mm}^2$$

Poisson's ratio, the ratio of the lateral strain to the longitudinal strain, under uniform axial stress. It is observed that the variations of longitudinal, lateral, and volumetric strains are depicted. At stress beyond 75%–80% of the compressive strength, the rate of volume reduction decreases; soon and then, the volume stops decreasing rather starts increasing due to major micro-cracking, leading to large lateral extensions. The values of Poisson's ratio have been generally ranging 0.10–0.30 and generally, 0.2 is considered for design purposes.

BEHAVIOR OF CONCRETE UNDER TENSION

In multiple states of stress, principal tensile stresses generally develop. Tensile stresses develop due to flexure, shrinkage, temperature changes, etc. For estimation of the moment at first crack, it is an important parameter to assess deflections and crack widths.

Modulus of Rupture

Generally, the flexure test is conducted as per IS 516 on a standard plain concrete prism under simply supported condition and subjected to third-point loading to destruction. Though linear stress distribution is assumed, the actual stress distribution is not at all linear. The modulus of rupture is found greater than the direct tensile strength.

The IS456 (clause 6.2.2) suggests an empirical formula as follows to assess cracking stress f_{cr}

$$f_{cr} = 0.7\sqrt{f_{ck}}, \text{both } E_c \text{ and } f_{ck} \text{ are in N/mm}^2$$

Stress-Strain Curve of Concrete in Tension

Normal concrete was found to be failed in tension ranging from strains 0.0001 to 0.0003. The stress-strain behavior in tension is generally considered linear up to failure. The modulus of elasticity in tension is considered the same as the modulus of elasticity in direct compression. Post-peak behavior is generally understood by using displacement-controlled tests. Fractures generally take place in different failure modes.

Shear Strength and Tensile Strength

Split cylinder test is conducted as per IS 5816. A standard plain concrete cylinder (generally 150 mm dia. and 300 mm long) is loaded under compression along a diametric plane. Splitting of the cylinder along the loaded plane occurs. Using the theory of elasticity, the following formula is used to assess splitting tensile strength f_{ct}. This is an indirect way of assessing the tensile strength of concrete.

$$f_{ct} = 2P/\prod DL, P = \text{Failure load}, D = \text{Diameter of Cylinder and}$$

$$L = \text{Length of the cylinder}$$

It is difficult to determine the resistance of concrete to pure shearing stresses experimentally, but assessment of the shear strength under combined stress condition is obtained reasonably acceptable. It has been reported that for plain concrete generally, shear strength is around 10%–25% of its direct compressive strength.

BEHAVIOR OF CONCRETE UNDER COMBINED STRESSES

It is observed that the strength of concrete in biaxial compression is greater than in uniaxial compression. The biaxial tensile strength is more or less equal to its uniaxial tensile strength. Failure modes suggest that tensile strains have tremendous influence on the failure criteria and failure mechanism of concrete for both uniaxial and biaxial states of stress. Compressive strength and the tensile strength of concrete are reduced in the presence of shear stress. However, the shear strength of concrete is enhanced under direct compression. When concrete is subject to compression in three orthogonal directions, its strength and ductility are increased. This effect provides confinement of concrete, which reduces the tendency for cracking and volume increase prior to failure.

EFFECT OF CONFINEMENT

The benefit from confinement of concrete i.e. providing transverse reinforcement in the form of stirrups/ties/spirals. Continuous spirals are effective in increasing the ductility and also to some extent, the compressive strength of concrete member. However, square or rectangular ties are less effective. The yielding of the confining steel contributes to the ability to undergo large deformations prior to failure, i.e., ductile behavior. IS13920 provides recommendation regarding ductility, in particular to the design and detailing of reinforced concrete structures subject to seismic and wind loads.

CREEP OF CONCRETE

Concrete creep is defined as the deformation of a structure under sustained load. Basically, long-term pressure or stress on concrete can make it change shape. This deformation usually occurs in the direction the force is being applied. Like a concrete column getting more compressed or a beam bending. Creep is indicated when strain in a solid increases with time while the stress producing the strain is kept constant. In more practical terms, creep is the increased strain or deformation of a structural element under a constant load. Depending on the construction material, structural design, and service conditions, creep can result in significant displacements in a structure. Severe creep strains can result in serviceability problems, stress redistribution, pre-stress loss, and even failure of structural elements. Concrete is a composite with properties that change with time. During service, the quality of concrete provided by initial curing can be improved by subsequent wetting as in the cases of foundations or water retaining structures. However, concrete can also deteriorate with time due to physical and chemical attacks. Structures are often removed when they become unsafe or uneconomical. Lack of durability has become a major concern in construction for the past 20–30 years. In some developed countries, it is not uncommon to find large amount of resources, such as 30%–50% of total infrastructure budget, applied to repair and maintenance of existing structures. As a result, many government and private developers are looking into lifecycle costs rather than the first cost of construction. Durability of concrete depends on many factors including its physical and chemical properties, the service environment and design life. As such, durability is not a fundamental property. One concrete that performs satisfactory in a severe environment may deteriorate prematurely in another situation where it is considered as moderate. This is mainly due to the differences in the failure mechanism from various exposure conditions. Physical properties of concrete are often discussed in terms of permeation, the movement of aggressive agents into and out of concrete. Chemical properties refer to the quantity and type of hydration products, mainly calcium silicate hydrate, calcium aluminate hydrate, and calcium hydroxide of the set cement. Reactions of penetrating agents with these hydrates produce products that can be inert, highly soluble, or expansive. It is the nature of these reaction products that control the severity of chemical attack. Physical damage to concrete can occur due to expansion or contraction under restraint (e.g., drying shrinkage cracking, frost action, cyclic wetting, and drying), or resulting from exposure to abrasion, erosion, or fire during service. It is generally considered that the surface layer or cover zone plays an important role in durability as it acts as the first line of defense against physical and chemical attacks from the environment.

Although durability is a complex topic, some of the basic fundamentals are well understood and have been documented. Many premature failures in recent years are due mainly to ignorance in design, poor specification, or bad workmanship.

Concrete creep is defined as deformation of structure under sustained load. Basically, long-term pressure or stress on concrete can make it change shape. This deformation usually occurs in the direction the force is being applied. Like a concrete column getting more compressed or a beam bending. Creep does not necessarily cause concrete to fail or break apart. Creep is factored in when concrete structures are designed. When in influence on aggregates, they undergo very little creep. It is really the paste that is responsible for the creep. However, the aggregate influences the creep of concrete through a restraining effect on the magnitude of creep. The paste that is creeping under load is restrained by aggregate which does not creep. The stronger the aggregate, the more is the restraining effect and hence the less is the magnitude of creep. The modulus of elasticity of aggregate is one of the important factors influencing creep. In case of mix proportions, the amount of paste content and its quality is one of the most important factors influencing creep. A poorer paste structure undergoes higher creep. Therefore, it can be said that creep increases with an increase in water/cement ratio. In other words, it can also be said that creep is inversely proportional to the strength of concrete. Broadly speaking, all other factors which are affecting the water/cement ratio are also affecting the creep. In view of age at which a concrete member is loaded will have a predominant effect on the magnitude of creep. This can be easily understood from the fact that the quality of the gel improves with time. Such gel creeps less, whereas a young gel under load being not so strong creeps more. What is said above is not a very accurate statement because the moisture content of the concrete is different at different ages, which also influences the magnitude of creep.

Elastic deformations occur immediately when concrete is loaded. Non-elastic deformations under sustained loading increase with time. Consequently, since concrete is frequently subjected to dead loading, which is sustained loading, it usually is subjected to both types of deformations. The non-elastic deformation, rightly called creep, increases at a decreasing rate during the period of loading. It has been shown that a significant amount of creep takes place during the first seconds after loading and yet creep may carry on for 25 years Approximately, one-fourth to one-third of the total creep takes place in the first month of sustained loading, and about one-half to three-fourths of the total creep occurs during the first half year of sustained loading in concrete sections of moderate size. When the sustained load is removed, there is some recovery but the concrete does not usually return to its original state. The elastic and creep recoveries are less than the deformation under load because of the increased age of the concrete at the time of unloading. Creep may be due to the giving-in of internal voids, viscous flow of the cement-water paste, crystalline flow in aggregates, and flow of water out of the cement gel due to external load and moisture. The last cause is generally believed to be the most important. The magnitude and rate of creep for most concrete structures are intimately related to the drying rate, but creep is also important in massive structures where little or no drying of the concrete takes place. In these structures, most of the creep is believed due to the flow of the absorbed water from the gel (seepage) caused by external pressure. `Creep strain, being caused by sustained load (or if the rate of loading is slow) unlike shrinkage which is independent of load, is expressed as a strain per unit compressive stress causing it. In order to visualize the effects of both elastic and creep deformation at a given time, the sustained modulus of elasticity, defined as to the sum of the elastic and creep deformation at a given time, may be used. Tests have shown that the modulus of elasticity of concrete after 2 years of sustained loading varied between one-half of the initial secant modulus of elasticity. A reduced value of short-time secant or chord modulus of elasticity is frequently used in design to take creep into account. Creep, unlike shrinkage, which is generally undesirable, may be either desirable or undesirable depending on the circumstances. It is desirable in that it generally promotes better distribution of stresses in reinforced concrete structures. It is undesirable if it causes excessive deformation and deflections that may necessitate costly repairs or if it results in large lasses of pre-stressed-concrete members. Young's modulus of elasticity of concrete decreases with time, which in turn increases long-term deflections. It is at best only an intelligent guess as to the combined effects of creep, humidity, temperature, and a host of other variables (unavoidable in the making of concrete) on the reduction of modulus of elasticity of concrete. However, under temperate weather conditions and using normal materials and construction practices, the finally reduced elastic modulus value may be effectively taken as 40%–50% of the initial (secant) modulus value. Where concrete is partly cast in situ, the differential creep effect usually nearly balances the differential shrinkage effect. Where a concrete structure is built part by part, even if each stage is cast in situ (e.g. span-by-span bridge–deck construction), the actual long-term distributions of dead load bending moment and shear equalize themselves by creep and finally very nearly agree with the distributions that would result if the structure was continuously cast in situ in one single operation. Therefore, the dead load moment and shear value at any section, in a span-by-span construction, may be taken as corresponding to the actually constructed scheme and the other as if the whole structure were cast in situ in one continuous operation. This point should be noted very carefully by the designer if he wants to understand what creep does to his initially assumed moment and shear distributions.

Creep affects strain and defluxions and often also stress distribution, but the effects vary with the type of structure. The influence of creep on the ultimate strength of a simply supported reinforced concrete beam subjected to a sustained load is not significant, but the deflection increases considerably and may, in many cases, be a critical consideration in design. According to Glanville and Thomas, there are two distinct "neutral surfaces" in a beam subjected to sustained loading, one of zero stress and the other of zero strain. This arises from the fact that an increase in the strain in concrete leads to an increased stress in the steel and a consequent lowering of the natural axis and an increasing depth of concrete is brought into compression. As a result, the elastic strain distribution changes, but the creep strain distribution changes, but the creep strain is not canceled out, so at the level of the new stress-neutral-axis, a residual tensile strain will remain. At some level, above this axis, there is a line of zero strain at any time although there is a stress acting. This is an interesting example of the influence of stress history on strain at any time. In reinforced concrete columns, creep results in a gradual transfer of load from the concrete to the reinforcement. Once the steel yields, any increase in load is taken by the concrete, so that the full strength of both the steel and the concrete is developed is developed before failure takes place. In statically indeterminate structures creep may relieve stress concentration induced by shrinkage, temperature changes, or movement of support in all concrete structures creep reduces internal stresses due to non-uniform shrinkage so that there is a reduction in cracking. In calculating creep effects in structures, it is important to realize that the actual time-dependent deformation is not the "free" creep of concrete, but a value modified by the quantity and position of reinforcement. On the other hand, in mass concrete, creep in itself may be a cause of cracking when a restrained concrete mass undergoes a cycle of temperature changes due to the development of the heat of hydration and subsequent cooling. Creep relieves the compressive stress induced by the rapid rise in temperature so that the remaining compression disappears as soon as some cooling has taken place. On further cooling of concrete, tensile stresses develop and, since the rate of creep is reduced with age, cracking may occur even before the temperature has dropped to the initial (placing) value. For this reason, the rise in temperature in the interior of a large concrete mass must be controlled by the use of low cement content precooling of mix ingredients, limiting the height of concrete lifts, and cooling of concrete by pipes embedded in the concrete mass (as is done in the case of dams).

Likewise, in very tall buildings, differential creep also has structural effects on beams and slabs. The loss of pre-stress due to creep is well known and, indeed, accounts for the failure of all early attempts of pre-stressing. It was only the introduction of high tensile steel, whose elongation is several times the contraction of concrete due to creep and shrinkage, which made pre-stressing a successful proposition. Concrete made with low-heat cement creep more than concrete made with normal cement, probably because of its influence on the degree of hydration. This desirable characteristic explains, at least in part, the relative freedom from cracking of mass concrete structures as they cool to normal temperatures. The effect of cement fineness on creep appears to be variable. Pozzolanic additions to cement generally increase creep. Approved air-entraining agents added in the proper amounts appear to have no appreciable effect on creep. As creep is an important factor, proprietary compounds should not be used unless their effects have been previously evaluated. Under comparable conditions, creep and shrinkage generally decrease when well-graded aggregates with low void contents are used and when the maximum size of the coarse aggregate is increased. Hard, dense aggregates with low absorption and a high modulus of elasticity are desirable when low shrinkage and creep are wanted. The mineral composition of the aggregate is important, and generally increasing creep may be expected with aggregates in the following order: limestone, quartz, granite, basalt, and sandstone. Creep of concrete increases as the water-cement ratio increases. In addition, it appears that if two concrete mixtures have the same water-cement ratio, the mixture having the greater volume of cement paste will creep the most. In general, lean concrete mixture exhibits considerably greater creep than rich concrete mixture because the water-cement ratio effect of increasing creep for the lean mixture is more important than the paste effect of decreasing creep. The tendency of concrete to creep decreases as cement hydration increases, and consequently, water-cured concrete should creep less than air-cured concrete. However, the shrinkage or swelling produced during the initial curing period also influences creep. Size effects are also important in curing because small specimens respond more rapidly to moisture changes than large concrete members. The rate and ultimate magnitude of creep increases as the humidity of the atmosphere decreases. The relation between relative humidity and creep is not linear concrete under sustained load in air at 70% relative humidity will have an ultimate about twice as large as concrete in air at 100% relative humidity. The ultimate creep in air at 50% relative humidity will be about three times as large. Protection of concrete members from rapid drying is beneficial in reducing the rate and ultimate amount of creep. The rate and magnitude of creep generally decrease as the size of concrete increases. It has been estimated that the creep of mass concrete may be about one-fourth of that obtained with small specimens stored in moist air. With a given material and sustained load, the rate of magnitude of creep decreases with age, as long as hydration continues. Creep is also approximately proportional to sustained stress within the range of usual working stresses. Sustained stresses above the working stresses produce creep that increases progressively faster rate as the magnitude of the sustained stress is increased. Most creep information for plain concrete has been obtained for sustained compressive loading. However, test information for creep under sustained tension, bending, torsion, and biaxial and tri-axial-stress conditions generally shows the same behavior pattern. A test method to determine the creep characteristics of molded concrete cylinders subjected to sustain longitudinal compressive load is provided in ASTM standard C512.

- Creep is smaller if the age at loading is higher. This effect, called aging (or maturing), is important even for the many-year-old concrete. It is caused by the gradual hydration of cement.
- At constant water content in sealed specimens and temperature *T*, creep is linearly dependent on stress up to about 0.4 of the strength and obeys the principle of superposition, provided that large strain reversals not stress reversals, and cyclic strains are exceeded. In contrast with polymers and metals, the diametric and volumetric creep are equally important. At constant water content and temperature, the poisson's ratio due to creep strains is about constant and equals its elastic value.
- The tensile creep is about the same as the compressive creep. After unloading, creep is partly irreversible. Creep recovery of fully unloaded sealed specimens is less than that predicted when the principle of superposition is applied. This is a nonlinear effect. Creep recovery is almost independent of age and is linearly dependent on the stress drop even if the previous stress has been high at 65%. Creep-recovery curves tend to straight lines in the logarithmic time scale. The additional creep and elastic strain due to a stress increment after a long creep period are less than those for the same stress increment on a virgin specimen of the same age. The creep properties for such increments seem to be anisotropic.
- At constant water content as well as temperature, the creep is less for smaller water content. The drop of elastic modulus due to incomplete drying is only moderate, not more than 10%. After complete drying, a hysteresis on rewetting occurs. When concrete is drying simultaneously with creep, creep is accelerated drying creep effect. The acceleration occurs not only in compression but also in shear and bending. This effect is also manifested in the dependence of creep on the size and shape of the specimen. Furthermore, under simultaneous drying, the nonlinearity of creep versus stress is more pronounced and the additional creep due to drying is irrecoverable.
- Creep is considerably accelerated by any rapid change in water content, both negative and positive, and by its cycling. Stationary permeation of water through concrete at constant water does not affect creep appreciably.

  When a dried specimen is rewetted which produces swelling and subsequently loaded in compression, the creep that follows may be substantially larger than the previous swelling. When concrete under load is drying, the poisson's ratio due to creep strains is decreased and the lateral creep in a uniaxial test is unaffected by drying.

- As compared with the prediction of the principle superposition, pulsating loads considerably accelerate creep of concrete, even at low-stress levels, cyclic creep. When pulsation occurs after a long period under constant load, cyclic creep is negligible as compared with a virgin specimen. Poisson's ratio decreases with the number of cycles. In cement paste at low stress, cyclic creep is not observed.
- Aging, in terms of cement hydration, is decelerated by a drop in pore humidity and accelerated by a rise in temperature. Creep properties are significant even in the many-year-old concrete, in which the amount of cement still undergoing hydration is negligible, and neither elastic modulus nor strength changes appreciably.
- Creep rate grows with temperature. Rapid heating as well as rapid cooling accelerates creep.

  An appreciable part of shrinkage, as well as creep acceleration due to drying, seems to be delayed with regard to the change in pore humidity. Specimens continuously immersed in water swell. Sealed specimens show autogenous shrinkage, normally small, and also gradual self-desiccation if the water-cement ratio is low.
- Instantaneous thermal dilatation is followed by delayed thermal dilatation. Under stresses exceeding about 0.4 of the strength, creep becomes progressively nonlinear with stress. The additional creep due to nonlinearity is largely irreversible and is caused mainly by gradual micro-cracking. The apparent Poisson's ratio in the uniaxial test rises with the stress and exceeds 0.5 prior to failure, which indicates incremental anisotropy. Diffusivity rises with temperature and decreases with aging. A number of further complex phenomena are observed in creep of frozen concrete, at high temperatures over 100°C and at very low temperature.

The standard compression test is usually the loading being gradually applied at a uniform strain rate of 0.001 mm/mm per minute. When the load is applied at a faster strain rate or an impact load is suddenly applied, it is observed that both the modulus of elasticity and the strength of concrete increase but the failure strain decreases. Again, when the load is applied at a slow strain rate, as much as 1 year or more, there is a slight reduction in compressive strength and a huge reduction of modulus of elasticity. Under long-term sustained loading at a constant stress level creeping of concrete occurs. The internal movement of adsorbed water, viscous flow or sliding between the gel particles, moisture loss, and the growth in micro-cracks, are responsible for deformation due to creep. Deformation keeps increasing with time, but the stress level is not changed. Creep is a time-dependent component of the total strain. Creep coefficient may be defined as the ratio of the creep strain at a particular time to the instantaneous strain (strain immediately after loading). Creep coefficient (θ) is recommended in IS456 for design purposes. There is an instantaneous recovery of strain by an amount equal to the elastic due to the removal of load at a particular age, termed as creep recovery.

As recommended by IS 456, effective modulus of elasticity (E_{ce}), is expressed as

$$E_{ce} = E_c/1+\theta$$

Creep generally depends on cement content, water-cement ratio, aggregate content, relative humidity, temperature, loading, age of loading, etc.

SHRINKAGE OF CONCRETE

Ettringite is a normal reaction product of the hydration of tricalcium aluminate with gypsum. Gypsum is included in cement mixtures to prevent flash setting. Most of it is converted to a mono-sulfoa-luminate later in the hydration process within a day or two and does not cause a problem. However, if additional sulfate is supplied (perhaps with contaminated mixing water), the ettringite can reform with a significantly greater volume and cause cracking. Ettringite can also form when lime or cement is used to stabilize high sulfate soils and cause heaving of the soil and damage to pavement that may have already been placed. **Decrease in either length or volume of a material resulting from changes in moisture content or chemical changes**.- according to ACI concrete terminology. Drying shrinkage is the contraction in the concrete caused by moisture loss from drying concrete. **Contraction of hardened concrete due to the loss of capillary water in the concrete** is identified as drying shrinkage in concrete. Drying shrinkage occurs in 3–4 days or even after several months after concreting. Thin members having a large surface area are susceptible to drying shrinkage cracks in concrete. Concrete shrinks in the hardened state due to loss of moisture due to evaporation, termed as drying shrinkage. Time-dependent strains occur in concrete. Shrinkage and creep are not independent phenomena, but their effects are additive. Shrinkage strains are independent of the stress conditions. It is reversible in alternating dry and wet conditions. Tensile stresses develop if shrinkage is restrained, leading to cracking. Internal stresses, curvature, and deflections occur due to un-symmetrical placement of reinforcements. Shrinkage can be minimized by maintaining water content in the mix as low as possible and total aggregate content as high as possible. Linear strain (mm/mm) is due to shrinkage. In the absence of reliable data, the IS code recommended designing an ultimate shrinkage strain equal to 0.0003 mm/mm. This literature review focuses on factors influencing the drying shrinkage of concrete. Although the factors are normally interrelated, they can be categorized into three groups: paste quantity, paste quality, and other factors. Concrete shrinkage can be separated into different forms, depending on the mechanisms causing it. These include plastic shrinkage, chemical shrinkage (autogenous shrinkage), and drying shrinkage. In addition, shrinkage can be considered over early and later ages that together contribute to the total shrinkage. Early age of shrinkage is normally defined as that occurring during the first day after batching, while long term refers to concrete at 24 hours and older. Plastic shrinkage occurs before setting due to moisture loss. Chemical shrinkage is also an early age behavior, especially in the first hour after mixing. It occurs because the volume of the hydrated system is lower than the volume of the ingredients.

DURABILITY OF CONCRETE

Many reinforced concrete structures built in the past in aggressive environments have shown signs of structural distress, due to chemical attacks, causing deterioration of concrete and corrosion of reinforcing steel leading to reduction of life of the structure. IS code has provisions pertaining to durability. Relative humidity and chemical attacks are basically responsible for degradation of concrete structures. Ingress of water, oxygen, carbon dioxide, chlorides, sulfates, and other harmful chemicals present in sea water and soil in humid atmosphere are responsible for rapid degradation of concrete and steel reinforcements. Concrete members have inadequate cover to reinforcement are vulnerable. Lack of good drainage of water and cracks in concrete also lead to the ingress of aggressive agents, which causes deterioration of concrete. Workmanship and maintenance are extremely important in this regard. Repairing and rehabilitating concrete structure may be difficult and expensive. Reduction of permeability need to be ensured at the construction stage. The IS code categorized environmental exposure conditions as mild, moderate, severe, very severe, and extreme for enforcing minimum criteria aimed at establishing the desired performance. Minimum grade of concrete, minimum clear cover to reinforcement, minimum cement content, maximum water-cement ratio, acceptable limits of surface crack width are recommended in IS456. Different structural members may be subject to different categories of exposure.

The ability of concrete to withstand the conditions for which it is designed without deteriorating or getting damaged for a long period of time is known as durability of concrete. The durability of concrete may also be defined as the ability of concrete to resist weathering action, environmental forces, chemical attack, and abrasion while simultaneously maintaining its desired engineering properties. Durability of concrete normally refers to the duration of a life span of the trouble-free performance of concrete. Different types of concrete require different degrees of durability depending on the exposure environment and properties desired. For example, concrete exposed to tidal seawater conditions will have different requirements than indoor concrete in dry areas. The cement paste structure is dense and of low permeability type. Under extreme exposure conditions, it has entrained air to resist freeze–thaw cycle. It is made with graded aggregate that is strong and inert-type material. The ingredients in the concrete mix contain minimum impurities such as alkalis, chlorides, and silt. Various factors that affect the durability of concrete are as follows: cement content, aggregate quality, water quality, concrete compaction, curing period, permeability, moisture, temperature, abrasion, carbonation, wetting and drying cycles, freezing and thawing, alkali-aggregate reaction, sulfate attack, and organic acids. Quantity of cement used in concrete mix will also be a factor affecting the durability of concrete. If cement content used is lower than the required, then water-cement ratio becomes reduced and workability also reduced. Adding more water to this mix results in formation of capillary voids, which will make concrete as permeable material. If excess cement content is used, problems like drying shrinkage, alkali-silica reaction may occur which finally effects the durability of concrete. Use of good quality aggregates in concrete mix will surely increase the durability of hardened concrete. The shape of aggregate particles should be smooth and round. Flaky and elongated aggregates affect the workability of fresh concrete. For better bond development between ingredients, rough textured angular aggregates are recommended, but they require more cement content. Aggregate should be well graded to achieve dense concrete mix. Aggregates should be tested for its moisture content before using. Excess moisture in aggregate may lead to highly workable mix.

Quality of water used in concrete mixing also affects the durability of concrete. In general, potable water is recommended for making concrete. pH of water used shall be in the range of 6–8. Water should be clean and free from oils, acids, alkalis, salts, sugar, organic materials, etc. The presence of these impurities will lead to corrosion of steel or deterioration of concrete by different chemical reactions while placing concrete, it is important to compact the placed concrete without segregation. Improperly compacted concrete contain number of air voids in it, which reduces the concrete strength and durability. Proper curing in initial stages of concrete hardening result in good durability of concrete. Improper curing leads to formation of cracks due to plastic shrinkage, drying shrinkage, thermal effects, etc. thereby durability decreases.

Concrete durability gets affected when there is a chance of penetrability of water into it. Permeability of water into concrete expand its volume and lead to formation of cracks and finally disintegration of concrete occurs. Generally, concrete contains small gel pores and capillary cavities. However, gel pores do not allow penetration of water through them since they are of very small size. But, capillary cavities in concrete are responsible for permeability, which are formed due to high water-cement ratio. To prevent permeability, lowest possible water-cement ratio must be recommended. The use of pozzolanic materials also helps to reduce permeability by filling capillary cavities. Moisture present in the atmosphere will also affect durability of concrete structures. Efflorescence in concrete occurs due to moisture, which will convert salts into soluble solutions, and when it evaporates, salts become crystallized on the concrete surface. This will definitely damage the concrete structure, and durability will be reduced.

Concrete is heterogeneous material, when fresh concrete is subjected to high temperature rate of hydration gets affected and strength and durability becomes reduced. Concrete ingredients have different thermal coefficients, so at higher temperatures, spalling and deterioration of concrete happen. Deterioration of concrete also occurs due to severe abrasion. When concrete is subjected to rapidly moving water, steel tires, floating ice continuously wearing of surface occurs and durability gets affected. The higher the compressive strength, the higher will be the abrasion resistance. When moist concrete is exposed to atmosphere, carbon dioxide presents in atmosphere reacts with concrete and reduces pH of concrete. When pH of concrete reaches below 10, reinforcement presents in the concrete starts corroding. Corrosion of reinforcement causes cracks in concrete, and deterioration takes place. When concrete is exposed to alternate wetting and drying conditions such as tidal waves from sea, etc., secondary stresses are developed in concrete. Due to these stresses, cracks will form and reinforcement is exposed to atmosphere. When chlorides or sulfates from sea water meets reinforcement, corrosion occurs and durability of concrete is reduced. Use of low-permeable concrete and proper cover for reinforcement can prevent this type of problems. When fully saturated concrete is exposed to repeat cycles of freezing and thawing, it is deteriorated by the action of freezing and softening of water in it. It causes cracking on concrete surface in the form of maps, which is called map cracking and affects durability of concrete. The coarse aggregates present in the concrete are also affected by freeze and thaw cycles; spalling of concrete may occur. In this case, durability of concrete can be achieved by adding air-entraining admixtures to the mix and also reduce the maximum size of coarse aggregate. Alkali-aggregate reaction, or alkali-silica reaction, that takes place between alkali content of cement and silica content of aggregates is also a major factor affecting durability of concrete. Due to this reaction, concrete expansion occurs, which finally leads to severe cracking and concrete gets deteriorated. Use of cement with less alkali content, non-reactive aggregates, pozzolanic materials like fly ash or slag cement, lithium-based admixture in concrete will help to overcome this problem. When concrete structures are attacked by sulfates like sodium sulfate, magnesium sulfate, etc., concrete disintegration happens. This reaction is due to the chemical reaction between hydrated cement products and sulfate solutions. Sulfate attack generally happens when water used for concrete mix contains high sulfate content. Due to unwashed aggregates, when soil around the concrete structure contains sulfates in it. This can be prevented by using sulfate resisting cement, by adding slag cement, by decreasing permeability, etc. When concrete surface is subjected to organic acids like acetic acid, lactic acid, butyric acid, etc., concrete durability gets severely affected. Formic acid on concrete surfaces can lead to corrosion of concrete.

Durability of concrete may be defined as the ability of concrete to resist weathering action, chemical attack, and abrasion while maintaining its desired engineering properties. Different concretes require different degrees of durability depending on the exposure environment and properties desired. Durability is defined as the capability of concrete to resist weathering action, chemical attack, and abrasion while maintaining its desired engineering properties. It normally refers to the duration or life span of trouble-free performance. Different concretes require different degrees of durability depending on the exposure environment and properties desired. For example, concrete exposed to tidal seawater will have different requirements than indoor concrete.

Concrete will remain durable if:

- The cement paste structure is dense and of low permeability.
- Under extreme condition, it has entrained air to resist freeze–thaw cycle.
- It is made with graded aggregate that are strong and inert
- The ingredients in the mix contain minimum impurities such as alkalis, chlorides, sulfates and silt.

2 Design Philosophy and Codes

PREAMBLE

Design of a structure fulfills the intended purpose of the structure during its lifetime with adequate safety in terms of strength and stability and adequate serviceability in terms of deflection, crack width, durability, and economy. **Safety** implies partial or total collapse of the structure, which is not at all acceptable under service loads, but under uncertain situations such as due to earthquake or extreme wind, partial secondary damage not leading to any accident is acceptable to some extent. Collapse may take place due to exceeding the load-bearing capacity, overturning, sliding, buckling, fatigue fracture, etc. **Serviceabilit**y implies satisfactory performance of the structure under service loads without any discomfort to the user due to excessive deflection, cracking, vibration, etc. Considerations may be included under the purview of serviceability are durability, impermeability, etc. A thin slab may be safe against collapse but may undergo excessive deflections, crack widths, etc. Leakage and the exposed steel become vulnerable to corrosion. Safety and serviceability can be enhanced by increasing the design margins of safety, but this increases the cost of the structure. While considering overall **economy**, the increased cost associated with increased safety margins needs to be judiciously examined before recommendation.

DESIGN CONSIDERATION

Design Philosophies

The RCC design, based on linear elastic theory, is useful in case of design of bridges, machine foundation, etc. Probabilistic concepts of design are extremely important. Different uncertainties in design more rationally handled by theory of probability. The risk can be assessed in terms of a probability of failure. A deterministic form involving a number of partial safety factors has been introduced in different codes. Latest codes IS 456, BS 8110, ACI 318, etc. have introduced Limit State Design (LSM) methodology, over a long period of time.

Uncertainties in Design

To safeguard against the risk of failure (collapse or unserviceability), safety margins are normally provided in the design. In the aforementioned designs, these safety margins were assigned (in terms of 'permissible stresses' in WSM and 'load factors' in the ultimate load design) primarily on the basis of past

DOI: 10.1201/9781003208204-2

experience and engineering judgment. Structures designed according to these methods were found, in general, to be safe and reliable. However, the safety margins provided in these methods lacked scientific basis. Hence, reliability-based design methods were developed with the objective of obtaining rational solutions, which provide adequate safety.

Variables such as loads, material strength, and member dimensions are subject to varying degrees of uncertainty and randomness. The deviations in the dimensions of members or strength of material, even though within acceptable tolerance, can result in a member having less than computed strength. Hence, the design should take into account the possibility of overload or under strength. Further, some idealization and simplifying assumptions are often used in the theories of structural analysis and design. There are also several unforeseen factors that influence the prediction of strength and serviceability. They include construction methods, workmanship and quality control, intended service life of the structure, human errors, possible future change of use, and frequency of loading. These uncertainties make it difficult for the designer to guarantee the absolute safety of the structure. Hence, in order to provide reliable safety margins, the design must be based on the probabilistic methods of design.

LIMIT STATES

In the limit states design, the term 'limit states' is preferably used instead of the term 'failure'. Thus, a limit state is a state of impeding failure, beyond which a structure ceases to perform its intended function satisfactorily.

The limit states usually considered relevant for RC are normally grouped into the following three types

1. **Ultimate (safety) limit states**, which correspond to the maximum load-carrying capacity and are concerned with the following:
   a. Loss of equilibrium (collapse) of a part or the whole structure when considered as a rigid body;
   b. Progressive collapse;
   c. Transformation of the structure into a plastic mechanism collapse;
   d. Rupture of critical sections due to the stress exceeding material strength (in some cases reduced by repeated loading) or by deformations;
   e. Loss of stability (buckling, overturning, or sliding); and
   f. Fracture due to fatigue.
2. **Serviceability limit states**, which deal with the discomfort to occupancy and/or malfunction, caused by excessive deflection, excessive crack width, undesirable vibration (e.g., wind induced oscillations, floor vibration), and so forth.
3. **Special limit states**, which deal with the abnormal conditions or abnormal loading such as damage or collapse in extreme earthquakes, damage due to fire, explosions, or vehicle collisions, damage due to corrosion or

deterioration (and subsequent loss of durability), elastic, plastic or creep deformation, or cracking leading to a change of geometry, which necessitates the replacement of the structure.

The attainment of one or more ultimate limit states be regarded as the inability to sustain any increase, whereas the serviceability limits states denote the need for remedial action or some loss of utility. Hence, ultimate limit states are conditions to be avoided, whereas serviceability limit states are conditions that are undesirable. Thus, it is clear that any realistic, rational, and quantitative representation of safety must be based on statistical and probabilistic analysis, which caters for both overload and under strength. The design for the ultimate limit state may be conveniently explained with reference to the type of diagram shown in Figure 2.1. This figure shows the hypothetical frequency distribution curves for the effect of loads on the structural element and the resistance (strength) of the structural element. When the two curves overlap, shown by the shaded area, effect of the loads

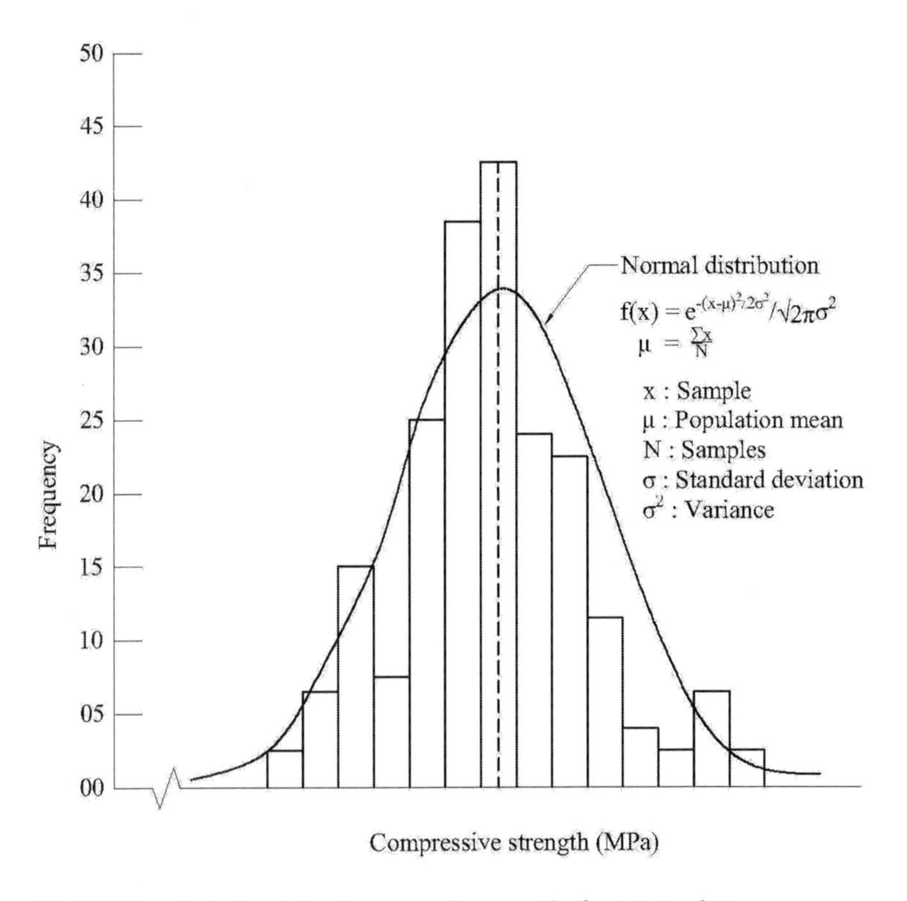

FIGURE 2.1 Typical variation in compressive strength of concrete cubes.

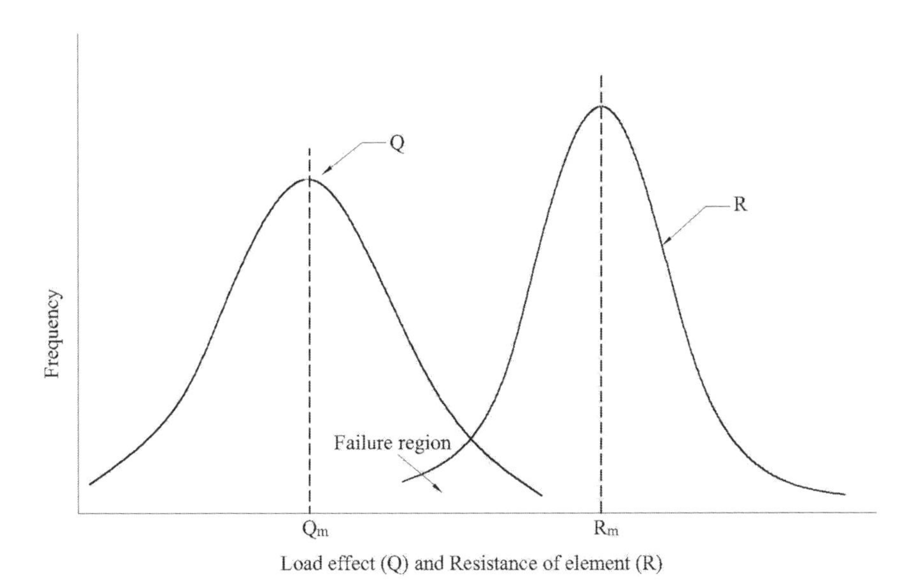

FIGURE 2.2 Frequency distribution cubes.

is greater than the resistance of the element, and the element will fail. Thus, the structure and its element should be proportioned in such a way that the overlap of the two curves (Figure 2.2) is small, which means that the probability failure is within the acceptable range.

Levels of Reliability Methods

There exists a number of levels of reliability analysis. These are differentiated by the extent of probabilistic information that is used. A full-scale probabilistic analysis is generally described in a level III reliability method, which uses the probability of failure P_f to evaluate the risk involved. It is highly advanced, mathematically involved, and generally used as a research tool. It is not suitable for general use in design offices.

The problem may be simplified by limiting the probability information of the basic variables to their 'second moment statistics' (i.e., mean and variance). Such a method is called a level II reliability method. In this method, the structural failure (the achievement of a limit state) is examined by comparing the resistance R with the load effect Q in a logarithmic form observing $\ln(R/Q)$ as shown in Figure 2.3. In this figure, the hatched region shows the failure. The distance between the failure line and the mean value of the function. $\ln f(R/Q)_m$. is defined by β called the reliability index, the concept of which was first proposed in 1956. The larger the value of β, the greater is the margin of safety of the system. The expression for β may be written as

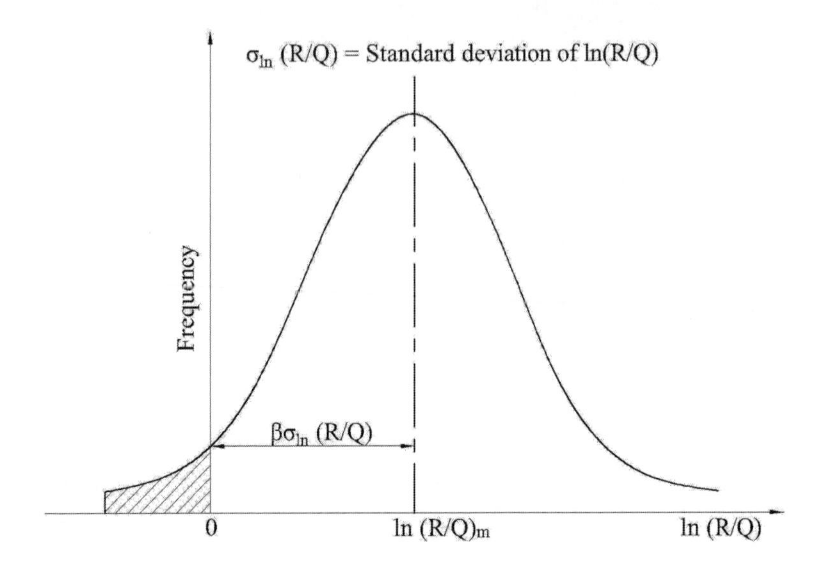

FIGURE 2.3 Level II reliability method.

$$\beta = \frac{\ln\left(R_m/Q_m\right)}{\sqrt{\left(V_r^2 + V_q^2\right)}}$$

where $V_R = (\sigma_R/R_m)$, $V_Q = (\sigma_Q/R_m)$, R_m and Q_m are the mean values of resistance and load, respectively, and σ_R and σ_Q are the standard deviations of resistance and load, respectively.

However, even such a 'simplified method' is unsuitable for use in a design office, as the determination of β requires an iterative procedure; it may require special software and is therefore time-consuming. The values of the reliability index β corresponding to various failure probabilities P_f can be obtained from the standardized normal distribution function of the cumulative densities and are given in Table 2.1.

It has to be noted that the values given in Table 2.1 are valid only if the safety margin is normally distributed. For code use, the method must be as simple as possible using deterministic rather than probabilistic data. Such a method called level I reliability method or first-order second moment reliability method, which is used in the code to obtain a probability-based assessment of structural safety.

TABLE 2.1
Reliability Index for Various Failure Probabilities

β	2.32	3.09	3.72	4.27	4.75	5.2	5.61
$P_f = \varphi(\beta)$	10^{-2}	10^{-3}	10^{-4}	10^{-5}	10^{-6}	10^{-7}	10^{-8}

CHARACTERISTIC LOAD AND CHARACTERISTIC STRENGTH

In normal design calculations, a single value is usually used for each load and for each material property, with a margin to take care of all uncertainties. Such a value is termed the characteristic strength (or resistance) or characteristic load. The characteristic strength of a material (such as steel, concrete, or wood) is defined as the value of strength below where more than a prescribed percentage of test results will fall. This prescribed percentage is normally taken as 95. Thus, the characteristic strength, f_{ck}, of concrete is the value of cube/cylinder strength, below which not more than 5% of the test values may fall (Figure 2.4). Similarly, the characteristic load. Q_c, is defined as the load that is not expected to be exceeded with more than 5% probability during the lifespan of a structure. Thus, the characteristic load will not be exceeded 95% of the time. The design values are derived from the characteristic values through the use of partial safety factors, both for materials and for loads. The acceptable failure probability, P_f for particular classes of structures is generally derived from experiences with past practice, consequences of failure, and cost considerations. Having chosen P_f and β, the determination of partial safety factors is an iterative process.

Sampling and Acceptance Criteria

The relationship between mean and characteristic strength is given by

$$\varphi_c^{-1}\left(P_f\right)=\left(f_{ck}-f_m\right)/s=-k$$

where φ_c is the cumulative normal distribution function, f_m and s are the mean value and standard deviation of the normal distribution, respectively, f_{ck} is the characteristic strength, k is an index that signifies the acceptable probability of failure.

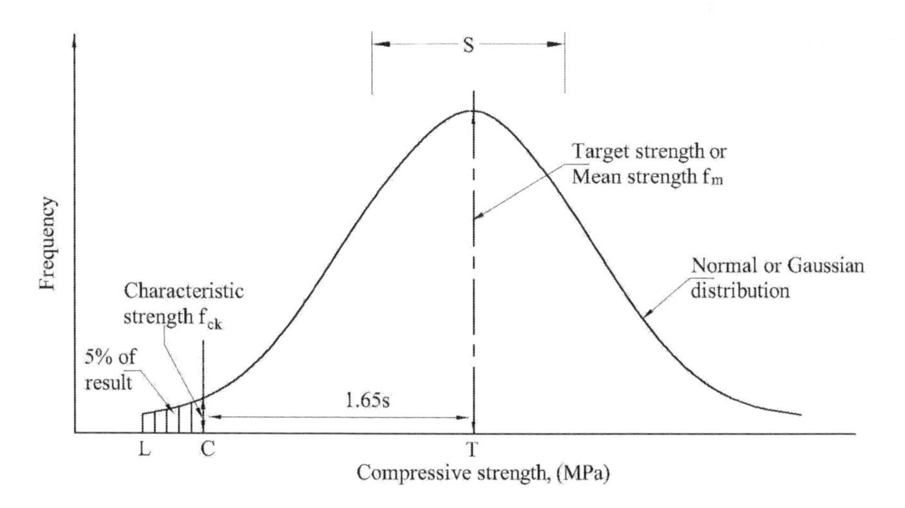

FIGURE 2.4 Normal frequency distribution of concrete strengths.

DUCTILITY REQUIREMENTS

If the structure can stay elastic under the action of the maximum expected earthquake in the respective area, then there is a need for ductile reinforcement. But increasing the elastic property with the demand was found as an uneconomical method of construction method even not appropriate for emerging with architecturally viable design ideas. For having economy and safety for building structures under earthquake motions that are unexpected, the method opted is to let the structure undergo damage either by means of plasticity, fracture, crushing, etc., keeping its strength to carry the vertical load as it is undergoing deformations during damage. For example, we will consider columns that must be designed for ductility. The method is to provide confined concrete with reinforcement to avoid buckling of longitudinal reinforcement. This would enable the column to continue taking vertical loads even if it is subjected to cracking, concrete crushing, or yielding of steel reinforcement. Now this arrangement would cause the material to become compliant, hence the stiffness force values in the structures and related components would drop. That is if the structure had remained elastic, then there would be higher internal forces and total base shear. Here the total base shear can be explained as the sum of internal shear forces in all the vertical load-carrying structural elements. So, incorporation of ductility in the material means, we have allowed the structure to get damaged, thus less internal forces. This provision is based on dynamic response modification factor or ductility factor, which is provided in building codes worldwide. The ductility factor depends on the lateral structural system implemented. For systems that have to undergo larger deformations without collapse when damage occurs would have a higher ductility factor. Thus, a structural system that has to undergo only small deformations before collapsing will gain only a small ductility factor.

The above-explained criteria are defined as a capacity design method, the phenomenon of ductility incorporation in building elements. This considers the problem of determining the failure mechanism of members. The idea is to force the member to undergo failure in a ductile manner by making the capacity of the member in other possible failure modes greater. Consider two bars of the same length and cross-sectional area. Here one of them forms a ductile material and the other one forms brittle. Pulling these two bars on either side will make them break under extreme load. It is observed that the ductile material elongates for a larger amount and the brittle material breaks for a small elongation. Hence among the materials used steel is more ductile than concrete which is brittle. Ductile Chain Design concept in building as per capacity design. When a brittle chain alone is pulled on either side, it would break suddenly. Among the chain, the weakest link would break first. If we make the weakest link as the ductile one, we can gain more elongation and more ductility. In buildings, we implement the same concept in the form of a ductile chain. The seismic inertia forces are transmitted from the floors to the beams and then to the columns. The failure of the column would affect the stability of the building than the beam failure. Hence it is appreciable to make beams as weak ductile links than columns. This design method is called as strong column-weak beam concept.

Ductility of concrete can be improved by confining it to spirals or stirrups/lateral ties. Confined concrete in compression higher ductility as well as strength. Ductility can be increased by decreasing the percentage of tension steel to a particular limit. Ductility can be defined as the ability of material to undergo large deformations without rupture before failure. Member or structural ductility is generally defined as the ratio of maximum deformation to the yield deformation. Ductility can be defined with respect to strains, rotations, curvature, or deflections. Strain-based ductility definition depends almost on the material, while rotation or curvature-based ductility definition also includes the effect of shape and size of the cross-sections. Each design code recognizes the importance of ductility in design because if a structure is ductile its ability to absorb energy without critical failure increases. Ductility behavior allows a structure to undergo large plastic deformations with little decrease in strength. In general, the ductility is increased steel under compression increases. If ductile members are used to form a structure, the structure can undergo large deformations before failure. This is beneficial to the users of the structures, as in case of overloading, if the structure is to collapse, it will undergo large deformations before failure and thus provides warning to the occupants. This gives a notice to the occupants and provides sufficient time for taking preventive measures. This will reduce loss of life. Structures are subjected to unexpected overloads, load reversals, impact, and structural movements due to foundation settlement and volume changes which are generally ignored in the analysis and design. The limit state design procedure assumes that all the critical sections in the structure will reach their maximum capacities at design load for the structure. Brittleness is a property of material that will fail suddenly without undergoing noticeable deformations. Brittle structures do not give notice before failure. Concrete is a brittle material. By suitably anchoring the reinforcement, the ductility of a structure can be increased to a greater extent. Reinforced concrete structures, unlike steel structures, tend to fracture or fail in a relatively brittle fashion as the ductility or deformation capacity of conventional concrete is limited. In such structures, the brittle failure as a result of inelastic deformation can be avoided only if the concrete is made to behave in a ductile manner so that the member can absorb and dissipate a large amount of energy. Hence in the case of reinforced concrete members subjected to inelastic deformation, not only strength but also ductility plays a vital role in the design. A ductile material is the one that can undergo large strains while resisting loads. Ductile detailing as per IS13920 is provided in structures so as to give them adequate toughness and ductility to resist severe earthquake shocks without collapse. The performance criteria in most earthquake code provisions require that a structure be able to resist earthquakes of minor intensity without damage. A structure would be expected to resist such frequent but minor/moderate shocks within its elastic range of stresses. It is believed that structural damage due to the majority of earthquakes will be limited to repairable damage. Ductile structures resist major catastrophic earthquakes without collapse. If excessive reinforcement is provided the concrete will crush before the steel yields, leading to brittle failure. A beam should be designed as preferably under-reinforced. The code recommends a value of 0.0035 for bending tensile strain. Ductility increases with the increase in the characteristic strength of concrete and decreases with the characteristic strength of steel. The presence of an enlarged compression flange in a T-beam reduces the depth of the compression zone at collapse and thus increases the ductility. Lateral reinforcement tends to improve ductility by preventing premature shear failures and by confining the compression zone, thus increasing the deformation capability of a reinforced concrete beam. The structural layout should be simple and regular avoiding offsets of beams to columns, or offsets of columns from floor to floor. Changes in stiffness should be gradual from floor to floor. The amount of tensile reinforcement in beam should be restricted and more compression reinforcement should be provided and should be tied up by stirrups to prevent reinforcement buckling and provide more ductility. Beams and columns in a reinforced concrete frame should be designed in such a manner that inelasticity is confined to beams only and the columns should remain elastic, i.e. indirectly **Strong column Weak beam concept**. Sum of the moment capacities of the columns for the design axial loads at a beam-column joint should be greater than the moment capacities of the beams, to ensure this behavior. The shear reinforcement should be adequate to ensure that the strength in shear exceeds the strength in flexure and thus, prevent a non-ductile shear failure before the fully reversible flexure strength of a member has been developed. Closed stirrups or spirals should be used to confine the concrete at sections of maximum moment to increase the ductility of members. Such sections include the upper and lower ends of columns and within beam-column joints, which do not have beams on all sides. Splices and bar anchorages must be adequate to prevent bond failures. The reversal of stresses in beams and columns due to the reversal of the direction of earthquake force must be taken into account in the design by appropriate reinforcement. Beam-column connections should be made monolithic. The factored axial stress on the member under earthquake loading shall not exceed $0.1f_{ck}$. The member shall have preferably had a width-to-depth ratio of more than 0.3. The width of the member shall not be less than 200mm. The depth of the member shall preferably be not less than ¼ of the clear span. The top and bottom reinforcement shall consist of at least two bars throughout the member length. The tension steel ratio on any face, at any section, shall not be less than $0.24\{(f_{ck}/f_y)\}^{0.5}$ as per IS code. The maximum steel ratio on any face at any section, shall not exceed 0.025. In an external joint, both the top and the bottom bars of the beam shall be provided with anchorage length, beyond the inner face of the column, equal to the development length in tension plus ten times the bar diameter minus the allowance for 90° bend. In an internal joint, both face bars of the beam shall be taken continuously through the column. Flexure members of lateral force resisting ductile frames are assumed to yield at the design earthquake load. The longitudinal bars shall be spliced, only if hoops are provided over the entire splice length, at spacing not exceeding 150mm. For confining the concrete and to support longitudinal bars. The lap length shall not be less than the bar development length in tension. Lap splices shall not be provided within a joint. To avoid the possibility of spalling of concrete cover under large reversed strains. When a column terminates into a footing or mat, special confining reinforcement shall extend at least 300mm into the footing or mat.

Thus ductility of a structure is one of the most important factors affecting its seismic performance. The prevailing Indian code IS 13920 ensures the overall ductility of the structure by providing extra reinforcement at critical locations of the structure like junctions. Improving the properties of concrete such as ductility, tensile strength, and energy dissipation capacity by introducing different types of admixtures/fibers may be used. A ductile concrete, with substantially higher tensile ductility compared to normal concrete, can contribute to higher structural resiliency and environmental sustainability, the latter by virtue of the need for less frequent repairs. Modification of the structure of C-S-H at the nanoscale to create hybrid, organic, cementitious nano-composites lately has received attention due to the interest in more sustainable concrete structures. Nano-reinforcements: carbon nano-tubes/nano-fibers are potential materials for use as nano-reinforcements in cement-based materials, may be used.

In seismic design, reinforced concrete structures must perform satisfactorily under severe load conditions. To withstand large lateral loads without severe damage, structures need strength and energy dissipation capacity. It is commonly accepted that it is uneconomical to design reinforced concrete structures for the greatest possible earthquake ground motion without damage. Therefore, the need for strength and ductility has to be weighed against economic constraints. Ductility is an essential property of structures responding inelastically during severe earthquakes. Ductility is defined as the ability of sections, members, and structures to deform inelastically without excessive degradation in strength or stiffness. The most common and desirable sources of inelastic structural deformations are rotations in potential plastic hinge regions. An energy dissipation mechanism should be chosen so that the desirable displacement ductility is achieved with the smallest rotation demands in the plastic hinges. The development of plastic hinges in frame columns is usually associated with very high rotation demand and may result in total structural instability (globalized failure). While for the same maximum displacement in a structural frame system, the rotation demand in the plastic hinges would be much smaller if they developed in the beams. For getting an efficient performance of beam at beam-column joint we need to give proper anchorage which will provide proper dissipation of energy and ductility to the structure. Otherwise, the failure may occur due to the poor anchorage at the joint by pulling out of the beam longitudinal bars from the joint.

Current design philosophy requires that beam-column joints have sufficient capacity to sustain the maximum flexural resistance of all the attached members. The mechanism of force transfer within beam-column joint of a rigid frame during seismic events is known to be complex involving bending in beams and columns, shear and bond stress transfer in the joint core. To provide proper anchorage of beam at the joint, various countries provide special detailing on and near hinged zones. Figure 2.1 shows the forces acting on joint core under lateral load. Indian Standard code recommends continuing the transverse loops around the column bars through the joint region. The length of anchorage is about $L_d + 10_{\mathrm{db}}$ inside the joint. The primary aim of joint design must be to suppress a shear failure. This often necessitates a considerable amount of joint shear reinforcement, which may result in construction difficulties. Current seismic code details for reinforced concrete structures are often considered impracticable by construction and structural engineers because of its installation and the difficulties in placing and consolidating the concrete in the beam-column joint regions For high seismic zones, load reversals in the joint can lead to significant bond deterioration along straight bar anchorages; therefore, American Concrete Institute (ACI) and Indian Standard(IS) requires that standard hooks be used to anchor longitudinal reinforcement terminated within an exterior joint. The use of standard hooks results in steel congestion, making the fabrication difficult to construct.

Ductility for earthquake-resistant design is important for buildings, structures, and building materials. Ductility and its importance in design are discussed. To understand the importance of ductility in influencing building performance, it is essential to know what ductility is. Ductility in general gains a definition in material engineering science as the ratio of ultimate strain to yield strain of the material. In a broader view, we must understand ductility as the ability of a structure to undergo larger deformations without collapsing. As per special provisions recommended in codes, the detailing of the structure that lets the structure gain a larger ductility other than the contributions of material ductility is called as ductility detailing or ductile detailing. When a structure undergoes dynamic forces (which is considered as the seismic demand) the structure cannot remain elastic anymore, and the next stage is damage. It can go through a plastic stage or fracture or damage, where stiffness will decrease appreciably and deformations will drastically increase even for a small load. These situations must be expected by an engineer, and we should ensure that our design sustains these loads without undergoing larger deformations or no collapse. For this target to be achieved, we should incorporate **Minimum flexural ductility.**

In the flexural design of RC beams, in addition to providing adequate strength, it is often necessary to provide a certain minimum level of ductility. For structures subjected to seismic loads, the design philosophy called strong column-weak beam is adopted, which is supposed to guarantee the following behavior: The beams yield before the columns and have sufficient flexural ductility such that the potential plastic hinges are in the beam maintain their moment resistant capacities until the columns fail. To ensure the ductile mode of failure, all beams should be designed as under-reinforced. More stringent reinforcing detailing like the provision of confining reinforcement in the plastic hinge zones is also generally imposed. In addition, traditionally limits were also imposed on either the tension steel ratio (which should not be more than 0.75 of the balanced steel ratio or the strain in steel should be greater than 0.005) or the neutral axis depth. This kind of limitation may result in a variable level of curvature ductility depending on the concrete grade and yield strength of steel.

3 Basic Understanding of Elastic Design Approach

PREAMBLE

It is essential for the reader to have relevant IS code of practices and handbooks for RCC design etc. as ready reference while going through this chapter.

The Working Stress Method (WSM) basically assumes that the structural material behaves in a linear elastic, and it is ensured by restricting the stresses in the material induced. Restricted stresses are specified in code of practices as permissible/allowable stresses, which are well below the material strength where the assumption of linear elastic behavior is justifiable. In general, the material is considered homogeneous, isotropic and follows Hook's law. Factor of safety is around three. The carrying load is termed as working loads/service loads. Perfect bond between steel reinforcement embedded in concrete is assumed, and strain compatibility, i.e., the strain in the reinforcing steel, is assumed to be equal to that in the adjoining concrete to which it is embedded. Hook's law is considered valid and the stresses in concrete and steel are calculated based on simple bending theory considering properties of the composite section based on the modular ratio concept. It may be noted here that modular ratio is defined as the ratio of the modulus of elasticity of steel to that of concrete. However, many factors related to long-term effects, i.e. creep and shrinkage, stress concentrations, other secondary effects, etc., are to be considered separately, and as a result, local changes in stresses may take place. It may be stated that most structures designed in accordance with WSM suggested by different codes of practices (IS 456 etc.) have been performing satisfactorily for many years, especially bridges, machine foundations, chimneys, etc. WSM is generally recommended for vibration-prone structures/special structures.

This was the traditional method of design not only for reinforced concrete but also for structural steel and timber design. The conceptual basis of WSM is simple. The method basically assumes that the structural material behaves in a linear manner, and that adequate safety can be ensured by suitably restricting the stress in the material induced by the expected 'working loads' (service loads) on the structure. As the specified permissible stresses are kept well below the material strength, the assumption of linear elastic behavior is considered justifiable. The ratio of the strength of the material to the permissible stress is often referred to as the factor of safety. The stresses under the applied loads are analyzed by applying the methods of 'strength of materials' such as the simple bending theory. In order to apply such methods to a composite material like

DOI: 10.1201/9781003208204-3

reinforced concrete, strain compatibility (due to bond) is assumed, whereby the strain in the reinforcing steel is assumed to be equal to that in the adjoining concrete to which it is bonded. Furthermore, as the stresses in concrete and steel are assumed to be linearly related to their respective strains, it follows that the stress in steel is linearly related to that in the adjoining concrete by a constant factor (called the modular ratio), defined as the ratio of the modulus of elasticity of steel to that of concrete.

However, the main assumption of linear elastic behavior and the assumption that the stresses under working loads can be kept within the 'permissible stresses' are not found to be realistic. Many factors are responsible for this, such as the long-term effects of creep and shrinkage, the effects of stress concentrations, and other secondary effects. All such effects result in a significant redistribution of the calculated stresses. Moreover, WSM does not provide a realistic measure of the actual factor of safety underlying a design. WSM also fails to discriminate between different types of loads that act simultaneously, but have different degrees of uncertainty. This can, at times, result in very uncomfortable designs, particularly when two different loads (say, dead loads and wind loads) have counteracting effects. Nevertheless, in defense against these and other shortcomings leveled against WSM, it may be stated that most structures designed in accordance with WSM have been generally performing satisfactorily for many years. The design usually results in relatively large sections of structural members (compared to ULM), thereby resulting in better serviceability performance (less deflections, crack widths, etc.) under the usual working loads. The method is also notable for its essential simplicity concept, as well as application.

BEHAVIOR OF REINFORCED CONCRETE UNDER FLEXURE

Linear Elastic Cracked Behavior

Assumptions

- Plane sections before bending remain plane after bending.
- The tensile strength of concrete may be generally ignored.
- The stress-strain relationships for Concrete and steel are linearly elastic.

However, creep needs to be considered using an effective modulus of elasticity of concrete.

Analysis of Bending Stress

Working stress method is based on the following important aspects:

1. Hooke's law is valid.
2. Material is homogeneous and isotropic.
3. Strain and stress profile are linear.
4. Theory of pure elastic bending is valid.

Several logical questions may be raised against the above considerations but became very popular as the structures designed according to WSM are performing successfully and for its simplicity. WSM is generally recommended for vibration-prone structures like bridges, machine foundations, etc./special structures.

From "Theory of pure elastic bending", we know

$$\sigma = My/I$$

σ = Bending stress at any point at a distance y from the neutral axis
$M \equiv$ Bending moment at the section, I = Moment of inertia about neutral axis

Transformed Section

Modular ratio concept transforms a composite section into equivalent homogeneous section so that the theory of simple bending for homogeneous isotropic material can be applied, either entirely in terms of concrete or in terms of steel. However, this transformation must not alter the magnitude, direction, and line of action of the resultant forces in the materials due to the bending stresses.

It is generally steel area is transferred into equivalent concrete by multiplying steel area with modular ratio, i.e., $m = E_s/E_c$ and considered at the steel reinforcement level along with the rest of the concrete. Although the modular ratio is $m = E_s/E_c$, but for WSM, IS 456 recommended to calculate modular ratio by an equation $m = 280/3\sigma_{cbc,}$ for transferring steel area in tension to equivalent concrete, where σ_{cbc} = allowable bending stress of concrete depending on its grade in N/mm^2. Modular ratio is recommended to transfer steel area in compression to equivalent concrete equal to $1.5 \times [280/3\sigma_{cbc}]$ (Figures 3.1 and 3.2).

Uncracked Section

*Area of equivalent concrete in compression

$$= (bx - A_{sc}) + 1.5m\, A_{sc} = bx + 1.5(m-1)A_{sc}$$

Area of equivalent concrete in tension

$$= b(d-x) - A_{st} + m\, A_{st} = b(d-x) + (m-1)A_{st}$$

where

x = depth of neutral axis from compression face
d = Effective depth = $D - d'$
d' = Effective cover to longitudinal reinforcements
A_{st} = Area of longitudinal steel in tension
A_{sc} = Area of longitudinal steel in compression
m = Modular ratio (m) = $280/3\sigma_{cbc}$
σ_{cbc} = Allowable bending stress of concrete as per IS456 in N/mm^2

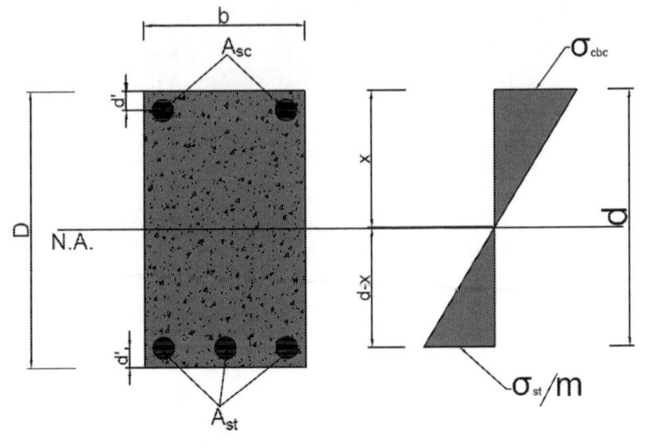

FIGURE 3.1 Stress distribution of an uncracked section.

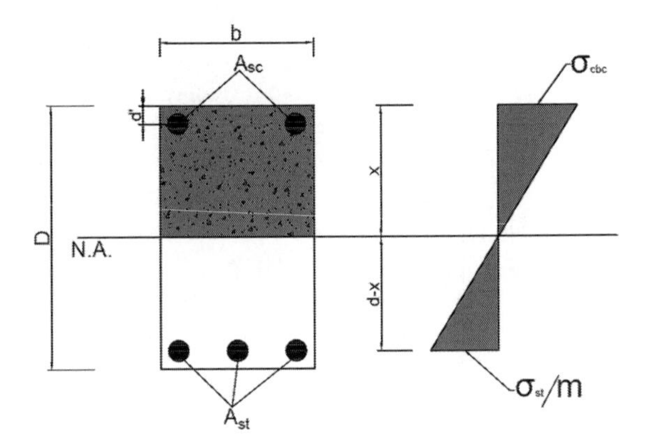

FIGURE 3.2 Stress distribution of a cracked section.

To get neutral axis depth (x) →

$$bx(x/2) + 1.5(m-1)A_{sc}\left(x - d'\right) = b(d-x)(d-x/2) + (m-1)A_{st}(d-x)$$

I = Uncracked moment of inertia

$$= bx^3/3 + b(d-x)^3/3 + m\,A_{st}(d-x)^2 + 1.5(m-1)A_{sc}\left(x-d'\right)^2$$

Cracked Sections

The effect of concrete in tension is neglected

To get neutral axis depth (x) →

$$bx(x/2) + 1.5(m-1)A_{sc}(x-d') = m\ A_{st}(d-x)$$

I_r = Cracked moment of inertia

$$= bx^3/3 + m\ A_{st}(d-x)^2 + 1.5(m-1)A_{sc}(x-d')^2$$

Singly Reinforced Cracked Sections

If we neglect the effect of steel in compression and concrete in tension and consider the effect of steel in tension and the effect of concrete in compression (Figure 3.3).

Equating moment of areas above and below neutral axis about neutral axis depth (x) → $b \times (x/2) = m\ A_{st}\ (d-x)$

I_r = Cracked moment of inertia = $bx^3/3 + m\ A_{st}\ (d-x)^2$

Gross Moment of Inertia as per IS 456

$I_g = bD^3/12$ → for rectangular section

Gross moment of inertia concept may be used for uncracked section i.e. when bending tensile stress is less than cracking stress of concrete (σ_{cr}) as given in codes of practices.

$$\sigma_{cr} = 0.7\sqrt{f_{ck}} \rightarrow \text{As per IS456}$$

where f_{ck} = Characteristic strength of concrete in N/mm^2 and σ_{cr} in N/mm^2

$$\text{Cracling Moment}(M_{cr}) = \sigma_{cr}\ (D/2)/I_g \rightarrow \text{as per IS456}$$

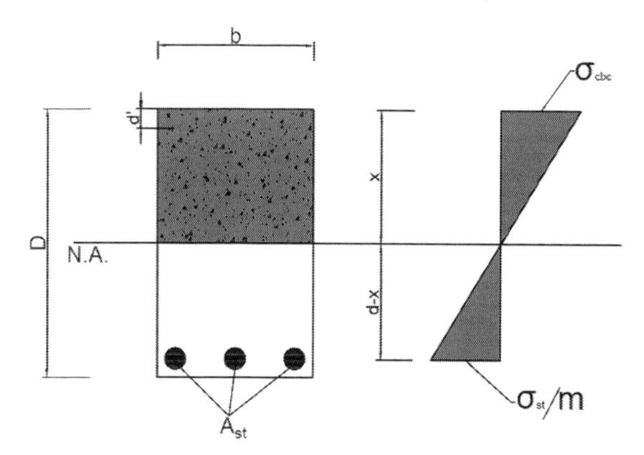

FIGURE 3.3 Stress distribution of a singly reinforced cracked section.

If $M < M_{cr,}$ there is no tension crack

Short-Term Static Modulus of Elasticity of Concrete

$$E_c = 5000\sqrt{f_{ck}}$$

f_{ck} = Characteristic strength of concrete in N/mm² and E_c in N/mm²

Permissible Stresses

From IS 456, Tables 21 and 22, Permissible σ_{cbc} and σ_{st}, respectively, are recommended (Figure 3.4).

Moment of resistance

Singly reinforced cracked rectangular sections

From similarity of triangles of stress diagram, it can be written as follows:

$$\boldsymbol{\sigma_{cbc}/x = (\sigma_{st}/m)/\ (d - x)}$$

Substituting $x = k\,d$ ie. $k = x/d \rightarrow \boldsymbol{\sigma_{cbc}\ /\ kd = (\sigma_{st}/m)\ /\ (d - kd)}$

$$\textbf{ie. } \boldsymbol{\sigma_{cbc}\ /\ k = (\sigma_{st}/m)/(1 - k)}$$

A value of 'k' can be obtained by substituting allowable bending compressive stress in concrete and tensile stress in tension steel as per codes of practices. Then $x = kd$

This condition is known as **'Balanced condition'** i.e. when bending compressive stress in concrete (σ_{cbc}) and tensile stress in steel (σ_{st}) are reaching their full allowable value as recommended by codes of practices simultaneously, the corresponding 'x' value is termed as 'Balanced neutral axis depth'(x_b) and $k_b = x_b/d$

The following equation can be used

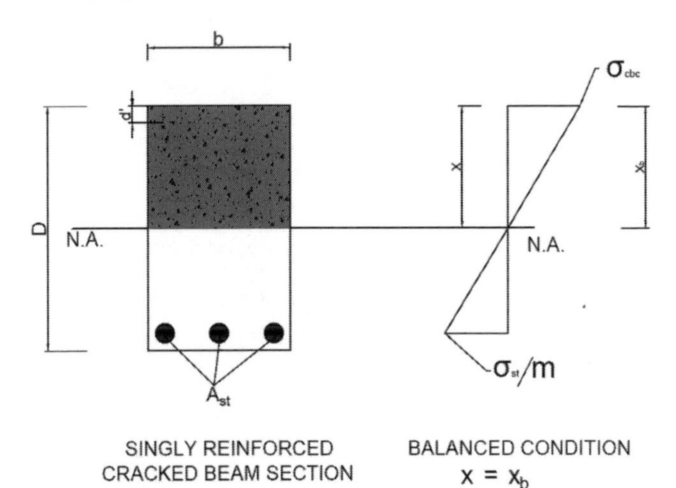

FIGURE 3.4 Singly reinforced cracked balanced sections.

$$\sigma_{cbc}/x_b = \sigma_{st}/m/(d-x_b) \rightarrow x_b \text{ and } k_b$$

Both σ_{cbc} and σ_{st} are allowable values of stresses as per IS 456

i.e., if $x = x_b$ or $k = k_b \rightarrow$ It is a **Balanced section**

It may be further noted that when the tensile stress in steel (σ_{st}) is reaching to its full allowable value but the bending compressive stress in concrete (σ_{cbc}) is less than its allowable value, the section is termed as an **Under-reinforced** one

In this case, $x < x_b$ or $k < k_b \rightarrow$ It will be an **Under-reinforced** section

It may be further noted that when the bending compressive stress in concrete (σ_{cbc}) is reaching to its full allowable value but the tensile stress in steel (σ_{st}) is less than its full allowable value, the section is an **Over-reinforced** one (Figures 3.5 and 3.6).

In this case, $x > x_b$ or $k > k_b \rightarrow$ **Over-reinforced section**

It is important to assess the allowable bending moment for a particular beam section, considering the above aspects, i.e. how much maximum bending moment may be allowed for a particular beam section, so that bending compressive stress in concrete and stresses in steel reinforcements are within permissible stresses recommended by IS 456

Referring to Force diagram shown in Figure 3.7, it can be written as follows:

$$C = \tfrac{1}{2}\sigma_{cbc} \times b,\ T = A_{st}\,\sigma_{st} \text{ and } C = T$$

An internal couple will be formed i.e. $M = C$ xa

where a = Distance between C and T i.e., Lever arm $= d - x/3 = d - kd/3$

$$\text{Allowable moment } (M) = T\,(a) = A_{st}\,\sigma_{st}\,(d - x/3)$$

$$\text{Or } M = C(a) = \tfrac{1}{2}\,\sigma_{cbc} \times b\,(d - x/3)$$

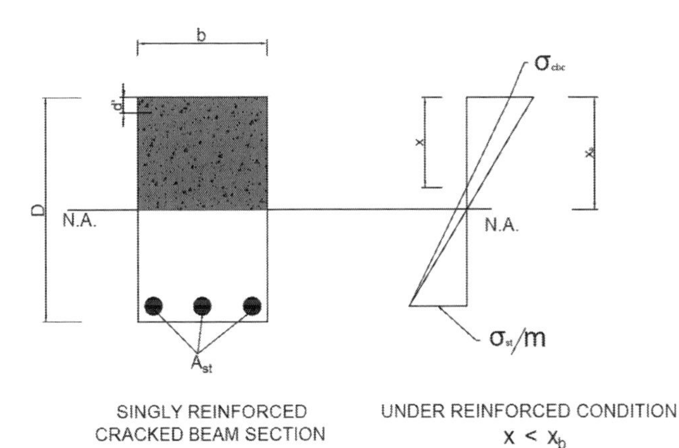

FIGURE 3.5 Singly reinforced cracked under-reinforced sections.

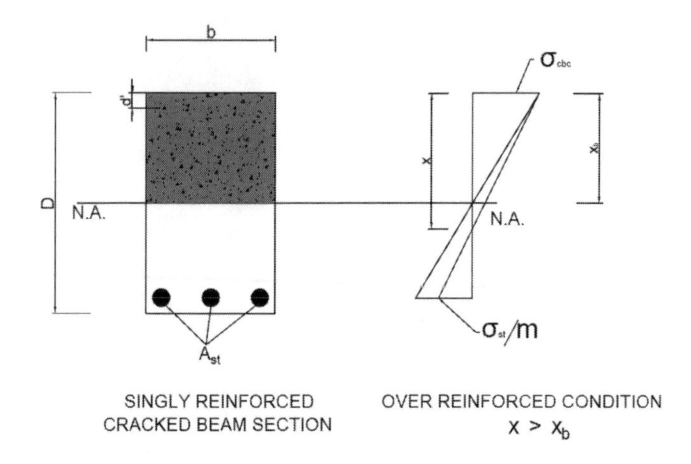

FIGURE 3.6 Singly reinforced cracked over-reinforced sections.

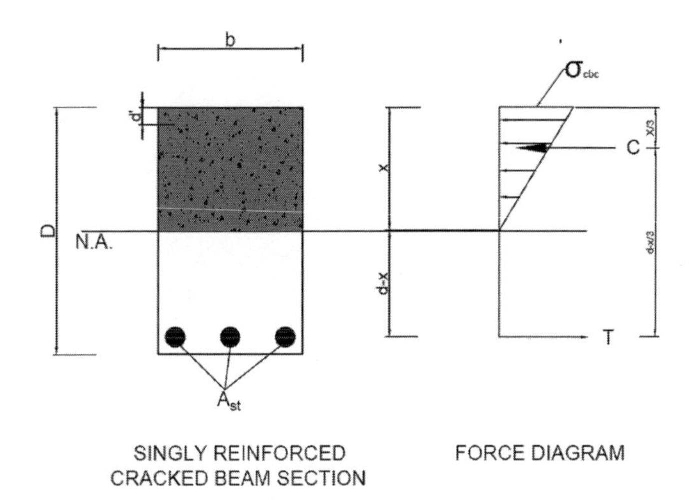

FIGURE 3.7 Force diagram of singly reinforced cracked section.

Substituting $x = kd \rightarrow M = [\frac{1}{2}\,\sigma_{cbc} \cdot k \cdot (1 - k/3)]\, bd^2 \rightarrow M = [\frac{1}{2}\,\sigma_{cbc}\, k\, j]\, bd^2$ where $J = 1 - k/3 \rightarrow \mathbf{M = Q\ bd^2}$, where $Q = [\frac{1}{2}\,\sigma_{cbc}\, k\, j]$

Referring to Figure 3.5 for a balanced condition, it can be written as follows:

$$M_b = A_{st}\,\sigma_{st}\,(d - x_b/3) \text{ where, } x_b \text{ can be obtained as discussed earlier}$$

$$\text{Expressing, } p_b = [A_{st} / bd] \times 100, \text{ ie. } A_{stb} = p_b bd / 100$$

where p_b = Balanced percentage of steel in tension

$$M_b = (p_b bd/100)(\sigma_{st})\,(d - k_b d/3) = [(p_b\,\sigma_{st}/100)\,(1 - k_b/3)]bd^2 \qquad (3.1)$$

Again,

$$M_b = \tfrac{1}{2}\,\sigma_{cbc}\,x_b\,b(d - x_b/3) = [\tfrac{1}{2}\,\sigma_{cbc}\,k_b(1 - k_b/3)]bd^2 \quad (3.2)$$

Therefore, $Q_b = \tfrac{1}{2}\,\sigma_{cbc}\,k_b\,j_b$, where $j_b = 1 - k_b / 3$

Equating the expressions of M_b, i.e., Eqs. 3.1 and 3.2, it can be written as follows:

$$(p_b\sigma_{st}/100)\,(1 - k_b/3) = [\tfrac{1}{2}\,\sigma_{cbc}\,k_b(1 - k_b/3)]$$

$$(p_b\sigma_{st}/100) = [\tfrac{1}{2}\,\sigma_{cbc}\,k_b]$$

$$p_b = 100\,\sigma_{cbc}\,k_b/2\sigma_{st} \quad (3.3)$$

Considering M20 concrete and Fe 415 steel

From IS 456, Table 21 and 22, $\sigma_{cbc} = 7$ N/mm^2 and $\sigma_{st} = 230$ N/mm^2, respectively

Therefore, $p_b = 100\,(7)\,(0.2887) / 2(230) = 0.44\%$

It can be stated that if $p = p_b$, it is a balanced section

$$j_b = (1 - K_b / 3) = (1 - 0.2889 / 3) = 0.9 \rightarrow Q_b = \tfrac{1}{2}(7)(0.2889)(0.9) = 0.9 \text{ N/mm}^2$$

It may be noted as follows

Moment factor (Q) increases with the percentage of steel 'p', but the rate of increase of moment factor (Q) decreases in over-reinforcement zone (Figure 3.8).

Moment of resistance (considering the effect of steel in compression)

When reinforcement in compression face is provided in addition to tension, it may be as hanger bars of nominal diameter for holding stirrups, which are not normally considered to assess moment of resistance. The concrete area for compression bars (A_{sc}) is accounted for by taking the effective transformed area of steel as $(1.5m - 1)\,A_{sc}$

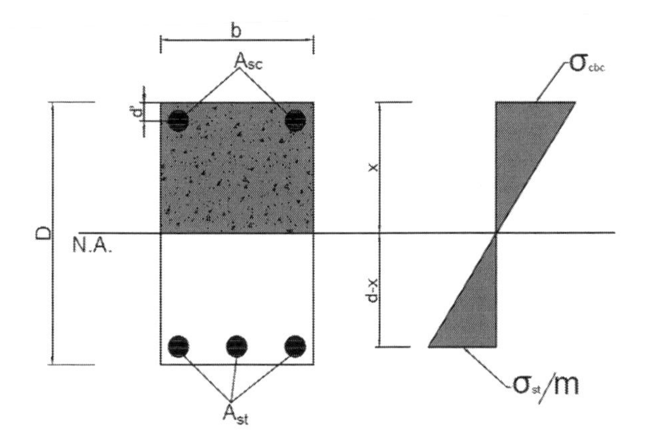

FIGURE 3.8 Stress distribution (considering the effect of steel in compression).

Neutral Axis Can Be Obtained from the Equation below

$$\frac{b\,(x)^2}{2} + (1.5m - 1)A_{sc}\,(x - d') = m\;A_{st}\,(d - x)$$

$$C_c + C_s = T$$

$$C_c = 0.5\;\sigma_{cbc}\;b\,(x)$$

$$C_s = (1.5m\;A_{sc})(\sigma_{cbc1}) - \sigma_{sc}\;A_{sc} = \;(1.5m - 1)\;A_{sc}\sigma_{sc}$$

$$\sigma_{cbc1} = \sigma_{cbc}\left(\frac{x - d'}{x}\right)$$

$$T = A_{st}\;\sigma_{st}$$

$$M = C_c\left(d - \frac{x}{3}\right) + C_s\,(d - d')$$

Typical Numerical Examples

Example 1

Considering a beam section having $b = 300\,\text{mm}$ and $D = 600\,\text{mm}$ and $d = 600 - 50 = 550\,\text{mm}$

Considering the grade of concrete is M20 and the grade of steel reinforcement is Fe 415. 4 - 25 Φ at the bottom.

$$A_{st} = 4\times\left[\pi(25)^2 / 4\right] = 1963\ \text{mm}^2$$

As per WSM IS 456, allowable stresses are $\sigma_{cbc} = 7$ N/mm² from Table 21, $\sigma_{st} = 230$ N/mm² from Table 22 and m = Modular ratio = 280/3 σ_{cbc} = 280 / 3 (7) = 13

Considering it as a cracked, singly reinforced section and using the expression,

$$\sigma_{cbc}/x_b = \sigma_{st}/m/(d{-}x_b) \rightarrow K_b = 0.2887,\ \text{ie. } Q_b = 0.9\ \text{N/mm}^2$$

Using the expression, $b \times (x/2) = m\,A_{st}\,(d - x) \rightarrow 300 \times (x/2) = 13\,(1963)\,(550 - x) \rightarrow x = 232.44$

$k = x/d = 0.423 > k_b$, Therefore, It is an over-reinforced section

Therefore, $\sigma_{cbc} = 7$ N/mm² (full allowable value)

M = Moment of resistance or Allowable Bending moment

$$= \tfrac{1}{2}\,\sigma_{cbc}\,b \times (d{-}x/3) = \tfrac{1}{2}\,(7)\,(300)\,(232.44)\,(550 - x/3) = 116\ \text{kNm}$$

$$Q = M / bd^2 = 116\times10^6 / (300)(550)^2 = 1.27\ \text{N/mm}^2 > 0.9\ \text{N/mm}^2$$

It is observed that for over-reinforced section $Q > Q_b$

Numerical Example 2

Considering a beam section having $b = 300$ mm and $D = 600$ mm and $d = 600 - 50 = 550$ mm

Considering the grade of concrete is M20 and grade of steel reinforcement is Fe 415. 2 - 20 Φ at the bottom.

$$A_{st} = 2\times\left[\pi(20)^2/4\right] = 628\ \text{mm}^2$$

As per IS 456, WSM, allowable stresses are $\sigma_{cbc} = 7$ N/mm² from Table 21, $\sigma_{st} = 230$ N/mm² from Table 22 and m = Modular ratio = 280/3 σ_{cbc} = 280/3 (7) = 13

Considering as a cracked, singly reinforced section and using the expression,

$$\sigma_{cbc}/x_b = \sigma_{st}/m/(d-x_b) \rightarrow k_b = 0.2887,\ \text{ie}\ Q_b = 0.9\ \text{N/mm}^2$$

Using the expression, $b \times (x/2) = m\,A_{st}\,(d - x) \rightarrow 300 \times (x/2) = 13\,(628)\,(550 - x)$ $\rightarrow x = 147.93$

$k = x/d = 0.27 < k_b \rightarrow$ Therefore, it is an under-reinforced section

Therefore, $\sigma_{st} = 230$ N/mm² (full allowable value)

M = Moment of resistance or Allowable bending moment

$$= A_{st}\sigma_{st}\,(d - x/3) = 628 \times 230 \times (550 - x/3) = 72.32\ \text{kNm}$$

$$Q = M/bd^2 = 72.32\times10^6/(300)(550)^2 = 0.797\ \text{N/mm}^2 < Q_b = 0.9\ \text{N/mm}^2$$

It is observed that for under-reinforced section $Q < Q_b$

It may be noted as follows:

If $p < p_b \rightarrow$ Under-reinforced section, If $p > p_b \rightarrow$ Over-reinforced section

If $Q > Q_b \rightarrow$ Over-reinforced section, $Q = Q_b \rightarrow$ Balanced section,

If $Q < Q_b \rightarrow$ Under-reinforced section

Numerical Example 3

Considering a beam section having $b = 300$ mm and $D = 550$ mm and $d = 550 - 50 = 500$ mm

Considering grade of concrete is M20 and grade of steel reinforcement is Fe 415. 3 - 16 Φ at bottom. Grade of concrete is M20 and the grade of steel reinforcement is Fe 415.

For M20 concrete and Fe 415 steel, $\sigma_{cbc} = 7$ N/mm² and $\sigma_{st} = 230$ N/mm² respectively as per WSM of IS456.

Using Eq. 3.3, ie. $p_b = 100\,\sigma_{cbc}\,k_b / 2\,\sigma_{st}$

$$p_b = 100(7)(0.2887)/2(230) = 0.44\%$$

For this problem, $A_{st} = 3 \times [\pi\,(16)^2/4] = 603$ mm²

$$\text{ie.}\ p = 100\,A_{st}/bd = 100(603)/(300)(500) = 0.40\% < p_b = 0.44\%$$

Therefore, it is an under-reinforced.

This is the easiest way to identify whether it is an over-reinforced or under-reinforced section.

Numerical Example 4

Calculate the flexural longitudinal reinforcement required for a simply supported beam of span

$l = 5.3$ m subjected to an imposed uniformly distributed load $w_L = 45$ kN/m over its entire span, using M20 concrete and Fe 250 steel. Diameters of available longitudinal bars are greater than 20 mm diameter.

From Table 22, IS 456, allowable $\sigma_{st} = 130$ N/mm² (for longitudinal bars are greater than 20 mm diameter and Fe 250 grade mild steel), and from Table 21, as per WSM of IS 456, $\sigma_{cbc} = 7$ N/mm² (for M20 concrete).

Therefore, $m = 280 / (3)(7) = 13$, $k_b = 0.418$ (calculated by using methods described earlier), $J_b = (1 - k_b/3) = (1 - (0.418/3)) = 0.86$

$$Q_b = ½\, \sigma_{cbc}\, k_b\, j_b = ½(7)\,(0.418)\,(0.86) = 1.25 \text{ N/mm}^2$$

Considering, a beam section 300 mm × 600 mm, ie. $D = 600$ mm, $d' = 50$ mm and $d = 550$ mm

w_d = Self-weight of beam = $b\, D\, \gamma_{conc} = (0.3)(0.6)(25) = 4.5$ kN/m

$$w = \text{Design load } = w_d + w_L = 45 + 4.5 = 49.5 \text{ kN/m}$$

M = Design bending moment = $wl^2/8 = (49.5)\,(5.3)^2 / 8 = 175$ kNm

For a balance section, $M_b = Q_b\, bd^2 \rightarrow 175 \times 10^6 = (1.25)\,(300)\,(d_b)^2 \rightarrow d_b = 683$ mm

$$d = 550 \text{ mm} < d_b = 683 \text{ mm}$$

Therefore, the beam section is to be designed as an over-reinforced one or doubly reinforced one

Let us go for over-reinforced design concept

For an over-reinforced section,

$$\sigma_{st} < 130 \text{ N/mm}^2 \text{ (allowable value) but } \sigma_{cbc} = 7 \text{ N/mm}^2 \text{ (allowable value)}$$

We know, $M = ½\, \sigma_{cbc}\, b\, x\, (d - x/3) \rightarrow 175 \times 10^6 = ½\,(7)\,(300) \times (550 - x/3) \rightarrow x = 400$ mm

Again, we know,

$$\sigma_{cbc} / x = (\sigma_{st}/m) / (d - x) \rightarrow 7/x = (\sigma_{st}/m) / (550 - x)$$

Substituting the value of x and $\sigma_{cbc} = 7$ N/mm² (allowable value), $\sigma_{st} = 34.125$ N/ mm²

We know $M = A_{st}\ \sigma_{st}\ (d - x/3) \rightarrow 175 \times 10^6 = A_{st}\ \sigma_{st}\ (550 - x/3) \rightarrow A_{st} = 12{,}307.69$ mm²

Total area of longitudinal steel = A_{st} + 2 nos. hanger bars (minimum 12 mm dia. As per IS 456) at the compression face = 12,307 +2 × [π (12)^2 / 4] = 12,533 mm^2

If we now decide to go under-reinforced design concept,

Providing, a beam section 300 mm × 750 mm i.e. b = 300 mm, D = 750 mm, d' = 50 mm

D = 750 – 50 = 700 mm > d_b = 683 mm, Therefore, it is an under-reinforced section.

$$w_d = \text{Self weight of beam} = b\,D\,\gamma_{\text{conc}} = (0.3)(0.75)(25) = 5.625 \text{ kN/m}$$

$$w = \text{Design load} = w_d + w_L = 5.625 + 45 = 50.625 \text{ kN/m}$$

$$M = \text{Design moment} = 50.625(5.3)^2 / 8 = 178 \text{ kNm}$$

We know $M_b = Q_b b d_b^2 \rightarrow 178 \times 10^6 = (1.25)(300)\,d_b^2 \rightarrow d_b = 689$ mm < 700 mm

Therefore, it is confirmed that it is an under-reinforced section

We know, for an under-reinforced section,

$$\sigma_{st} = 130 \text{ N/mm}^2 \text{ (allowable value) but } \sigma_{cbc} < 7 \text{ N/mm}^2 \text{ (allowable value)}$$

We know $\sigma_{cbc} / x = (\sigma_{st}/m) / (d - x) \rightarrow \sigma_{cbc} = (\sigma_{st}\, x) / (m\,(d - x))$

We have $M = ½\, \sigma_{cbc} \times b\,(d - x/3)$

Substituting the expression of σ_{cbc} in the above equation, we get

$$178 \times 10^6 = ½\,(\sigma_{st} x)/\,(m(d - x)) \times b(d - x / 3)$$

ie. $178 \times 10^6 = ½[(130x) / (13(700 - x))] \times (300)(700 - x/3) \rightarrow x = 697.25$ mm

We know $M = A_{st}\ \sigma_{st}\ (d - x/3) \rightarrow 178 \times 10^6 = A_{st}\ (230)\ (700 - x/3) \rightarrow A_{st} = 1654$ mm^2

Total area of longitudinal steel = A_{st+} 2 nos. hanger bars (minimum 12 mm dia.As per IS 456) at the compression face = 1654 + 2 × [π (12)^2 / 4] = 1880 mm^2

Doubly Reinforced Sections

If there is some restriction of depth or width of headroom consideration or some other reasons and demand of the same considering singly reinforced section or economic considerations, doubly reinforced concept is proposed. It is a balanced section having higher moment of resistance than singly reinforced balanced section. By providing additional steel in tension face as well as providing steel in compression face, the moment capacity is increased as required. A typical numerical example is shown in Example 5 (Figure 3.9).

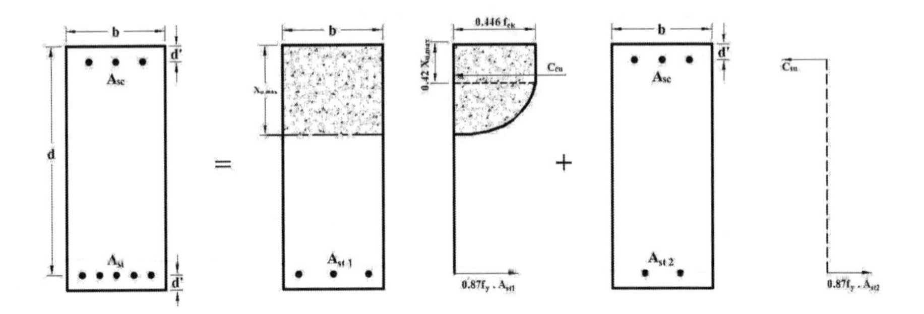

FIGURE 3.9 Stress distribution and force diagram of rectangular section (considering the effect of steel in compression).

Numerical Example 5

Considering, $l = 5.3$ m, $w_L = 45$ kN/m, and beam section 300 mm × 600 mm (same as before)
i.e., $b = 300$ mm, $D = 600$ mm, $d' = 50$ mm, As per WSM of IS456, Allowable. $\sigma_{sc} = 130$ N/mm² and $\sigma_{cbc} = 7$ N/mm² (same as before)

Referring to calculations of previous numerical example, $d = 600 - 50 = 550$ mm $> d_b = 683$ mm

$$w_d = \text{Self weight of beam} = (0.3)(0.6)(25) = 4.5 \text{ kN/m}$$

$$w = 45 + 4.5 = 49.5 \text{ kN/m} \rightarrow M = 49.8(5.3)^2 / 8 = 175 \text{ kNm}$$

Doubly reinforced section is a modified balanced section. Therefore, neutral axis depth is equal to the balanced depth x_b.

Referring to calculations of the previous numerical example, $k_b = 0.418$ and $d = 550$ mm,

$$x_b = k_b d$$

$$M_b = ½\, \sigma_{cbc}\, x_b b(d - x/3) = Q_b bd^2 \text{ where, } Q_b = 1.25 \text{ N/mm}^2,\ x_b = 0.418 \times 550 = 230 \text{ mm}$$

$$M_b = (1.25)(300)(550)^2 = 114 \times 10^6 \text{ N-mm}$$

$$A_{st1} = M_b / \sigma_{st}\,(d - x_b/3) = 114 \times 10^6/(130)(550 - 230/3) = 1857 \text{ mm}^2$$

$$M - M_b = A_{st2}\, \sigma_{st}\,(d - d')$$

$$A_{st2} - (M - M_b) / \sigma_{st}\,(d - d') = (175 \times 10^6 - 114 \times 10^6)/(130)\,(550 - 50) = 935 \text{ mm}^2$$

$$A_{st1} + A_{st2} = 1857 + 935 = 2792 \text{ mm}^2$$

Stress at compression steel level $\sigma'_{cbc} = (x_b - d' / x_b)\,\sigma_{cbc}$

$$C_{cs} = (1.5m - 1)\,A_{sc}\sigma'_{st}\;(1.5m - 1)\,A_{sc}\,(x_b - d'/x_b)\sigma_{cbc}$$

$$A_{st2}\,\sigma_{st} = (1.5m - 1)\,A_{sc}\,(x_b - d'/x_b)\,\sigma_{cbc}$$

$$(935)(130) = \left(1.5(13) - 1\right) A_{sc}\left(230 - 50 / 230\right)(7) \rightarrow A_{sc} = 1168 \text{ mm}^2$$

Total area of longitudinal steel = $A_{st} + A_{sc}$ = 2792 + *1168 = 3960 mm^2
*It may be noted here that this area is more than the cross-sectional area of 2 nos.–12 mm round bars (Minimum requirement).

Singly Reinforced Non-rectangular Sections

Analysis of Singly Reinforced Flanged Sections

A portion of the slab acts integrally with the beam and bends in the longitudinal direction of the beam. This slab portion is termed as flange of the T or L beam. The beam portion below the flange is termed as web. While using theory of flexure, a uniform stress distribution across the width of the section is considered and IS456 defined effective flange width b_f (cl. 23.1.2(c)), where b_w is the width of the web, D_f is the thickness of the flange (i.e., Slab thickness), and l_o is the distance between points of zero moments in the beam, assumed as 0.7 times the effective span in case of continuous beams and frames (Figures 3.10 and 3.11)

Flange Width as per IS 456

$$b_f = l_o / 6 + b_w + 6D_f, \text{ for T-beam}$$

$$= l_o / 12 + b_w + 3D_f, \text{ for L-beam}$$

$$= l_o / \left(l_o / b + 4\right) + b_w, \text{ for Isolated T-beam}$$

$$= 0.5 l_o / \left(l_o / b + 4\right) + b_w, \text{ for Isolated L-Beam}$$

If kd > D_f

$$\left(b_f - b_w\right) D_f \left(kd - D_f / 2\right) + b_w (kd)^2 / 2 = mA_{st}(d - kd)\frac{1}{2}\sigma_{cbc}(kd)$$

$$-\frac{1}{2}\sigma_{cbc1}\left(b_f - b_w\right)\left(kd - D_f\right) = A_{st}\sigma_{st}$$

$$\sigma_{cbc1} = \sigma_{cbc}\frac{\left(kd - D_f\right)}{kd}$$

$$M = \frac{1}{2}\sigma_{cbc} b_f (kd)\left(d - \frac{kd}{3}\right) - \frac{1}{2}\sigma_{cbc1}\left(b_f - b_w\right)\left(kd - D_f\right)\left(d - D_f - \frac{\left(kd - D_f\right)}{3}\right)$$

(a)

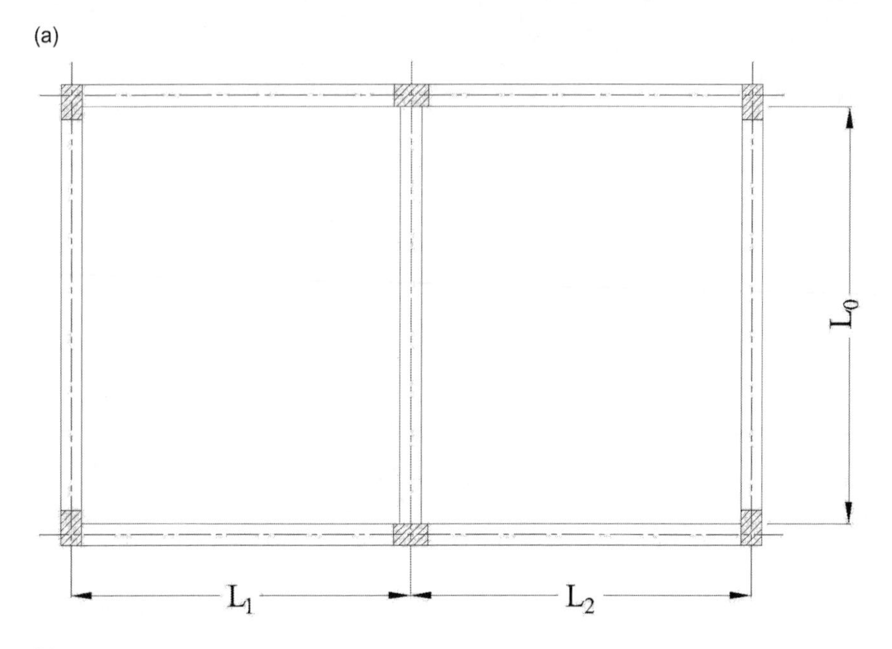

(b)

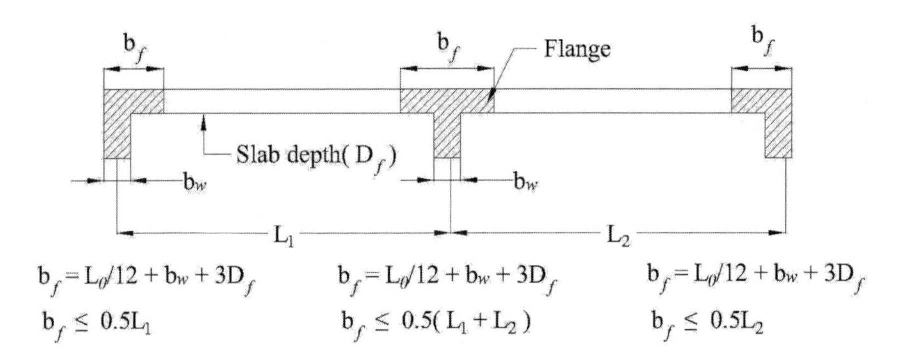

FIGURE 3.10 T-beam and L-beam action.

If kd < D_f

$$(b_f - b_w)D_f(kd - D_f/2) + b_w(kd)^2/2 + (1.5m - 1)A_{sc}(kd - d') = mA_{st}(d - kd)$$

$$C_c = 0.5\sigma_{cbc}\left[b_f(kd) - (b_f - b_w)\frac{(kd - D_f)^2}{kd}\right]$$

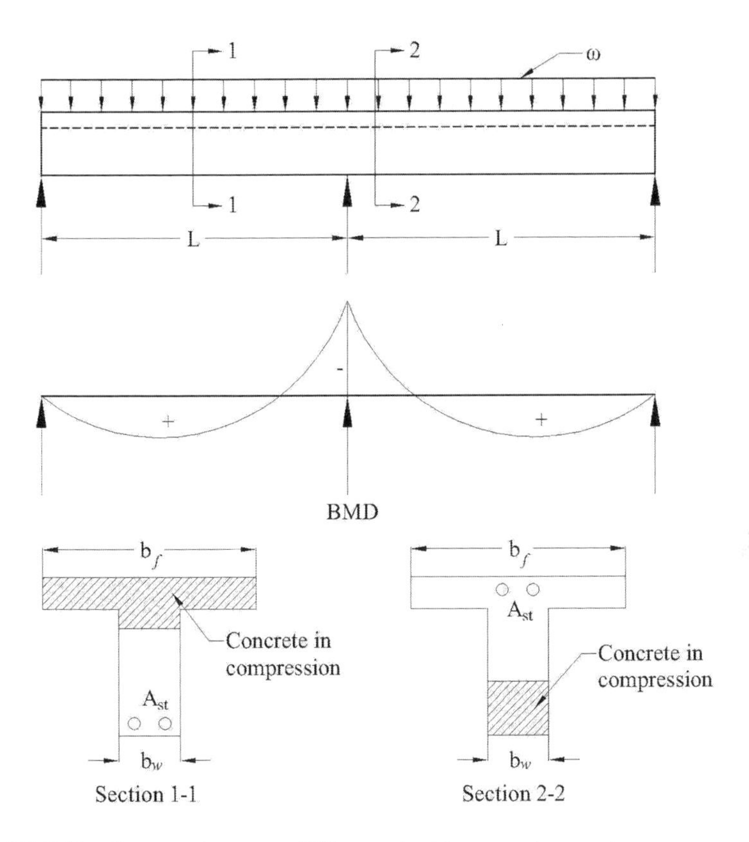

FIGURE 3.11 Compression zone of T-beam at mid-span and support.

$$M = \frac{1}{2}\sigma_{cbc}\left[b_f(kd)\left(d - \frac{kd}{3}\right) - (b_f - b_w)\frac{(kd - D_f)^2}{kd}\left(d - D_f - \frac{(kd - D_f)}{3}\right)\right]$$

$$+ (1.5m - 1)A_{sc}\sigma_{cbc}\frac{(kd - d')}{kd}(d - d')$$

Doubly Reinforced Flanged Beams

Sometimes the dimensions of the web are restricted from architectural or other considerations. If the beam is expected to carry more bending moment than it can take as a balanced section, compression reinforcement is provided to increase its moment of resistance.

Such a beam is known as doubly reinforced T-beam.

A_{sc} = area of compressive reinforcement
d_c = cover for the compressive reinforcement
c' = compressive stress in concrete surrounding compressive steel.

The design of such a beam is done in the following steps:

- Find the moment of resistance M_1 of the T-beam as the balanced section.

$$M_1 = b_f \cdot D_f \cdot \frac{c}{2}\left[1+\frac{kd-D_f}{kd}\right]\left[d-\frac{D_f}{2}+\frac{D_f^2}{6\left(2kd-D_f\right)}\right]$$

- Calculate the moment of resistance (M_2) to be provided by the compressive steel.

$$M_2 = M - M_1, \text{ where } M = \text{Design bending moment}$$

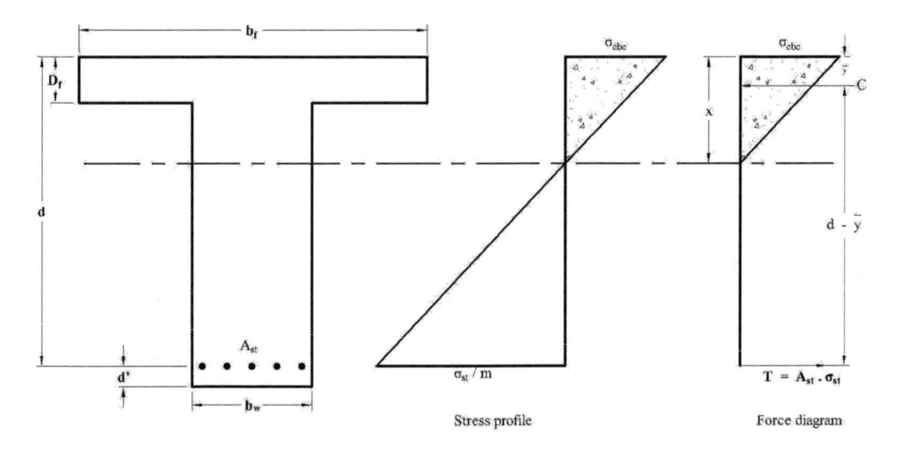

FIGURE 3.12 Stress distribution and force diagram of singly reinforced flange section.

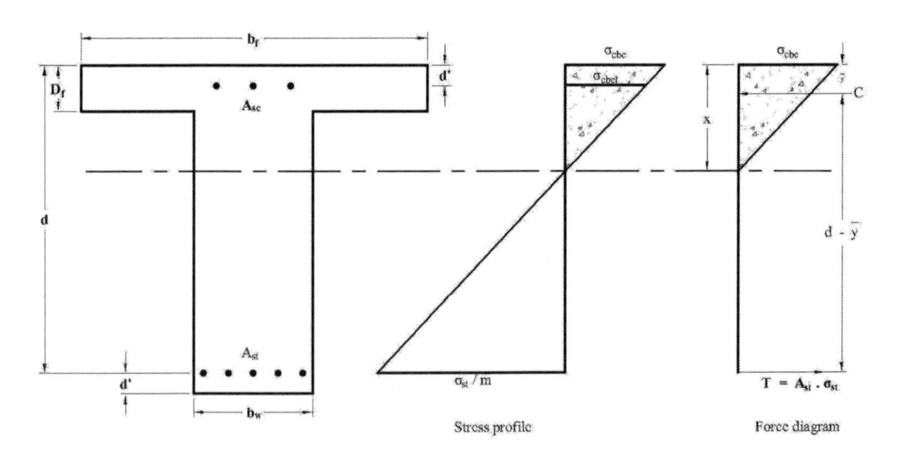

FIGURE 3.13 Stress distribution and force diagram of flange section (considering the effect of steel in compression).

$$\text{But } M_1 = (m_c - 1)A_{sc} \times c'(d -$$

where $c' = \dfrac{kd - d}{kd} \cdot c$ and $m_c = 1.5\,\text{m}$

$$M_2 = (m_c - 1)A_{sc} \times c \times \frac{kd - d_c}{kd}(d - d_c)$$

Thus, A_{sc} is known from this equation.

- Find the tensile reinforcement as under

$$A_{sc} = \frac{M_1}{\sigma_{st} \times a}\left(\text{ where } t = \sigma_{st}\right)$$

$$a = d - \frac{D_f}{2} + \frac{D_f^2}{6(2kd - D_f)}$$

$$A_{st2} = \frac{M_2}{\sigma_{st}(d - d_c)}$$

Total area of steel $A_{st} = A_{st} + A_{st2}$

For a doubly reinforced T-beam of given dimensions, the neutral axis can be located by equating the moments of equivalent areas about the neutral axis.

$$\text{Thus}: b_f \cdot D_f\left(n - 1\frac{D_f}{2}\right) + (m_c - 1)A_{sc}(n - d_c) = m \cdot A_{st}(d - n)$$

(Assuming the neutral axis to lie in the web)

The moment of resistance of such a beam, neglecting the compressive force in the rib, is given by

$$M_r = b_f \cdot D_f\left(\frac{c + c_1}{2}\right) \times a + (m_c - 1)A_{sc} \times c\frac{(n - d_c)}{n}(d - d_c)$$

Taking 'a' equal to $d - \dfrac{D_f}{2} + \dfrac{D_f^2}{6(2n - D_f)}$, the above expression reduces to:

$$M_r = b_f \cdot D_f \cdot \sigma_{cbc}$$

$$\left[\frac{2n - D_f}{2n}\right]\left[d - \frac{D_f}{2} + \frac{D_f^2}{6(2n - D_f)}\right. + (m_c - 1)A_{sc} \times c\frac{(n - d_c)}{n}(d - d_c)$$

where $c = \sigma_{cbc}$ = permissible compressive stress in concrete.

Example 6

An isolated T-beam carries a uniformly distributed load of 45 kN/m run, inclusive of its own weight, over an effective span of 6 m. The beam has the following dimensions: thickness of slab = 100 mm; effective depth of the beam = 480 mm and width of beam = 300 mm. Determine required tensile and compressive reinforcement. M20 grade concrete and Fe415 steel.

As per WSM of IS456, $\sigma_{cbc} = 7$ N-mm^2 $\sigma_{st} = 230$ N/mm^2, $m = 280/3\ \sigma_{cbc} = 13$.
The effective width of an isolated T-beam as per IS456

$$b_f = \frac{l_o}{\left(\frac{l_0}{b}\right)+4} + b_w = \frac{6000}{\frac{6000}{800}+4} + 300 = 822 \text{ mm. say } b_f = 800 \text{ mm}$$

$$\text{Design Bending moment} = \frac{wL^2}{8} = \frac{45 \times 6^2}{8} = 202.5 \text{ kN-m} = 202.5 \times 10^6 \text{ N-mm}$$

$$\text{For balanced section, } n = kd = \frac{m\sigma_{cbc}}{m\sigma_{cbc} + \sigma_{st}} d = \frac{13 \times 7}{13 \times 7 + 230} \times 480 = 136 \text{ mm.}$$

The moment of resistance of the balanced section is

$$M_r = b_f \cdot D_f \frac{\sigma_{cbc}}{2}\left[\frac{2kd - D_f}{kd}\right]\left[d - \frac{D_f}{2} + \frac{D_f}{6(2kd - D_f)}\right]$$

$$= 800 \times 100 \times \frac{7}{2}\left[\frac{2 \times 136 - 100}{136}\right]\left[480 - 50 + \frac{100 \times 100}{6(2 \times 136 - 100)}\right]$$

$$= 354117.65 \times [439.69] = 155.7 \times 10^6 \text{ N-mm}$$

$$A_{st} = \frac{M_1}{\sigma_{st} \times a} = \frac{155.7 \times 10^6}{230 \times 439.69} = 1539.6 \text{ mm}^2$$

$$M_2 = M - M_1 = (202.5 - 155.7)10^6 = 46.8 \times 10^6 \text{ N-mm}$$

$$46.8 \times 10^6 = (m_c - 1) A_{sc} \times c\left[\frac{kd - d}{kd}\right](d - d_c)$$

Let, $d_c = 40$ mm,

$$A_{sc} = \frac{\left(46.8 \times 10^6\right) \times 136}{(1.5 \times 13 - 1) \times 7(136 - 40)(480 - 40)} = 116.4 \text{ mm}^2$$

and

$$A_{st}{}^2 = \frac{M_2}{\sigma_{st}(d - d_c)} = \frac{46.8 \times 10^6}{230(480 - 40)} = 462 \text{ mm}^2$$

Total tensile steel $A_{st} = 1539.6 + 462 = 2002\,mm^2$ and compression steel $A_{sc} = 116.4\,mm^2$

Example 7

A reinforced T-beam section, reinforced with 5 nos-25ϕ Fe 415 bars at tension face 3 nos-25ϕ Fe 415 bars at compression face. Determine the moment of resistance of the section. The beam is used over an effective span of 6 m. Considering M20 grade concrete and Fe415 steel.

As per WSM of IS456, $\sigma_{cbc} = 7$ N-mm^2 $\sigma_{st} = 230$ N/mm^2, $m = 280/3\ \sigma_{cbc} = 13$.

As per IS 456, the effective width of flange

$$b_f = \frac{l_o}{\left(\frac{l_o}{b}\right)+4} + b_w = \frac{6000}{\frac{6000}{1200}+4} + 300 = 966.67 \text{ mm}$$

Let the neutral axis lie in the web at a distance (n) below the top of the flange. Neglecting the contribution of web. Taking the moment of equivalent areas about neutral axis, it may be written as follows

$$b_f \cdot D_f\left(n - \frac{D_f}{2}\right) + (m_c - 1)A_{sp}(n - d_c) = m \cdot A_{st}(d - n)$$

where $A_{sc} = 3\frac{\pi}{4}(25)^2 = 1473 \text{ mm}^2$ and $A_{st} = 5\frac{\pi}{4}(25)^2 = 2454 \text{ mm}^2$

$$966.67 \times 100(n - 50) + (1.5 \times 13 - 1)1473(n - 40) = 13 \times 2454(500 - n)$$

or,

$$(966.67 \times 100)n - (966.67 \times 100 \times 50) + (18.5 \times 1473)n - (18.5 \times 1473 \times 40)$$

$$= 13 \times 2454 \times 500 - 13 \times 2454 \times n$$

$$155{,}819.5\ n = 21{,}874{,}370 \rightarrow n = 140.38 \text{ mm}$$

Depth of critical N.A. is given by

$$n_c = \frac{m\sigma_{cbc}}{m\sigma_{cbc} + \sigma_{st}} d = \frac{13 \times 7}{13 \times 7 + 230} \times 500 = 141.74 \text{ mm}$$

Since the actual N.A. falls above the critical N.A., the section is under-reinforced and the stress in steel will reach its maximum value first. Hence, $\sigma_{st} = 230$ N/mm^2.

$$C = \frac{\sigma_{st}}{m}\frac{n}{d-n} = \frac{230}{13}\times\frac{140.38}{500-140.38} = 6.91\ \text{N/mm}^2$$

$$\text{Moment of resistance } = M_r = b_f \cdot D_f\frac{c}{2}\left[1+\frac{n-D_f}{n}\right]\left[d-\frac{D_f}{2}+\frac{D_f^2}{6(2n-D_f)}\right]+$$

$$(m_c-1)A_{sc}\times c\frac{(n-dc)}{n}(d-d_c)$$

$$= 966.67\times 100\times\frac{6.91}{2}\left[1+\frac{140.38-100}{140.38}\right]\left[500-\frac{100}{2}+\frac{(100)2}{6(2\times 140.38-100)}\right]$$

$$+(1.5\times 13-1)1473\times 6.91\frac{(140.38-40)}{140.38}(500-40) = 197{,}489{,}719.2+61{,}937{,}305.43$$

$$= 259{,}427{,}024.6\,\text{N-m} = 259.43\ \text{kN-m}.$$

Let us consider the contribution of web. Taking the moment of equivalent areas about N.A., it may be written as follows:

$$b_f\cdot D_f\left(n-\frac{D_f}{2}\right)+\frac{b_w(n-D_f)2}{2}+(m_c-1)A_{sc}(n-d_c) = m\cdot A_{st}(d-n)$$

$$966.67\times 100(n-50)+150\times(n-100)^2+(1.5\times 13-1)1473(n-40)$$

$$= 13\times 2454\times(500-n)$$

$$150n^2-2\times 100\times 150n+150\times 100^2+966.67\times 100n+18.5\times 1473n+13\times 2454n$$

$$-966.67\times 100\times 50-18.5\times 1473\times 40-13\times 2454\times 500 = 0$$

$$150n^2+125{,}819n-20{,}374{,}370 = 0 \rightarrow n = 138.93\ \text{mm, as found earlier, } n_c$$

$$= 141.74\ \text{mm}.$$

The section is therefore under-reinforced, so $\sigma_{st} = 230$ N/mm^2

The corresponding stresses are as follows:

$$\sigma_{cbc} = \frac{\sigma_{st}}{m}\frac{n}{d-n} = \frac{230}{13}\times\frac{138.93}{500-138.93} = 6.81\ \text{N/mm}^2.$$

$$\sigma'_{cbc} = \frac{\sigma_{st}}{m}\frac{n-d_c}{d-n} = \frac{230}{13}\times\frac{138.93-40}{500-138.93} = 4.85\ \text{N/mm}^2.$$

$$c_1 = \frac{\sigma_{st}}{m}\frac{n-D_f}{d-n} = \frac{230}{13}\times\frac{138.93-100}{500-138.93} = 1.91\ \text{N/mm}^2.$$

The moment of resistance of the beam will be equal to the sum of the moment of resistance due to (i) compressive force of flange, (ii) compressive force of rib, and (iii) compressive force of steel.

$$M_r = M_{r1} + M_{r2} + M_{r3}$$

Now M_{r1} = Moment of resistance due to compressive in flange = $b_f \cdot D_f \left(\frac{c+c_1}{2}\right)\left(d - y^-\right)$

$$\text{where } y^- = \frac{c+2c_1}{c+c_1} \times \frac{D_f}{3} = \frac{6.81+3.82}{6.81+1.91} \times \frac{100}{3} = 40.63 \text{ mm}$$

$$\mathrm{M_{r1}} = 966.67 \times 100 \times \left(\frac{6.81+1.91}{2}\right) \times (500 - 40.63) = 193.61 \times 10^6 \text{ N-mm}$$

M_{r2} = Moment of resistance due to compressive force in steel

$$= b_W \left(n - D_f\right) \frac{\sigma'_{cbc}}{2}\left[d - D_f - \frac{1}{3}\left(n - D_f\right)\right]$$

$$= 300(138.93 - 100)\frac{1.91}{2}\left[500 - 100 - \frac{1}{3} \times (138.93 - 100) = 4.32 \times 10^6 \text{ N-mm}\right.$$

M_{r3} = Moment of resistance due to compressive force in steel.

$$= \left(m_c - 1\right) A_{sc} . \sigma'_{cbc} \left(d - d_c\right) = \left(1.5 \times 13 - 1\right) \times 1473 \times 4.85 \times \left(500 - 40\right) = 60.8 \times 10^6 \text{ N-mm}$$

Total $M = (193.61 + 4.32 + 60.8)\ 10^6 = 258.73 \times 10^6$ N-mm = 258.73 kN-mm

Example 8

A simply supported T-beam having a span of 6m and cross-sectional dimensions $D = 600$ mm, $d = 600–50 = 550$ mm, $b_f = 1000$ mm, $D_f = 100$ mm, $b_w = 250$ mm is subjected to bending moment of 100 kNm. Compute the maximum stresses in concrete and steel, assuming M20 concrete and Fe 415 steel & reinforced with 2 nos. 25Φ. Find also moment of resistance.

As per WSM of IS456, $\sigma_{cbc} = 7$ N-mm^2 $\sigma_{st} = 230$ N/mm^2, $m = 280/3\ \sigma_{cbc} = 13$.

$$A_{st} = 2 \times (\pi / 4) \times 25^2 = 982 \text{ mm}^2$$

Assuming $kd \le D_f$ and equating moments of compression area and tension area about the neutral axis,

$$b_f \times (kd)^2 / 2 = m \times A_{st} \times (d - kd)$$

$$\Rightarrow 1000 \times (kd)^2 / 2 = 13 \times 982 \times (550 - kd)$$

$$\Rightarrow kd = 106.42 \text{ mm} > D_f = 100 \text{ mm}$$

Therefore, the assumption of $kd \le D_f$ made at the beginning is incorrect.

Considering, $kd > D_f$, the neutral axis is located in the web

$$\left(b_f - b_w\right) \times D_f \times \left(kd - D_f / 2\right) + b_w \times (kd)^2 / 2 = m \times A_{st} \times (d - kd)$$

$$\Rightarrow (1000-250)\times 100\times (kd-100/2)+250\times (kd)^2/2=13\times 982\times (550-kd)$$

$$\Rightarrow kd = 106.56 \text{ mm}$$

Compressive force $C = 0.5 \times \sigma_{cbc} \times [b_f \times (kd) - (b_f - b_w) \times (kd - D_f)^2 / (kd)]$

Taking moments of forces about tension steel centroid,

$$M = 0.5\times \sigma_{cbc}$$

$$\times\left[b_f\times(kd)\times(d-kd/3)-(b_f-b_w)\times(kd-D_f)^2/(kd)\times\left(d-D_f-(kd-D_f)/3\right)\right]$$

$$\Rightarrow 100\times 10^6 = 0.5\times\sigma_{cbc}\times\left[1000\times(106.56)\times(550-106.56/3)\right.$$

$$\left.-(1000-250)\times(106.56-100)^2/(106.56)\times\left(550-100-(106.56-100)/3\right)\right]$$

$$\Rightarrow \sigma_{cbc} = 3.66 \text{ N/mm}^2 < 7 \text{ N/mm}^2$$

We know that $C = T$

$$0.5\times\sigma_{cbc}\times\left[b_f\times(kd)-(b_f-b_w)\times(kd-D_f)^2/(kd)\right]=A_{st}\times\sigma_{st}\,0.5\times 3.66$$

$$\times\left[1000\times(106.56)-(1000-250)\times(106.56-100)^2/(106.56)\right]=982\times\sigma_{st}$$

$$\Rightarrow \sigma_{st} = 198 \text{ N/mm}^2 < 230 \text{ N/mm}^2$$

It is an under-reinforced section. So, $\sigma_{st} = 230$ N/mm²

Stress in concrete, $\sigma_{cbc} = kd/(d - kd) \times (1/m) = 106.56/(550 - 106.56) \times (1/13) = 4.25$ N/mm²

Hence, moment of resistance $(M) = 0.5\times f_c\times[b_f\times(kd)\times(d-kd/3)-(b_f-b_w)$

$$\times(kd-D_f)^2/(kd)\times(d-D_f-(kd-D_f)/3)]$$

$$= 0.5\times 4.25\times[1000\times(106.56)\times(550-106.56/3)-(1000-250)$$

$$\times(106.56-100)^2/(106.56)$$

$$\times(550-100-(106.56-100)/3)] = 116.21 \text{ N/mm}^2$$

Example 9

A triangular section having effective depth d and width b at the level of the reinforcement. Let the neutral axis be at a depth of n below the apex. At depths x below the apex, the width b_x and stress (Figure 3.14)

$$b_x = b / dx, \quad \sigma_{\text{cbc}\,x} = kd - x / kd \times \sigma_{\text{cbc}}$$

Hence, the compressive force dC of elementary strip of thickness dx is given by

$$dC_x = b_x \cdot \sigma_{\text{cbc}\,x} \cdot d_x$$

The total compressive force is given by

$$C = \int dC_x = \int_0^{kd} bx \cdot cx \cdot dx = \int_0^{kd} \frac{b}{d} x \cdot \frac{kd}{kd} - \frac{x}{kd} \sigma_{\text{cbc}\,x}\, dx = \sigma_{\text{cbc}\,x} d / kd^2 \int_0^{kd} x \cdot (kd - x)\, dx$$

$$= \sigma_{\text{cbc}\,x} d / kd^2 \left[kd \cdot x^2 / 2 - x^3 / 3 \right]_0^{kd} = k^2 bd / 6 \times \sigma_{\text{cbc}}$$

Taking $c = \sigma_{\text{cbc}}$ and $k = k_c$ for a balanced design

$$\sigma_{\text{cbc}\,x} = k_c^{\,2} bd / 6 \cdot \sigma_{\text{cbc}}, T = A_{\text{st}} \cdot \sigma_{\text{st}}$$

Equating

$$A_{\text{st}} \cdot \sigma_{\text{st}} = k_c^{\,2} \cdot bd / 6 \cdot \sigma_{\text{cbc}} / \sigma_{\text{st}} ---- A_{\text{st}} = k_c^{\,2} \cdot bd / 6 \cdot \sigma_{\text{cbc}} / \sigma_{\text{st}}$$

In determining reinforcement for a balanced section, where k_c is given by

$$k_c = m \cdot \sigma_{\text{cbc}} / m\sigma_{\text{cbc}} + \sigma_{\text{st}} = 1 / 1 + 0.0107\sigma_{\text{st}}$$

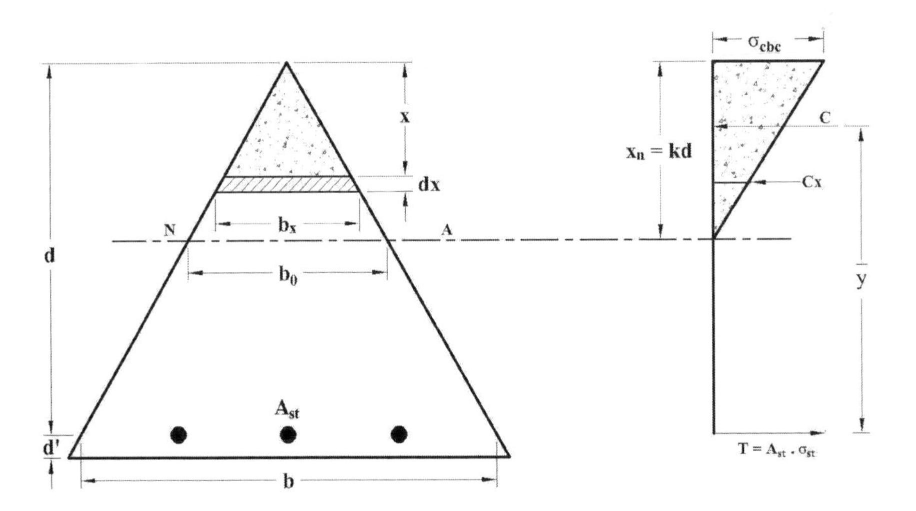

FIGURE 3.14 Section of triangular beam.

In order to find the moment of resistance of the section, taking the moment of compressive force about the center of tensile steel.

$$M_x = \int_0^{kd} (d-x)\,dCx = \int_0^{kd} (d-x)bx \cdot Cx\,dx$$

$$= \int_0^{kd} (d-x)\cdot \frac{b}{d} x \cdot \frac{kd-x}{kd} \cdot c\,dx = \frac{bc}{kd^2}\int_0^{kd} x\cdot (kd-x)(d-x)\,dx$$

$$= \frac{b\sigma_{\text{cbc}}}{kd^2}\left[kd^2 \cdot \frac{x^2}{2} - d(k+1)\cdot \frac{x^3}{3} + \frac{x^4}{4} \right]_0^{kd}$$

$$M_r = \frac{b\sigma_{\text{cbc}}}{kd^4}\left[\frac{kd^2}{2}\cdot (kd)^2 - \frac{d(k+1)}{3}(kd)^3 + \frac{1}{4}(kd)^4 \right]$$

$$M_r = \frac{\sigma_{\text{cbc}} \cdot k^2(2-k)}{12} bd^2$$

$$Q = \frac{kc^2(2-Kc)}{12}\cdot \sigma_{\text{cbc}} = K \cdot \sigma_{\text{cbc}}$$

$$K = \frac{1}{12} Kc^2(2-kc)$$

Example 10

TRAPEZOIDAL BEAM SECTION

The moment of resistance for a trapezoidal section can be found by taking the sum of M_r for the middle rectangular and $M_{r'}$ for the triangular section (Figure 3.15).

Thus, $M_r = M_{r1} + M_{r2}$

$$M_{r1} = \text{Moment of resistance of triangular portion} = \frac{ck^2(2-k)}{12}(b-b_1)d^2$$

$$M_{r2} = \text{Moment of resistsnce of triangular portion} = \frac{1}{2}c\left(1-\frac{k}{3}\right)kb_1 d^2$$

$$M_r = \left[\frac{ck^2(2-k)}{12}(b-b_1) + \frac{1}{2}c\frac{1-k}{3}kb_1 \right] d^2$$

For a balanced section $c = \sigma_{\text{cbc}}$ and $k = K_c$

$$M_r = \left[\sigma_{\text{cbc}} \frac{kc^2(2-k_c)}{12}(b-b_1) + \frac{1}{2}\sigma_{\text{cbc}}\left(1-\frac{k_c}{3}\right)k_c \cdot b_1 \right] d^2$$

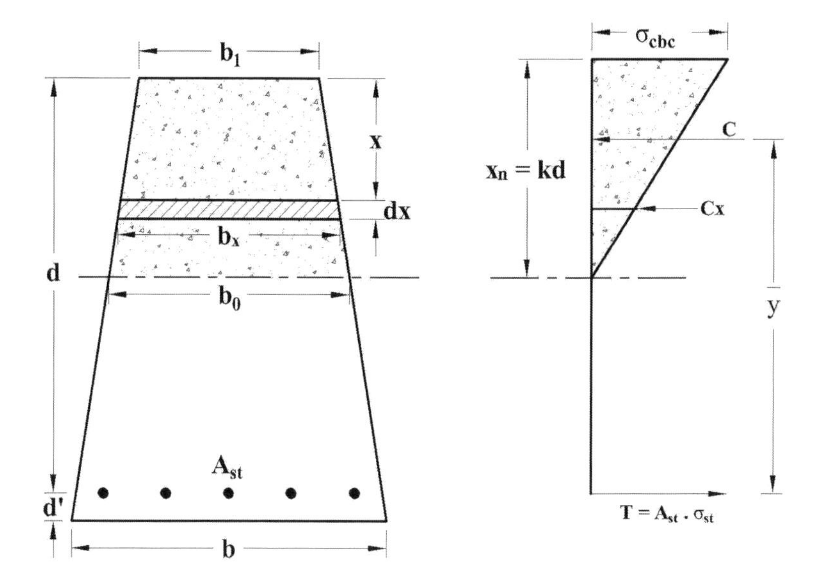

FIGURE 3.15 Section of trapezoidal beam.

or,

$$M_r = [Q(b - b_1) + Rb_1]d^2$$

where $Q = K_c^2\ (2 - k_c) / 12 \cdot \sigma_{cbc}$

$$R = \frac{1}{2}\sigma_{cbc} \cdot \left(1 - \frac{kc}{3}\right) \cdot kc$$

and

$$k_c = \frac{m\sigma_{cbc}}{m\sigma_{cbc} + \sigma_{st}} = \frac{1}{1 + 0.0107\sigma_{st}}$$

The total compressive force is given by $C = C_1 + C_2$

where C_1 = compressive force for triangular portion $= \frac{K^2}{6}c\ (b - b_1)d$

and C_2 = compressive force for rectangular portion $= \frac{1}{2}cb_1\ kd$

$$\text{Since, } c\ = \frac{k^2}{6}c(b - b_1)d + \frac{1}{2}cb_1kd$$

$$c = \frac{kc\,d}{2}\left[b_1 + \frac{kc(b - b_1)}{3} + b_1\right]c$$

For a balanced section, $c = \sigma_{cbc}$ and $k = k_c$

$$C = \frac{kc\,d}{2}\left[b_1 + \frac{kc(b-b_1)}{3}\right]\sigma_{cbc}$$

Total tensile force, for a balanced section is $T = A_{st} \cdot \sigma_{st}$

$$\text{Equating, } A_{st} = \frac{kc \cdot d}{2}\left[b_1 + \frac{kc}{3}(b-b_1)\right]\sigma_{cbc} / \sigma_{st}.$$

If the dimensions of the section are given, the M_r and actual stresses can be determined by first locating neutral axis. This can be done by equating the moment of compressive area about neutral axis to the moment of equivalent tensile area about neutral axis.

$$\text{Moment of compressive area about neutral axis} = \frac{\frac{1}{6}(b-b_1)}{d} \cdot n^3 + \frac{b_1 n^2}{2}$$

$$\text{Moment of equivalent tensile area about neutral axis} = m \cdot A_{st}(d-n)$$

$$\text{Equating, } \frac{1}{6}\frac{(b-b_1)}{d} \cdot n^3 + \frac{b_1 n^2}{2} - m \cdot A_{st}(d-n) = 0$$

This is a cubic equation – the solution of which (by trial) would yield the value of n.

Hence, $k = n/d$

The depth if critical neutral axis is given by $n_c = \frac{m\sigma_{cbc}}{m\sigma_{cbc} + \sigma_{st}} \cdot d = kc\,d$

i. If $n = n_c$, the section is balanced and M_r can be obtained
ii. If $n < n_c$, the section is under-reinforced and stress in steel will reach its maximum value first.

$$\sigma_{cbc} = \frac{\sigma_{st}}{m} \cdot \frac{n}{d-n} \quad \text{Knowing } c, M_r \text{ can be obtained}$$

iii. If $n > n_c$, the section is over-reinforced, and stress in concrete will reach its maximum value first, i.e., σ_{cbc}. The corresponding stress in steel $= \sigma_{cbc} \cdot m\,(d-n)/n$

Therefore,

$$M_r = (b-b_1) + \frac{1}{2}\left(1-\frac{k}{3}\right)kb_1]\sigma_{cbc} \cdot d^2 \quad (\text{where } k = n/d)$$

TRAPEZOIDAL SECTION WITH $b_1 > b$

Since the area of concrete in tension zone is not useful, it is more economical to keep b_1 greater than b. All the equations will be valid, except that the term containing $(b - b_1)$ factor will be negative. Alternatively, these equations can be rearranged in the following form:

$$M_r = \left[\frac{1}{2}c\left(1-\frac{k}{3}\right)kb_1 - \frac{ck^2(2-k)}{12}(b_1-b)\right]d^2$$

$$M_r = \left[Rb_1 - Q(b_1-b)\right]d^2$$

$$c = \frac{kd}{2}\left[b_1 - \frac{kc}{3}(b_1 - b)\right]c$$

$$c = \frac{kc\,d}{2}\left[b_1 - \frac{kc}{3}(b_1 - b)\right]\sigma_{cbc}$$

$$A_{st} = \frac{kc \cdot d}{2}\left[b_1 - \frac{kc}{3}(b_1 - b)\right]\sigma_{cbc} / \sigma_{st}$$

BEHAVIOR OF REINFORCED CONCRETE UNDER SHEAR AND BOND

The shear-resisting mechanisms are related to the bond between concrete and the embedded reinforcement as well as anchorage at the end. Shear capacity depends on the tensile and compression strength of the concrete. Shear failure is non-ductile. For earthquake-resistant RCC structures, sufficient attention was paid to ductility through IS13920 and IS 4326. The designer must ensure that a shear failure should not occur as well and that ductility is ensured. The analogy between the shear resistance of a parallel chord truss and a web-reinforced concrete beam is one of the concepts. The analogy, postulated implies that the web of the equivalent truss consists of stirrups acting as tension members and concrete struts running parallel to diagonal. The flexural concrete compression zone and the flexural reinforcement form the top and bottom chords of this analogous pin-jointed truss. The forces in the truss can be determined from considerations of equilibrium only. The slope of the compression diagonals has been traditionally assumed to be 45° to the beam axis. It has been observed, however, that the slope of the diagonal cracks at the boundaries of the struts varies along the beam. This is often the case, and design equations based on compression struts at 45° are conservative. On the other hand, the struts are steeper in the vicinity of point loads. However, in these areas, local arch action boosts the capacity of the other shear-carrying mechanisms. Generally, a beam having high concrete strength and low web steel content represents a less rigid tension system. The compression struts arc at an angle less than 45°, hence the stirrups arc is more effective than in a 45° truss. The transverse (external) shear force is denoted as V (and has a maximum value near the support, equal to the support reaction) (Figure 3.16).

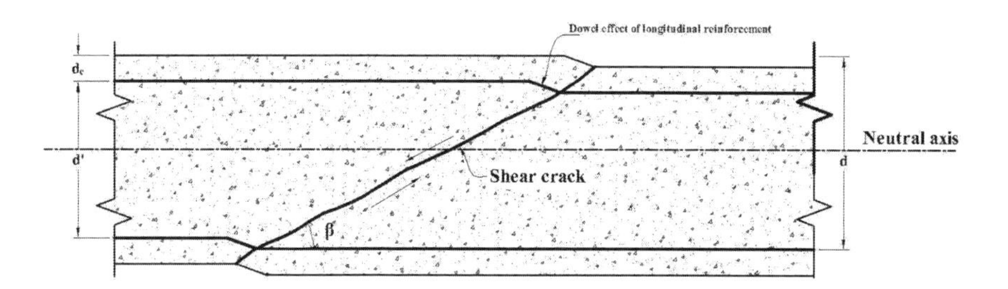

FIGURE 3.16 Shear failure in reinforced concrete section.

1. Shear resistance V_{cz} of the uncracked portion of concrete
2. Vertical component of the 'interface shear' (aggregate interlock) force V_a
3. Dowel force V_d in the tension reinforcement (due to dowel action)
4. Shear resistance V_s carried by the shear (transverse) reinforcement

The interface shear V_a is a tangential force transmitted along the inclined plane of the crack, resulting from the friction against relative slip between the interlocking surfaces of the crack. Its contribution can be significant if the crack width is limited. The dowel force V_d comes from 'dowel action' of longitudinal reinforcement.

When a beam is loaded with transverse loads, the bending moment varies from section to section. Shearing stresses in the beam are caused by this variation of bending moment along the span. Let us consider, two vertical sections 'dx' apart, of a homogeneous beam, the bending moments are M and $M + \delta M$. Hence, the bending stresses at section where bending moment is $M + \delta M$ is greater than those where the bending moment is M. Thus, at the same fiber, there are unequal bending stresses at two cross-sections distance 'dx' apart. This inequality of stresses produces a slip tendency in each fiber to slide over adjacent fiber in a horizontal plane, causing horizontal shear stress which is accompanied by the complementary shear stress in the vertical direction.

In the case of homogeneous beams, the shear stress q at any plane is given by

$$q = V(Ay) / I\ b$$

V = Shear force at section, I = Moment of inertia of the beam section
b = width of the section, Ay = moment of area above the section, about neutral axis.

Due to this variation of bending moment in a length ∂x, the compressive forces in concrete at these two sections will be C and $C + \partial C$, and the tensile forces in steel will be T and $T + \partial T$. Let us assume that the shear force V remains constant in the length dx. Let us assume that concrete does not take any tension. Let q be the intensity of shear at the plane. Hence total horizontal shear = $q{\cdot}b{\cdot}\partial x$ at a layer. Also, the total horizontal force that tends to slide this layer past the adjacent one is equal to $(T + \partial T) - (T) = \partial T$.

Hence, $q{\cdot}b{\cdot}\partial x = \partial T$

$$q = \frac{\partial T}{b\partial x}$$

$$\Sigma M = 0,$$

$$(T + \partial T)jd = Tjd + V \cdot \partial x \rightarrow \partial T = V \cdot \frac{\partial x}{jd} \rightarrow q = V / bjd$$

The shear stress distribution below the neutral axis (concrete in tension) may be considered rectangular.

The shear stress distribution above the neutral axis may be considered as usual parabolic, considering homogeneous. The intensity of shear stress S at any layer distant y above the neutral axis is given by

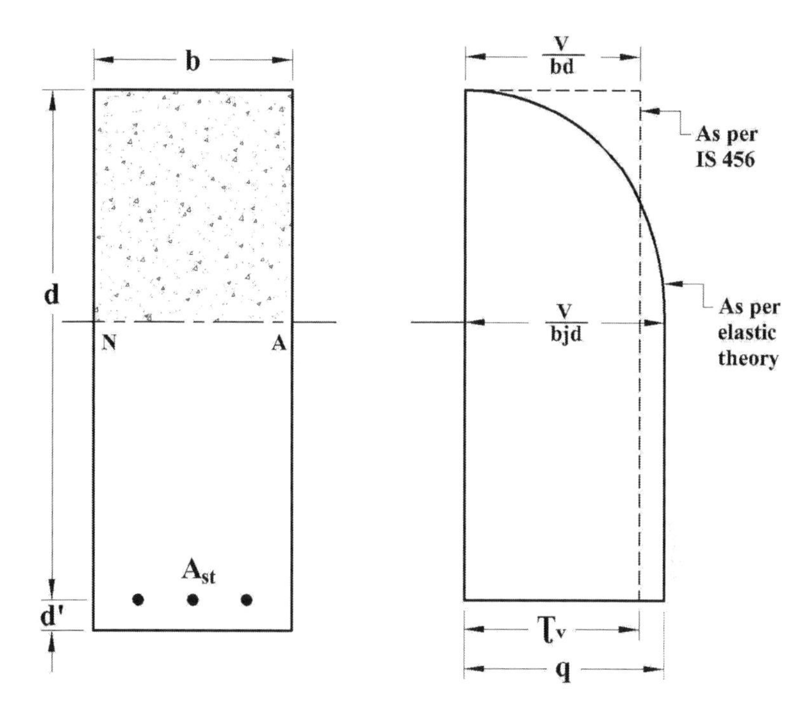

FIGURE 3.17 Shear stress distribution diagram.

$$s = \frac{v}{Ib}\int_{y}^{n} b \cdot d \cdot dy = \frac{v}{2I}\left(n^2 - y^2\right)$$

where I is the moment of inertia of the reinforced concrete beam section about the neutral axis and can be found from the expression: $I = 1/3\ b{\cdot}n^3 + A_{st}(d-n)^2$

The complete shear stress distribution diagram is shown in Figure 3.17. The intensity of maximum shear (q) can also be found by equating the external shear force to the shearing resistance of the beam as under:

$$V = (\text{Area of shear stress diagram})b = \{(2/3qn) + q(d-n)\}b$$

$$= \{2/3qn + qd - qn\}b = qb\{d - n/3\} = q \cdot bjd$$

$$q = V / b \cdot jd$$

From the stress-based approach, it gives the value of maximum shear stress q at neutral axis and the same for all layers below the neutral axis considering cracked reinforced concrete beam. Though in pure flexure, we neglect the concrete area below neutral axis, the same concrete between the cracks is needed for shear transfer between steel reinforcement and the compression zone. Also, the flexural cracks (assumed vertical in pure flexure) are not vertical in the shear-flexural zone under consideration. The shear deformations, which are really responsible for the mobilization of aggregate interlock at the cracks and dowel forces in the reinforcement, are neglected. Due to these reasons, the stress-based approach followed may not represent the actual behavior of the beam in shear and flexure.

Conventional shear stress formula $q = V/bjd$ has now been used. Indian Standard Code (IS: 456) recommends to use nominal shear stress given by the expression.

$$\tau_v = V / bd$$

τ_v = nominal shear stress, V = shear force at the section due to design loads

We know that the bending stress and the shearing stress vary across the cross-section of the reinforced concrete beam below the neutral axis the bending stress (f) is tensile while the shear stress curve is constant. Figure 3.18 shows a small element taken from the portion below the neutral axis the element is subjected to a longitudinal tensile stress (f) and horizontal shear stress q along with the vertical complementary shearing stress q.

$$p = \frac{f}{2} \pm \sqrt{\left(\frac{f}{2}\right)^2 + q^2}$$

The inclination of the principal plane with horizontal is given by $\tan 2\theta = -\dfrac{2q}{f}$

The major principal stress p_1 is tensile and given by $P_1 = \dfrac{f}{2} \pm \sqrt{\left(\dfrac{f}{2}\right)^2 + q^2}$

At the support where bending stress 'f' is practically zero the value of the principal tensile stress is equal to the share stress cube and it is inclined at 45° to the horizontal that is it acts diagonally. The diagonal tension which is caused by the combined action of longitudinal tension and the traverse shearing stresses is to be resisted by the provision of shear reinforcement or diagonal reinforcement or web reinforcement.

Beams with Inclined Stirrups

The inclined stirrups are assumed to be placed at angle α (usually, not less than 45°) with the axis of the beam and spaced 'S_v' apart. Let us consider a stirrup with a total area of legs as A_{sv} and with design yield stress equal to $0.87f_{yv}$. The horizontal length over which the bar is effective is given by

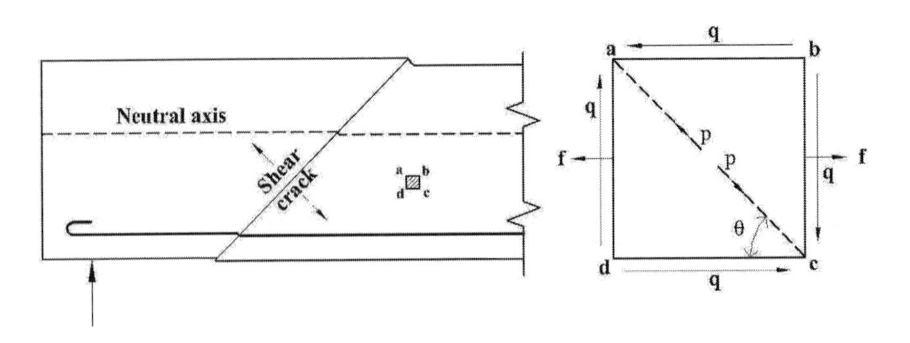

FIGURE 3.18 Diagonal tension due to shear.

$$L_h = (d - d')(\cot\theta + \cot\alpha)$$

The number of effective stirrups is given by $n = [(d - d')(\cot\theta + \cot\alpha)] / S_v$

Using truss analogy, the vertical component of shear carried by one stirrup = $0.87 f_{yv} \cdot A_{sv} \cdot \sin\alpha$

Hence, the shear resistance of all the stirrups intercepting the crack, as shown in figure 3.19, is given by

$$V_{us} = 0.87 f_{yv} \cdot A_{sv} \cdot \sin\alpha\left(\left[(d - d')(\cot\theta + \cot\alpha)\right] / S_v\right.$$

As per truss analogy, the crack angle θ is taken as 45° and $(d - d')$ may be taken approximately as d; using the aforementioned equation and simplifying, we get the formula given in Clause 40.4(b) of IS 456 as

$$V_{us} = \left[(0.87 f_{yv} \cdot A_{sv} \cdot d) \times (\sin\alpha + \cos\alpha)\right] / S_v$$

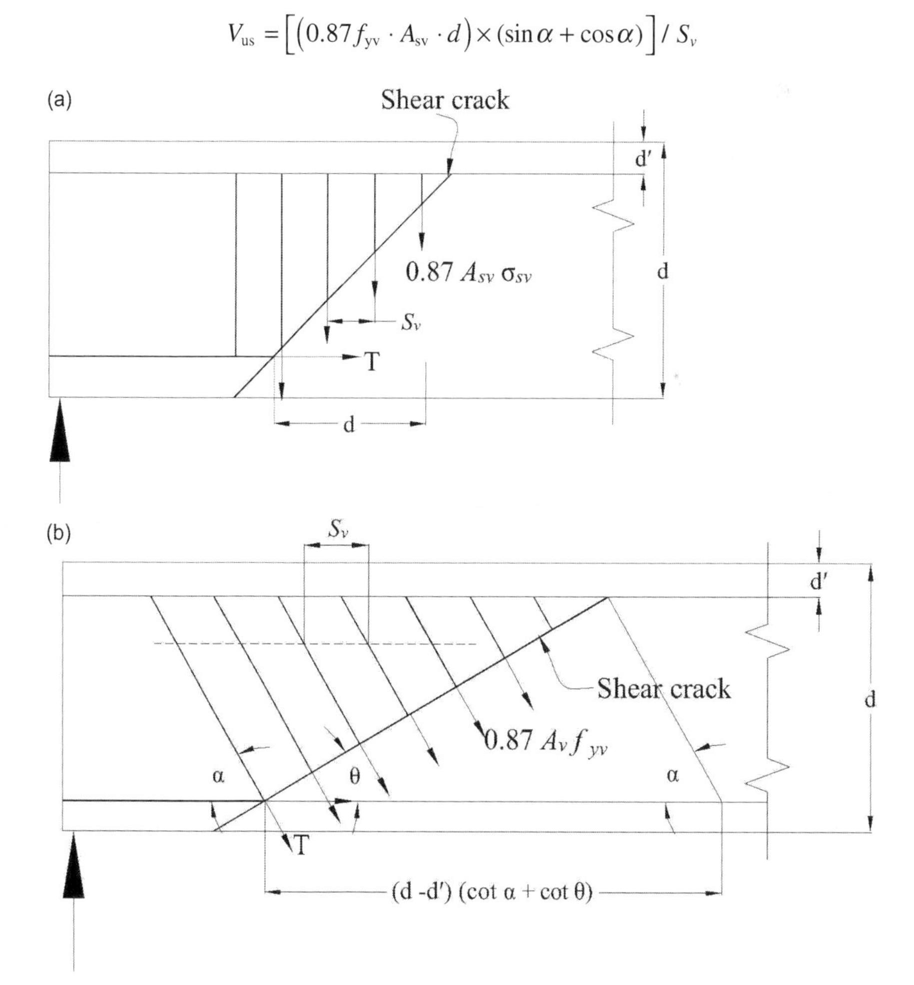

FIGURE 3.19 Design of stirrups: (a) vertical stirrups and (b) inclined stirrups.

The case of vertical stirrups may be considered as a special case of inclined stirrups, with $\alpha = 90°$.

Bent-Up Bars

The shear force V_{us} resisted by the bent-up bars inclined at an angle α is equal to the vertical component of the forces in the bars. Hence, we get the formula given in Clause 40.4(c) of IS 456 as

$$V_{us} = T \sin\alpha = 0.87 f_{yv} \cdot A_{sv} \cdot \sin\alpha$$

As stated earlier, bent-up bars alone cannot suffice as shear reinforcement.

where V_u is the design factored shear force acting at the section under consideration and V_n is the nominal shear capacity of beam with uniform depth, which is obtained by using the following equation (Clause 40.1 of the code):

$$V_n = \tau_y b_w d$$

As discussed, using the truss model, the nominal shear capacity V_s the beam with transverse reinforcement is considered as the sum of the contributions of concrete, V_c and that of the stirrups, V_n as follows:

$$V_n = V_c + V_s$$

where V_c is the nominal shear resistance provided by the concrete (V_c is calculated as given in Section 6.4 of IS456; $V_c = \tau_c\, b_w\, d$) and V_s is the nominal shear provided by the shear reinforcement.

Beams with Vertical Stirrups

If the crack is assumed to form at an angle 45° to the neutral axis, then the horizontal distance of the crack is approximately d. We consider a stirrup with the total area of legs as A_{sv} and the stirrups spaced at a distance S_v apart.

Number of stirrups across the crack, $n = d / S_v$

Assuming that the stress in the stirrups is equal to the design yield stress (= $0.87 f_y$), the following equation given in Clause 40.4(a) of IS 456 for the shear resistance of the vertical stirrups as follows:

$$V_s = 0.87 f_y \cdot A_{sv} \cdot (d / S_v)$$

$$V_n = \tau_v b_w d = \tau_v b_w d + 0.87 f_{sv}\, A_{sv} (d / S_v)$$

Simplifying, we get

$$\frac{A_{sv}}{S_v} = \frac{(\tau_v - \tau_c) b_w}{0.87\ f_{yv}}$$

where A_{sv} is the total cross-sectional area of a 'single' stirrup leg effective in shear, s_v is the stirrup spacing along the length of the member, τ_v is the calculated nominal shear stress (V_s/bd) in MPa, τ_c is the design shear strength of concrete in MPa, and f_{yv} is the yield stress of the stirrup steel. Thus, when τ_v exceeds τ_c (calculated using Table 19 of the code), shear reinforcement has to be provided. However, while deriving the formula it was assumed that all the links crossing the crack inclined at an angle of 45° are effective. To rectify this, the code (Clause 26.5.1.5) limits the spacing to $0.75d$.

It can be written

$$V_s = V_u - V_c = (\tau_v - \tau_c) b_w d$$

where V_u is the applied factored shear force due to the external loads and V_c is the shear strength provided by concrete.

It should be noted that d/s_v is used to the number of stirrups crossed by the crack, $n = d/s$, compute the number of stirrups crossing a shear crack forming at an angle of about 45°. This ratio seldom results in whole numbers, allowing for fractional shear contribution to shear strength. In reality, a crack cannot cross a fractional portion of a stirrup.

Design Procedure for Shear Reinforcement

The design of a reinforced concrete beam for shear using vertical stirrups involves the following steps:

1. Determine the maximum factored shear force V_u at the critical sections of the member.
2. Check the adequacy of the section for shear. Compute the nominal shear stress $\tau_v = V_u / (b_w d)$. Check whether, τ_v is less than the maximum permissible shear stress, $\tau_{c\ max}$, as given in Table 20 of the code. If τ_v is greater than $\tau_{c\ max}$. Increase the size of the section or the grade of concrete and recalculate steps 1 and 2.
3. Determine the shear strength provided by the concrete (for the percentage of tensile reinforcement available at the critical section) V_c using design shear strength τ_c given by Table 19 of the IS456; $V_c = \tau_c bd$.
4. If $V_u > V_c$ shear reinforcements have to be provided for $V_{us} = V_u - V_c$.
5. Compute the distance from the support beyond which only minimum shear reinforcement is required (where $V_u < 0.5 V_c$).
6. Design of stirrups needs to be made as per Clause 40.4 of IS 456.

Where stirrups are required, it is usually advantageous to select a bar size and type and determine the required spacing. The total cross-sectional area of stirrup, A_{sv} = Number of legs × Area of cross-section of each leg of stirrup. Minimum diameter of stirrup will be specified in IS456, but usually minimum diameter is 6 mm/8 mm, which is adopted in practice depending on availability in the market.

a. If vertical stirrups are chosen, calculate the spacing using

$$s_v = \frac{0.87\ f_y\ A_{sv}\ d}{V_{us}}$$

b. If inclined stirrups are chosen, calculate the spacing using

$$s_v = \frac{0.87\ f_y\ A_{sv}\ d}{V_{us}}(\sin\alpha + \cos\alpha)$$

c. In regions where only minimum stirrups are required, clause 26.5.1.6 of IS456 should be followed

$$s_v = \frac{0.87\ f_y\ A_{sv}}{0.4\ b_w}$$

The calculated spacing s_v should be smaller than the following maximum spacing as per clause 26.5.1.5 of IS 456.

a. For vertical stirrups, the lesser of 0.75*d* or 300 mm
b. For inclined stirrups, the lesser of *d* or 300 mm

7. Check anchorage requirements and details.

Although Clause 26.5.1.6 of the IS 456 states that shear reinforcement need not be provided in the regions of the beam where $\tau_v < 0.5\tau_c$, it is a better practice to provide nominal (minimum) shear reinforcement in such regions of the beam to improve ductility and to prevent failure due to accidental loading. It should be noted that while larger diameter stirrups at wider spacing are more cost-effective than smaller stirrup sizes at closer spacing (due to fabrication and placement costs), closely spaced stirrups of smaller diameter give better crack control and ductility. In wide beams with a large number of longitudinal rods and carrying heavy shear forces (such as those encountered in raft foundation beams), it is advisable to provide multi-legged stirrups, so that the longitudinal forces are evenly distributed among the longitudinal rods of the beam. From the tests conducted on wide members, it was that the effectiveness of the shear reinforcement decreases as the spacing of the web reinforcement legs across the width of the member increases.

For a given arrangement of vertical stirrup (with a given number of legs, diameter of bar, and spacing), the shear resistance V_{us}/d (N/mm) is constant.

$$\frac{V_{us}}{d} = \frac{0.87\ f_y\ A_{sv}}{S_v}$$

Design aids (SP16) can be generated using this equation (Table C.21 of Appendix C for the design aid of a two-legged vertical stirrup and Table C.22). Such design aids for vertical stirrups and bent-up bars may be found in SP 16, Tables 62 and 63, respectively

Shear design of beams with varying depth

Beams of varying depth are encountered in beams having haunches. In such members, it is necessary to account for the contribution of the vertical component of the flexural tensile force T_u, which is inclined at an angle β to the longitudinal direction, in the nominal shear stress, τ_v. The following two cases may arise in practice:

1. The bending moment M_u increases numerically in the same direction in which the effective depth d increases. In this case, as seen in figure 3.20, the net shear force is

$$V_{u,\text{net}} = V_u - (M_u / d)\tan\beta$$

The nominal shear stress (Clause 40.1.1 of IS 456) is obtained as

$$\tau_v = \frac{V_u - (M_u / d)\tan\beta}{b_w d}$$

2. The bending moment M, decreases numerically in the direction in which the effective depth d increases, the net shear force is

$$V_{u,\text{net}} = V_u + (M_u / d)\tan\beta$$

The nominal shear stress (Clause 40.1.1 of IS 456) is obtained as

$$\tau_v = \frac{V_{u,\text{net}}}{b_w d}$$

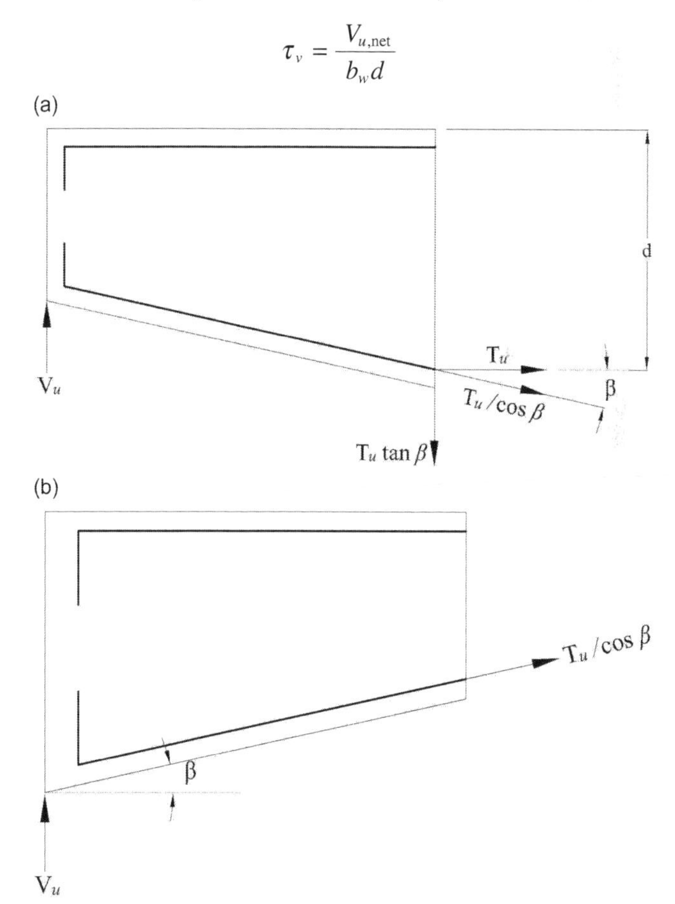

FIGURE 3.20 Beams of variable depth: (a) bending moment increases with increasing depth and (b) bending moment decreases with decreasing depth.

A similar situation arises in tapered base slabs or footings, where flexural compression is inclined to the longitudinal axis of the beam, since the compression face may be sloping. It should be noted that in the case of cantilever beams, the depth increases in the same direction as the bending moment and hence is used to check for reduced shear stress.

Shear Strength of Members with Axial Force

The beams in moment-resistant frames are often subjected to axial forces in addition to the bending moments and shears. Columns are also subjected to axial loads, bending moments, and shear forces. The compressive force acts like pre-stressing and delays the onset of flexural cracking; also, flexural cracks do not penetrate to a greater extent into the beam. However, tensile forces directly increase the stress and hence the strain in the longitudinal reinforcement. Axial tension increases the inclined crack width and reduces the aggregate interlock, and hence the shear strength provided by the concrete is reduced.

Clause 40.2.2 of IS 456 allows an increase in the design shear strength of concrete due to compressive forces, using the following factor, which should be multiplied with the values given in Table 19 of the code.

$$\delta = 1 + \frac{3P_u}{A_g f_{ck}} \leq 1.5$$

where P_u is the axial compressive force acting on the member in N, A_g is the gross area of concrete section in mm^2, and f_{ck} is the characteristic compressive strength of concrete in MPa. IS 456 does not give the guidelines for the detrimental effects of the tensile force, but may be considered with P_u taken negative for tension and using the following expression (NZS 3101, Part 1:2006):

$$\delta = 1 + \frac{12P_u}{A_g f_{ck}} \geq 0$$

Again, Clause 11.2.1.2 of ACI 318-08 gives the following expression for the design shear strength of members with shear and axial compression:

$$V_c = 0.15\left(1 + \frac{P_u}{14A_g}\right)\lambda\sqrt{f_{ck}}\,b_w d$$

In this equation, P_u/A_g is expressed in MPa.

For members subjected to significant axial tension, the following expression is suggested by ACI 318-08 (Clause 11.2.2.3):

$$V_c = 0.15\left(1 + \frac{0.29P_u}{A_g}\right)\lambda\sqrt{f_{ck}}\,b_w d \geq 0$$

where P_u is taken as negative for tension and P_u / A_g is expressed in MPa.

Numerical Examples

Examples 11

A reinforced concrete beam 250 mm and 450 mm effective depth, is subjected to shear force of 110 kN at the support. Percentage of tensile reinforcement at the support is 0.8%. Find the spacing of 12 mm diameter 2-legged stirrup to resist the shearing stress at supports. M25 grade concrete, using Fe415 HYSD bars, is used. Also, design the minimum shear reinforcement at the mid-span of the beam.

$$P = \frac{100 A_{st}}{bd} = 0.8\%$$

From Table 3.1 for M25 concrete, we get $\tau_c = 0.37$ N/mm²

From Table 3.3, maximum shear stress $\tau_{c\ max} = 1.9$ N/mm²

Nominal shear stress at support $\tau_v = \frac{V}{bd} = \frac{110 \times 1000}{250 \times 450} = 0.98\ \text{N/mm}^2 >$ $\tau_c < \tau_{c\ max}$, So ok.

Therefore, shear reinforcement is provided to carry a shear $V_s = V - \tau_c \cdot bd$.

$$V_s = V - \tau_c bd = 95{,}000 - 0.29 \times 250 \times 400 = 66{,}000\ \text{N}$$

$$S_v = \frac{\sigma_{sv} \cdot d \cdot A_{sv}}{V_s} = \frac{230 \times 450}{68{,}375} A_{sv} = 1.514 A_{sv}$$

Using 12 mm φ bars two-legged stirrups,

$$A_{sv} = 2 \times \frac{\pi}{4} \times (12)^2 = 226.2\ \text{mm}^2$$

$$S_v = \frac{\sigma_{sv} \cdot d \cdot A_{sv}}{V_s} = \frac{230 \times 450}{68375} A_{sv} = 1.514 A_{sv}$$

Maximum allowable spacing = 0.75 d = 0.75 × 450 = 337.5 mm.

Provide stirrups @ 300 mm c/c.

Mminimum shear reinforcement using 8 mm dia bars.

$$A_{sv} = 2 \times \frac{\pi}{4} \times (8)^2 = 100.53\ \text{mm}^2$$

$$\text{Spacing, } S_v = \frac{2.5 A_{xv} f_y}{\text{b}} = \frac{2.5 \times 100.53 \times 415}{250} = 417\ \text{mm.}$$

Providing 8 mm dia. 2-legged stirrups at 300 mm spacing at mid-span

Example 12

Design the stirrups of the beam carrying a UDL of 50,000 N/m over the entire span, using Fe 415 HYSD bars, $\sigma_{sv} = 230$ N/mm² and $f_y = 415$ N/mm².

$$V = \frac{wL}{2} = \frac{50,000 \times 6}{2} = 150,000 \text{ N}$$

$$\tau_v = \frac{V}{bd} = \frac{150,000}{300 \times 550} = 0.91 \text{ N/mm}^2$$

From Table 3.1, for $\frac{100As}{bd} = 0.8\%$, we get $\tau_c = 0.37$ N/mm² $< \tau_v = 0.91$ N/mm²

Hence, shear reinforcement is necessary at the support. Let us use 8 mm dia 2-legged vertical stirrups,

$$A_{sv} = 2\left(\frac{\pi}{4}(8)^2\right) = 100.53 \text{ mm}^2$$

Now $V_s = V - \tau_c\, bd = 150,000 - 0.37 \times 300 \times 550 = 88,950$ N

$$S_v = \frac{\sigma_{sv} \cdot A_{sv} \cdot d}{V_s} = \frac{230 \times 100.53 \times 500}{49,000} = 143 \text{ mm}$$

Maximum spacing of stirrups is given by

$$S_v = \frac{2.5 A_{sv} \cdot f_y}{b} = \frac{2.5 \times 100.53 \times 415}{300} = 348 \text{ mm} \approx 340 \text{ mm}$$

Maximum allowable spacing, $0.75d = 0.75 \times 550 = 412.5$ mm

Hence, providing spacing of stirrups 140 mm c/c at support and 300 mm at mid-span, where minimum shear reinforcement, is sufficient.

Shear force at a distance 'x' from support

$$V_x = \frac{V}{2}(x) = \frac{150,000}{2} x = 75,000x$$

$$V_s = V_x - \tau_c bd = 75,000x - 0.37 \times 300 \times 550 = 75,000x - 61,050$$

$$S_v = \frac{\sigma_{sv} \cdot A_{sv} \cdot d}{V_s} = \frac{230 \times 100.53 \times 550}{75,000x - 61,050} = 340$$

$$\frac{230 \times 100.53 \times 550}{75,000\text{x} - 61,050} = 340, \text{ ie. } x = 1.313 \text{ m}$$

Thus, spacing at support may be 140 mm c/c a and 300 mm c/c at 1.35m beyond support near mis-span.

Example 13

A reinforced concrete beam has an effective depth of 650 mm and width of 300 mm and is reinforced with 6 bars of 20 mm diameter at the center of the span. The maximum shear force at the support is 180 kN. Design the stirrup reinforcement required. Take $\sigma_{sv} = 230$ N/mm. Use M 25 concrete.

For M 25 concrete, $J = 0.904$ with MS. Bars, Lever arm $a = jd = 0.904 \times 650 = 587.6$ mm

Let us bend two bars at a distance of 800 mm from the support.

Shear resistance of the two bars bent (45°) is

$$V_s = A_{sv} \cdot \sigma_{sv} \sin 45° = 0.707 \cdot A_{sv} \cdot \sigma_{sv} = 0.707\left\{2 \times \frac{\pi}{4} \times (20)^2\right\} \times 230 = 102{,}171 \text{ N.}$$

However, maximum contribution toward shear resistance shall not be more than half the total shear reinforcement.

$$\text{Also,} \frac{100As}{bd} = \frac{100 \times 4\left(\frac{\pi}{4}\right)(20)2}{300 \times 650} = 0.64\%$$

Hence from Table 23, $\tau_c = 0.34$ N/mm^2

$$V_c = \tau_c \cdot bd = 0.34 \times 300 \times 650 = 66{,}300 \text{ N}$$

$$V_s = V - V_c = 180{,}000 - 66{,}300 \text{ N} = 113{,}700 \text{ N}$$

$$\text{Maximum contribution } V_s \text{ due to bent up bars} = \frac{1}{2} V_s = \frac{1}{2} \times 113{,}700 = 568{,}500 \text{ N}$$

$$V_{ss} \text{ from stirrups} = 568{,}500 \text{ N.}$$

Using 8 mm dia. 2-legged stirrups $A_{sv} = 2\left(\frac{\pi}{4}(8)^2\right) = 100.5 \text{ mm}^2$

$$S_v = \frac{\sigma_{sv} \cdot A_{sv} \cdot d}{V_{ss}} = \frac{230 \times 100.5 \times 650}{56{,}850} = 264.4 \text{ mm.}$$

This spacing should not be more than S_v corresponding to minimum reinforcement requirement.

$$\text{i.e. } S_v \leq \frac{2.5 \cdot A_{sv} \cdot f_y}{b} = \frac{2.5 \times 100.5 \times 415}{300} = 348 \text{ mm.}$$

Maximum allowable spacing = 0.75 d = 0.75 × 650 = 487.5 mm

Considering IS456 recommendations and practical considerations, let us provide 2-legged 8 mm dia. @ 250 mm c/c.

BEHAVIOR UNDER TORSION

Usually, torsion occurs in combination with flexure and transverse shear. The interactive behavior of torsion with bending moment and flexural shear in reinforced concrete beams is quite complex, considering non-homogeneity, non-linearity, and composite nature of the material in the presence of tension. For convenience in design, codes recommended simplified design procedures, which reflect a judicious blend of theoretical considerations and experimental results. In reinforced concrete design, the terms 'equilibrium torsion' and 'compatibility torsion' are commonly used to refer to two different torsion-inducing situations.

In 'equilibrium torsion', the torsion is induced by an eccentric loading with respect to the shear center at any cross-section, and equilibrium conditions alone suffice in determining the twisting moments. In 'compatibility torsion', the torsion is induced by the need for the member to undergo an angle of twist to maintain deformation compatibility, and the resulting twisting moment depends on the torsional stiffness of the member. This is associated with twisting moments that are developed in a structural member to maintain static equilibrium with the external loads and are independent of the torsional stiffness of the member. Such torsion has been considered in the design procedure recommended in IS456 (Code Cl. 41.1). The magnitude of the twisting moment does not depend on the torsional stiffness of the member, and is entirely determinable from statics alone. The member has to be designed for the full torsion, which is transmitted by the member to the supports. Moreover, the end of the member should be suitably restrained to enable the member to resist effectively the torsion induced. The type of torsion induced in a structural member by rotations (twists) applied at one or more points along the length of the member. The twisting moments induced are directly dependent on the torsional stiffness of the member. These moments are generally statically indeterminate, and their analysis necessarily involves (rotational) compatibility conditions; hence, the name 'compatibility torsion'. In statically indeterminate structures, the torsional restraints are 'redundant', and releasing such redundant restraints will eliminate the compatibility torsion. Thus, the Code states: "...where torsion can be eliminated by releasing redundant restraints, no specific design for torsion is necessary, provided torsional stiffness is neglected in the calculation of internal forces" [Cl. 41.1 of the Code].

Estimation of Torsional Stiffness

Behavior of reinforced concrete members under torsion shows that the torsional stiffness is little influenced by the amount of torsional reinforcement in the linear elastic phase, and may be taken as that of the plain concrete section. However, once torsional cracking occurs, there is a drastic reduction in the torsional stiffness. The post-cracking torsional stiffness is only a small fraction (less than 10%) of the pre-cracking stiffness and depends on the amount of torsional reinforcement, provided in the form of closed stirrups and longitudinal bars. Heavy torsional reinforcement can, no doubt, increase the torsional resistance (strength) to a large extent, but this can be realized only at very large angles of twist (accompanied by very large cracks). Hence, even with torsional reinforcement provided, in most practical situations, the maximum twisting moment in a reinforced concrete member under compatibility torsion is the value corresponding to the torsional cracking of the member.

Equivalent Shear

Equivalent shear V_e shall be calculated from the formula,

V_e = equivalent shear, V = Shear force, T = torsional moment, and b = breadth of beam.

The equivalent nominal shear stress (τ_{ve}) shall be calculated from the expression:

$$\tau_{ve} = V_e / bd$$

i. The value of τ_{ve} shall not exceed the value of $\tau_{c\,max}$ given in table (if it exceeds, the section should be redesigned by increasing the concrete area).
ii. If the equivalent nominal shear stress τ_{ve} does not exceed τ_c given in the table, minimum shear reinforcement shall be provided in the form of stirrups, such that:

$$\frac{A_{sv}}{b \cdot S_v} \geq \frac{0.4}{f_y}$$

The above equation can be rearranged in the following form to give the maximum spacing of stirrups:

$$S_v \leq \frac{2.5\, A_{sv}\, f_y}{b}$$

where A_{sv} is the total cross-sectional area of stirrups legs effective in shear, s_v is the stirrup spacing and f_y, is the characteristic strength of the stirrups reinforcement, in N/mm^2, which shall not be taken greater than 415 N/mm

iii. If τ_{ve} exceeds τ_c given in the table, both longitudinal and transverse reinforcement shall be provided in accordance with IS 456.

Longitudinal Reinforcement

As per IS 456, the longitudinal reinforcement shall be designed to resist an equivalent bending moment M_{e1}.

$$M_{e1} = M + M_r$$

where M = Design Bending moment

$$M_T = T\frac{1+\frac{D}{b}}{1.7}$$

where T is the torsional moment, D is overall depth of the beam and b is the breadth of the beam. If the numerical value of M_T defined above exceeds the numerical value of M_s, longitudinal reinforcement shall be provided on the flexural compression face, such that beam can also withstand an equivalent moment M_{e2} given by

$$M_{e2} = M_T - M$$

the moment M_{e2} being taken as acting in the opposite sense to the moment M.

Transverse Reinforcement

2-legged closed hoops enclosing the corner longitudinal bars shall have an area of cross-section A_{sv} given by

$$A_{sv} = \frac{\tau_{sv}}{b_1 d_1 \sigma_{sv}} + \frac{V_{sv}}{2.5 d_1 \sigma_{sv}}$$

However, the total transverse reinforcement shall not be less than

$$\text{Than } (\tau_{ve} - \tau_c) b \cdot s_v / \tau_{sv}$$

where

T = torsional moment
V = shear force
S_v = spacing of the stirrup reinforcement
b_1 = center to center distance between corner bars in the direction of width
d_1 = center to center distance between corner bars in the direction of depth
b = breadth of member
σ_{sv} = permissible tensile stress in shear reinforcement
τ_{sv} = equivalent shear stress
τ_c = shear strength of the concrete as specified in Table 31

When a member is designed for torsion, torsion reinforcement shall be provided as below.

a. The transverse reinforcement for torsion shall be rectangular closed stirrups placed perpendicular to the axis of the member. The spacing of the stirrups shall not exceed the least of x_1, $X_1 + Y_1 / 4$ and 300 mm, where x_1 and y_1 are respectively the short and long dimensions of the stirrup
b. When the cross-sectional dimension of the member 450 mm, additional longitudinal bars shall be provided to satisfy the requirements of minimum reinforcement and spacing given.Where the depth of the

web in a beam exceeds 450 mm the side face reinforcement shall be provided along the two faces. The total area of such reinforcement shall be not less than 0.1% of the web area and shall be distributed equally on two faces at a spacing not exceeding 300 mm or web thickness whichever is less.

Typical Numerical Examples

Numerical Example 14

A fixed ended beam AB as shown on the plan in Figure 3.21 is subjected to a uniformly distributed live load of 30 kN/m and a concentrated point live load *P*. The distance of the point of application of the concentrated load *P*, $a = 1$ m. Design the beam against torsional moment, bending moment, and shear force. Apply 'Limit State of Design methodology, as per IS 456.

Considering, width of the beam = 250 mm and overall depth = 600 mm

$$\text{Self-Weight} = (25 \times 0.25 \times 0.6)\,\text{kN/m} = 3.75\ \text{kN/m}$$

Bending moment at end support (M_1), due to the uniformly distributed load of 30 kN/m

And self-weight of the beam = $(3.75 + 30)\,(6)^2 / 12 = 101.25$ kNm

Bending moment at end support (M_2), due to the concentrated load $P = 15$kN:

$$= PL/8 = 15(6)/8\ \text{kNm} = 11.25\ \text{kNm}$$

Torsional moment at end (T) = Pa/2 = 15 × 1/2 = 7.5 kNm

Design bending moment (M) = $M_1 + M_2 = 101.25 + 11.25 = 112.5$ kNm

Design torsional moment (T) = 7.5 kNm

Factored design bending moment (M_u) = $1.5(M_1 + M_2)$

$$= 1.5(101.25 + 11.25) = 168.75\ \text{kNm}$$

Factored design torsional moment (T_u) = $(1.5 \times T) = 11.25$ kNm

According to IS 456, the longitudinal reinforcement shall be designed to resist an equivalent bending moment (M_{e1}) = $M_u + M_t$

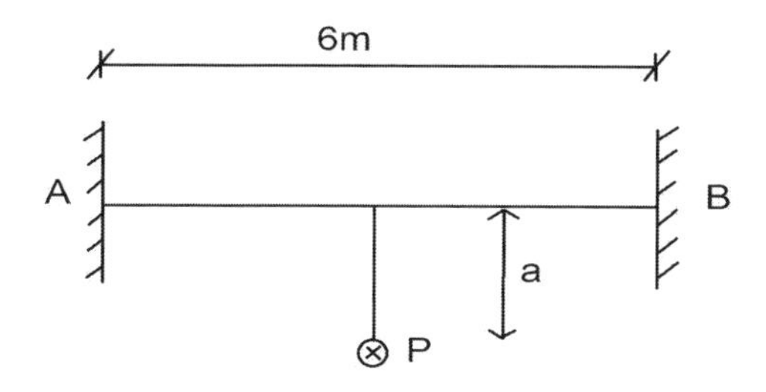

FIGURE 3.21 Plan of a fixed beam under eccentric vertical load.

where M_u = Factored design bending moment and
M_t = Factored equivalent bending moment of the torsional moment (T_u)

$$= T_u(1+D/b)/1.7 = 11.25[(1+600/250)/1.7] = 22.5 \text{ kNm}$$

Equivalent bending moment M_{e1} = (168.75+22.5) kNm = 191.25 kNm
Considering M25 grade concrete, Fe500 grade HYSD bars
Q_b = 3.34 N/mm² (from SP 16, page 10)
Let us consider 20Φ bars then, d = (600–20/2–25) mm = 565 mm

$$Q = M_{e1}/bd^2 = (191.25\times10^6)/(250(565)^2)$$

$$= 2.40 \text{ N/mm}^2$$

From SP 16, Page 49, p_t = 0.632%

$$A_{st} = (0.632/100)(250)(5/65)\text{mm}^2 = 892.7 \text{ mm}^2$$

According to IS 456, Equivalent Shear, $V_e = V_u + 1.6\ (T_u/b)$

$$V_u = \text{Factored design shear force } = 1.5[(33.75(6/2))+15/2]\text{kN} = 163.125 \text{ kN}$$

$$V_e = 163.125+1.6(11.25/0.250) = 235.125 \text{ kN}$$

Equivalent nominal shear stress (ζ_{ve}), as per clause 40.1(IS 456:2000)

$$V_e/bd = (235.125\times10^3/(250\times(600-25-20/2-8))) = 1.69 \text{ N/mm}^2$$

$\zeta_{ve} < \zeta_{c\,max}$ = 3.1 N/mm² from Table 20 of IS456
For p_t = 0.632%, ζ_c from Table 19, IS 456

$$\zeta_c = 0.49+(0.57-0.49)(0.632-0.49)/(0.75-0.50) = 0.535 \text{ N/mm}^2$$

$\zeta_{ve} > \zeta_c$, Therefore, transverse reinforcement is to be designed

TRANSVERSE REINFORCEMENT

2- legged closed hoops/stirrups, enclosing the corner longitudinal bars, shall have an area of cross-section A_{sv} given as;

$$A_{sv} = [T_u S_v/(0.87 f_y\, bd)] + [(V_u S_v)/2.5 d_1(0.87 f_y)]$$

Considering S_v = 200 mm c/c, b_1 = 250 – 25 – 25 – 20 / 2 – 20/2 = 180 mm and d_1 = 600 – 25 – 25 – 20/2 – 20/2 = 530 mm
Therefore,

$$A_{sv} = [(11.25\times10^6\times250)/(180\times530\times0.87 f_y)]$$

$$+[(163.125\times10^3\times200)/(2.5\times530\times0.87\times500)]$$

$$= 67.773+56.604 \text{ mm}^2 = 124.377 \text{ mm}^2$$

However, minimum transverse reinforcement shall not be less than

$$A_{sv,min} = \left[\left(\zeta_v - \zeta_c\right) \times bS_v\right] / (0.87 f_y)$$
$$= [(1.69 - 0.535) \times 250 \times 200] / (0.87 \times 500)\ \text{mm}^2 = 132.759\ \text{mm}^2$$

A few cases are designed, and some useful information is observed and furnished in Table 3.1

It is quite evident that when the concentrated load (*P*) increases, both shear force and torsional moment increase. If the beam section is kept unchanged, then the requirement for longitudinal reinforcement and transverse reinforcements increases with the increase of torsional moment due to the increase of concentrated load *P*. However, if the width is increased, keeping the depth unchanged, then the requirement for longitudinal and transverse reinforcements decreases. Torsional moment plays a major role in the design output and width plays a major role.

TABLE 3.1
Comparative result of different torsion-related design problems

Cases	b = Breadth (mm)	D = Depth (mm)	P (kN)	A_{st} required (mm²)	A_{sv} required (mm²)
1	250	600	15	892	132
2	250	600	20	963	154
3	250	600	25	1038	174
4	300	600	25	1000	154
5	350	600	25	979	138

FLEXURAL BOND STRESS

Due to this variation of bending moment in a length ∂x, the tensile forces in steel will be T and $T + \partial T$ (ref to Figure 3.22)

We know $M = T{\cdot}jd$, ie. $M + dM = (T + dT)\, jd$

Hence, $dM = dT{\cdot}jd$, or, $dT = \dfrac{dM}{jd}$

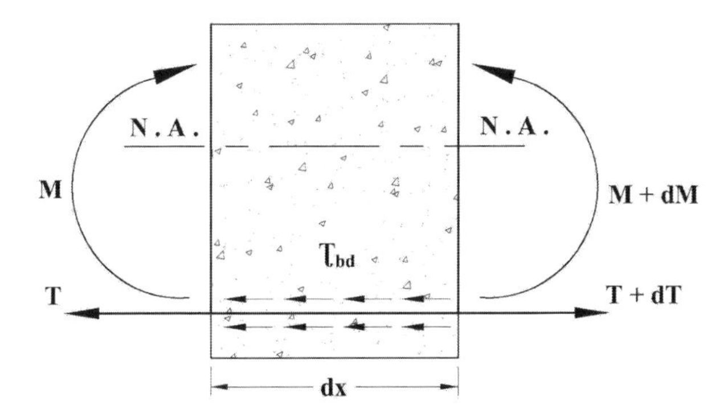

FIGURE 3.22 Bond mechanism of reinforced concrete.

The difference in the tensile force (dT) in steel bars between sections is resisted by bond stress developed in the length dx.

Let τ_{bx} = bond stress development in concrete around the steel reinforcement.

$\Sigma 0$ = sum of the perimeters of the steel bars resisting tension.

Total surface area of bars = $dx\ \Sigma 0$.

Hence $\tau_{bx}\ [dx\ \Sigma 0] = dT$

Substituting the value of dT in (i), we get

$$\tau_{bx}[dx\,\Sigma 0] = \frac{dM}{dx}$$

or

$$\tau_{bx} = \frac{dM}{dx} \cdot \frac{1}{jd\,\Sigma 0}$$

But, $\dfrac{dM}{dx}$ = shear force = V

Hence, $\tau_{bx} = \dfrac{V}{jd\,\Sigma 0} = \dfrac{V}{a\,\Sigma 0}$

The stress τ_{bx} at a particular section is called local bond stress.

Figure 3.23 shows a steel bar embedded in concrete and subjected to a tensile force T. Due to this force, there will be a tendency for the bar to slip out and this tendency is resisted by the bond stress developed over the perimeter of the bar, along its length of embedment. Let us assume that average, uniform bond stress is developed along the length. The required length, necessary to develop full resisting force, is called anchorage length in case of axial tension (or compression) and development length in the case of flexural tension, and is designated by symbol L_d.

Hence, if φ is the nominal diameter of the bar, we have

$$\frac{\pi}{4}\varphi^2 \cdot \sigma_{\text{st}} = t_{\text{bd}} \cdot \pi\varphi \cdot L_d$$

$$L_d = \frac{\varphi \cdot \sigma_{\text{st}}}{4t_{\text{bd}}}$$

The permissible bond stress is specified in the IS code (Table 3.4) for various grades of concrete. It shows that a bar must extend a length L_d beyond any section at which it is required to develop its full strength so that sufficient bond resistance is mobilized.

For M 15 concrete, $T_a = 0.6$ N/mm while for mild steel bars, $\sigma_{\text{st}} = 140$ N/mm.

$$L_d = \frac{140\varphi}{4 \times 0.6} = 58.3\varphi$$

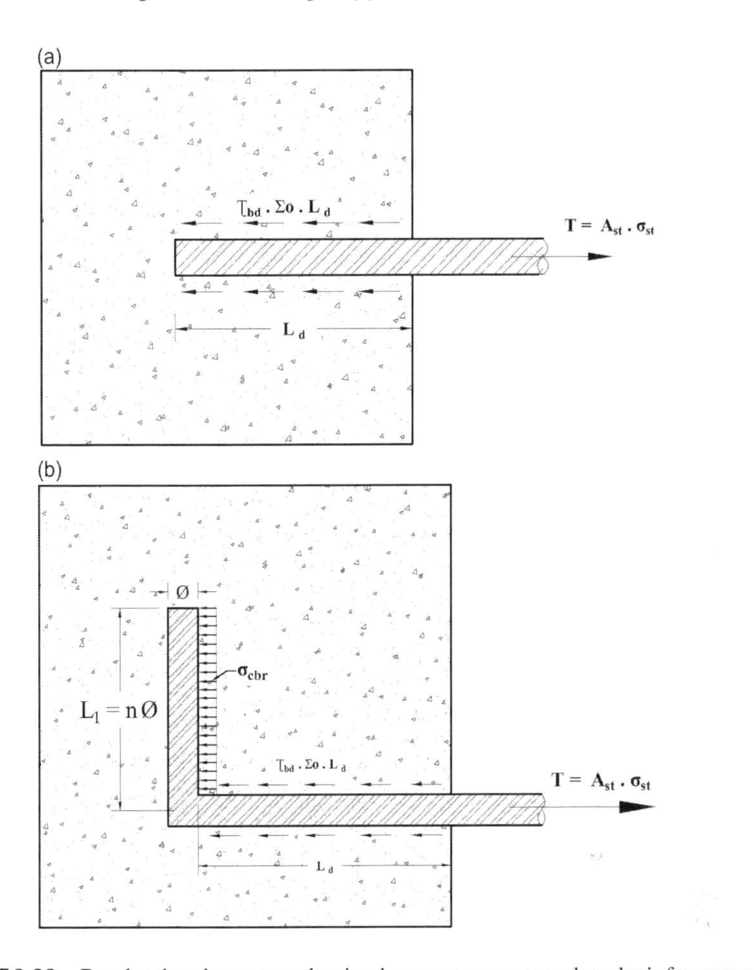

FIGURE 3.23 Bond and anchorage mechanism between concrete and steel reinforcement.

Quite often space available at the end of beam is limited to accommodate the full development length L_d. In that case, hooks or bends are provided. The anchorage value (L_e) of hooks or bend accounted for contributing to the development length L_d.

A semi-circular hook, fully dimensioned, with respect to a factor *K*. The value of *K* is taken as 2 (ref. to Figure 3.24) in the case of mild steel conforming to IS: 432-1966, (specifications for Mild-Steel and Medium Tensile Steel bars and Hard-Drawn steel wires for concrete reinforcement) or IS: 1139-1959, (specifications for 'Hot rolled mild steel and medium tensile steel deformed bars for concrete reinforcement'). The hook with $K = 2$ is shown in Figure 3.24 (aii) with equivalent horizontal length of the hook. For the case of Medium Tensile Steel conforming to IS: 432-1966 or IS: 1139-1959, *K* is taken as 3. In the case of cold worked steel conforming to IS: 1986-1961, (specifications for cold twisted steel bars for concrete reinforcement), *K* is taken as 4. In the case of bars above 25 mm, however, it is desirable to increase the value of *K* to 3, 4, and 6, respectively (ref. to Figure 3.24).

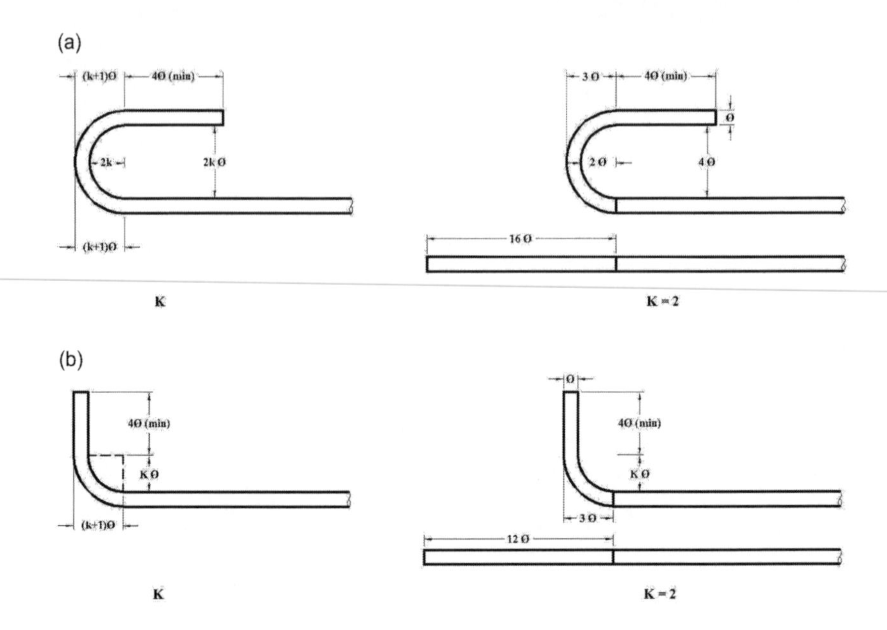

FIGURE 3.24 Different kinds of hooks.

In the case of deformed bars, the value of bond stress for various grades of concrete is greater by 40% than the plain bars. Hence deformed bars may be used without hooks, provided anchorage requirements are adequately met (Figure 3.25).

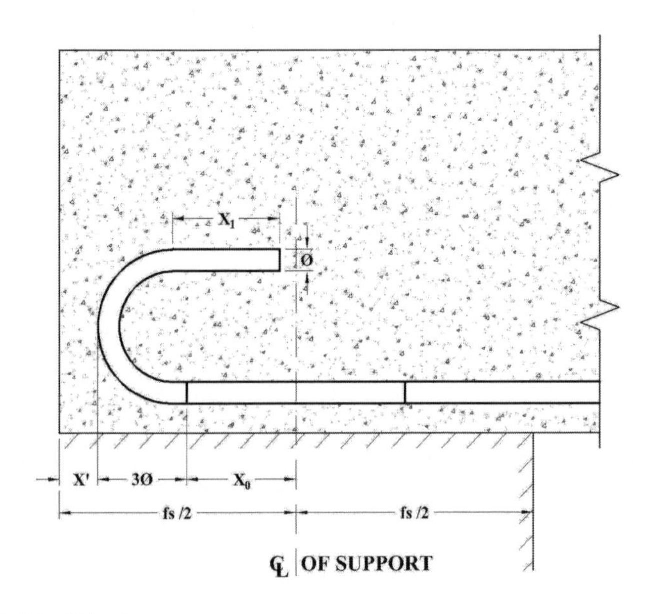

FIGURE 3.25 U-hook.

The anchorage value of a standard U-type hook shall be equal to 16 times the diameter of the bar. The anchorage length of straight bar in compression shall be equal to the development length of bars in compression. The projected length of hooks, bends and straight lengths beyond bends if provided for a bar in compression, shall be considered for development length.

Inclined bars. The development length shall be as for bars in tension; this length shall be measured as under: (1) in tension zone from the end of the sloping or inclined portion of the bar, and (2) in the compression zone, from mid-depth of the beam.

Stirrups. Notwithstanding any of the provisions of this standard, in case of secondary reinforcement, such as stirrups and transverse ties, complete development lengths and anchorage shall be deemed to have been provided when the bar is bent through an angle of at least 90° round a bar of at least its own diameter and is continued beyond the end of the curve for a length of at least eight diameters, or when the bar is bent through an angle of 135° and is continued beyond the end of curve for a length of at least six bar diameters or when the bar is bent through an angle of 180° and is continued beyond the end of the curve for a length at least four bar diameters.

The code stipulates that at the simple supports (and at the point of inflection), the positive moment tension reinforcement shall be limited to a diameter such that

$$L_d \leq \frac{M_1}{V} + L_o$$

where

L_d = development length computed for σ_{st}

M_1 = moment of resistance of the section assuming all reinforcement at the section to be stressed to σ_{st}

V = shear force at the section due to design loads

L_o = sum of anchorage beyond the center of support and the equivalent anchorage value of any hook or mechanical anchorage at the simple support (at the point of inflection, L_o is limited to d or 12Φ whichever is greater)

Note. For the computation of L_o the support width should be known. A beam with end support, in which is the side cover and x_o is the distance of the beginning of the hook from the center line of the support. Now L_o = sum of anchorage beyond the center of the support and the equivalent anchorage value

The dark portion shows the hook which has an anchorage value of 16Φ. The distance of the beginning of the hook from is apex of the semi-circle can be taken to be equal to 3Φ. Let L_s be the length of support

$$\text{So, } L_o = x_o + 16\Phi, \text{ where } x_0 = \frac{ls}{2} - x' - 3\Phi$$

$$\text{So, } L_o = \left(\frac{ls}{2} - x' - 3\Phi\right) + 16\Phi$$

$$L_o = \frac{ls}{2} - x' + 13\Phi$$

If no hook is provided, as in case of deformed bars,

$$L_o = \frac{ls}{2} - x'$$

The code further recommends that the value of M_1/V may be increased by 30% when the ends of the reinforcement are confined by a compressive reaction. This condition of '**confinement**' of reinforcing bars may not be available at all the types of simple supports.

As per IS456, at simple supports, where the compressive reaction confines the ends of reinforcing bars, we have

$$L_d \leq 1.3M_1 / V + L_o$$

DEFLECTION ANALYSIS

The performance of structures at the service loads is an important design consideration. If sections arc proportioned by strength requirements alone, there is a danger that although the degree of safety against collapse will be adequate, the performance of the structure at the sen ice loads may be unsatisfactory. For example, at the service loads the deflections of the members may be excessively large, or the cracking of the concrete may be unacceptably great.

Numerical Examples 15

Calculate the flexural longitudinal reinforcement required for a simply supported beam of span

$l = 6$ m subjected to an imposed uniformly distributed load $w_L = 40$ kN/m (Dl +LL) over its entire span, using M25 concrete and Fe 415 steel. Use SP16. Apply WSD. Check against deflection.

From Table 22, IS 456, allowable $\sigma_{st} = 230$ N/mm² and from Table 21, IS 456,

$$\sigma_{cbc} = 8.5 \text{ N/mm}^2 \text{ (for M25 concrete).}$$

As per Is456, m = 280 / (3 × 8.5) = 10.98, $k_b = 0.289$ (calculated by using methods described earlier), $J_b = (1 - k_b/3) = (1 - (0.289/3)) = 0.904$

$$Q_b = \tfrac{1}{2}\,\sigma_{cbc}\,k_b j_b = \tfrac{1}{2}(8.5)(0.289)(0.904) = 1.11 \text{ N/mm}^2$$

Considering a beam section 350 mm × 800 mm, i.e., $D = 800$ mm, $d' = 50$ mm, and $d = 750$ mm

$$w_d = \text{Self weight of beam} = b\,D\,\gamma_{\text{conc}} = (0.35 \times 0.8 \times 25) = 7 \text{ kN/m}$$

$$w = \text{Design load} = w_d + w_L = 7 + 40 = 47 \text{ kN/m}$$

$$M = \text{Design bending moment} = wl^2/8 = \left(47 \times 6^2\right)/8 = 211.5 \text{ kNm}$$

Considering $d = 750$ mm, $M/bd^2 = 211.5 \times 10^6 / (350 \times 750^2) = 1.074$ N/mm²

Using SP16, $P_t = 0.516$, $A_{st} = p_t / 100 \times b \times d = 0.516/100 \times 350 \times 750 = 13\ 54.5$ mm²

CHECK AGAINST DEFLECTION

Short-Term Deflection

As per IS456,

$$I_{\text{eff}} = \frac{I_r}{1.2 - \dfrac{M_r}{M}\dfrac{z}{d}\left(1 - \dfrac{x}{d}\right)\dfrac{b_w}{b}}$$

where I_r = moment of inertia of the cracked section and $M_r = f_{cr}\, I_{gr}/y_t$

f_{cr} = Modulus of rupture of concrete = $0.7\sqrt{\text{fck}} = 0.7\sqrt{25} = 3.5$ N/mm²

I_{gr} = Moment of inertia of the gross section about the centroidal axis, neglecting the reinforcement

$$= bD^3/12 = 350 \times 800^3/12 = 1.49 \times 10^{10} \text{ mm}^4$$

y_t = Distance from the centroidal axis of the gross section to the extreme fiber in tension

$$= D/2 = 800/2 = 400 \text{ mm}$$

Therefore, $M_r = f_{cr}\, I_{gr}/y_t = 3.5 \times 1.49 \times 10^{10} / 400 = 13.067 \times 10^7$ N-mm

M = maximum moment under service loads = 211.5×10^6 N-mm

d = effective depth = 750 mm, $p_t = 0.516$, $p_c = 0$

$$\frac{p_c(m-1)}{p_t m} = 0$$

$$p_t m = 0.516 \times 10.98 = 5.67$$

$$\frac{d'}{d} = \frac{50}{750} = 0.067$$

Referring to Table – 92 SP-16, $x/d = 0.285$ mm

z = lever arm = $d - x/3 = d\,(1 - x/3d)$

b_w = breadth of web, b = breadth of compression face

$$I_{\text{eff}} = \frac{I_r}{1.2 - \dfrac{M_r}{M}\left(1 - \dfrac{x}{3d}\right)\left(1 - \dfrac{x}{d}\right)\dfrac{b_w}{b}}$$

$$= \frac{I_r}{1.2 - \dfrac{13.067 \times 10^7}{211.5 \times 10^6}\left(1 - \dfrac{0.285}{3}\right)(1 - 0.285)}$$

$$\frac{I_{\text{eff}}}{I_r} = 1.25$$

Referring to Table 88, SP – 16, $\dfrac{I_r}{\left(\dfrac{bd^3}{12}\right)} = 0.44$

$$I_r = 0.44 \times \frac{bd^3}{12} = 0.44 \times 350 \times 750^3 / 12 = 5.414 \times 10^9 \text{ mm}^4$$

Therefore, $I_{\text{eff}} = 1.25 \times I_r = 1.25 \times 5.414 \times 10^9 \text{mm}^4 = 6.77 \times 10^9 \text{mm}^4$

Maximum short-term deflection (elastic) of a simply supported beam carrying udl $= \dfrac{5\ ML^2}{48EI_{\text{eff}}}$

$$E_c = 5000\sqrt{f_{\text{ck}}} = 5000\sqrt{25} = 25{,}000 \text{ N/mm}^2$$

Maximum short-term deflection = $5 \times 211.5\times106 \times 6000^2 / 48 \times 25{,}000 \times 6.77 \times 10^9 = 4.686$ mm

LONG-TERM DEFLECTION

Deflection due to Shrinkage

$$a_{\text{cs}} = k_3 \times \Psi_{\text{cs}} \times l^2$$

where $k_3 = 0.125$ for simply supported members

$$\Psi_{\text{cs}} = k_4 \times \frac{\varepsilon_{\text{cs}}}{D}$$

$$p_t - p_c = 0.516 - 0 = 0.516$$

$$k_4 = 0.72 \times \frac{(p_t - p_c)}{\sqrt{p_t}} \leq 1 \text{ for } 0.25 \leq p_t - p_c < 1$$

$$= 0.72 \times \frac{(0.516 - 0)}{\sqrt{0.516}} = 0.5172$$

$$\varepsilon_{\text{cs}} = 0.0003 \text{ (Cl.6.2.4 IS - 456)}$$

$$\Psi_{\text{cs}} = 0.5172 \times 0.0003 / 800 = 1.94 \times 10^{-7}$$

$$a_{\text{cs}} = 0.125 \times 1.94 \times 10^{-7} \times (6000)^2 = 0.873 \text{ mm}$$

Deflection due to Creep

As per IS456

$$a_{\text{cc(perm)}} = a_{i,\text{cc(perm)}} - a_{i,\text{(perm)}}$$

In the absence of data, the age at loading is assumed to be 28 days and the value of creep coefficient, θ is taken as 1.6 from clause 6.2.5- IS 456

$$E_{\text{ce}} = E_c / (1+\theta)$$

$$= 25{,}000 / (1+1.6) = 9.62 \times 10^3 \text{ N/mm}^2$$

$$m = E_s / E_{\text{ce}} = 2 \times 10^5 / 9.62 \times 10^3 = 20.79$$

$$p_t = 0.516,\ p_c = 0,\ \frac{p_c(m-1)}{p_t m} = 0,\ p_t m = 0.516 \times 20.79 = 10.82$$

Referring to Table 88, SP – 16

$$\frac{I_r}{\left(\frac{bd^3}{12}\right)} = 0.72$$

$$I_r = 0.72 \times \frac{bd^3}{12} = 0.72 \times 350 \times 750^3 / 12 = 8.86 \times 10^9 \text{ mm}^4$$

Therefore, $I_{\text{eff}} = 1.25 \times I_r = 1.25 \times 8.86 \times 10^9 \text{mm}^4 = 11.075 \times 10^9 \text{mm}^4 < I_{\text{gr}} = 1.4$ $9 \times 10^{10} \text{mm}^4$

$$a_{i,\text{cc(perm)}} = \frac{5ML^2}{48 E_{\text{ce}} I_{\text{eff}}} = 5 \times 211.5 \times 10^6 \times 6000^2 / 48 \times 9.62 \times 10^3$$

$$\times 11.075 \times 10^9 = 7.44 \text{ mm}$$

$$a_{i,\text{(perm)}} = \frac{5ML^2}{48 E_c I_{\text{eff}}} = 5 \times 211.5 \times 10^6 \times 6000^2 / 48 \times 25{,}000 \times 6.77 \times 10^9 = 4.686 \text{ mm}$$

$$a_{\text{cc(perm)}} = 7.44 - 4.686 = 2.754 \text{ mm}$$

Total deflection including deflection due to shrinkage and creep

$$= 4.686 + 0.873 + 2.754 = 8.313 \text{ mm.}$$

Permissible total deflection = Span/250 = 6000/250 = 24 mm. < 8.313 mm Ok.

Long-term deflection = 0.873 + 2.754 = 3.627 mm

Permissible long-term deflection = Span/350 = 6000/350 = 17.143mm < 3.627 mm Ok.

Crack Width Analysis

Numerical Examples 16

Calculate the flexural longitudinal reinforcement required for a simply supported beam of span

$l = 6$ m subjected to an imposed uniformly distributed load $w_L = 40$ kN/m (Dl +LL) over its entire span, using M25 concrete and Fe 415 steel. Use SP16. Apply WSD. Check against crack width.

From Table 22, IS 456, allowable $\sigma_{st} = 230$ N/mm² and from Table 21, IS 456,

$$\sigma_{cbc} = 8.5\ \text{N/mm}^2\ (\text{for M25 concrete}).$$

As per Is456, $m = 280 / (3 \times 8.5) = 10.98$, $k_b = 0.289$ (calculated by using methods described earlier), $J_b = (1 - k_b/3) = (1 - (0.289/3)) = 0.904$

$$Q_b = \tfrac{1}{2}\,\sigma_{cbc}\,k_b j_b = \tfrac{1}{2}(8.5)(0.289)(0.904) = 1.11\ \text{N/mm}^2$$

Considering, a beam section 350 mm × 800 mm, ie. $D = 800$ mm, $d' = 50$ mm and $d = 750$ mm

$$w_d = \text{Self weight of beam} = b\,D\,\gamma_{\text{conc}} = (0.35)(0.8)(25) = 7\ \text{kN/m}$$

$$w = \text{Design load} = w_d + w_L = 7 + 40 = 47\ \text{kN/m}$$

$$M = \text{Design bending moment} = wl^2/8 = (47)(6)^2/8 = 211.5\ \text{kNm}$$

Considering $d = 750$ mm, $M/bd^2 = 211.5 \times 10^6 / (350 \times 750^2) = 1.074$ N/mm²

Using SP16, $P_t = 0.516$, $A_{st} = p_t/100 \times b \times d = 0.516/100 \times 350 \times 750 = 135$ 4.5 mm²

$$W_{cr} = \frac{3a_{cr}\varepsilon_m}{1 + \dfrac{2(a_{cr} - C_{min})}{(h - x)}}$$

$$a_{cr} = 30\ \text{mm}$$

$$a = 800 - 50 - 30\ \text{mm} = 720\ \text{mm}$$

$$C_{min} = 25\ \text{mm}$$

$$h = 800\ \text{mm}$$

$$x = 0.285 \times 750 = 213.75\ \text{mm}$$

$$\begin{aligned}
\varepsilon_m &= \varepsilon - b(h-x)(a-x)/3E_s A_s(d-x) \\
&= 230/\left(2\times10^5\right) - \frac{350\times(800-213.75)\times(720-213.75)}{3\times2\times10^5\times1354.5\times(750-213.75)} \\
&= 1.15\times10^{-3} - 2.38\times10^{-4} \\
&= 9.12\times10^{-4}
\end{aligned}$$

$$w_{cs} = \frac{3\times 30\times 9.12\times 10^{-4}}{1+\dfrac{2(30-25)}{(800-213.75)}}$$
$$= 0.081 \text{ mm}$$

Permissible crack width = 0.3 mm (in normal environment) > 0.081 mm Ok.

4 Basic Understanding of Limit State Approach

It is essential for the reader to have relevant IS code of practices and handbooks for RCC design etc. as ready reference while going through this chapter.

PREAMBLE

Design Considerations

The aim of structural design is to fix up the sectional requirements of structural elements like slab, beams, columns, foundation, etc. along with other details, so that it fulfills its purpose during its proposed lifetime with adequate safety in terms of strength, stability, serviceability and gives an economical solution.

Safety implies that partial or total collapse of the structure is not acceptable under the normal expected loads, but partial damage (within recommendation of codes) may be allowed under abnormal conditions, such as due to earthquake or extreme wind or under overloading and later on retrofitting may be made. Collapse may occur due to exceeding the bearing capacity of soil underneath foundations, overturning, sliding, buckling, fracture due to fatigue, etc.

Serviceability implies satisfactory performance of the structure under service loads, excessive deflection, cracking, vibration, discomfort to the user, etc. Durability, impermeability, acoustic and thermal insulation, etc. are also coming under serviceability. Design may be safe but may not satisfy the 'serviceability' requirements. However, a balance between safety, serviceability, and cost are three pillars of economic solution.

Design Philosophies

Working stress method of design (WSM) is on linear elastic theory and is used mostly to design vibration-prone structures like machine foundation, RC bridges, etc. Around 75 years back, alternative design philosophy, ultimate load carrying capacity, was proposed and gained acceptance to some extent. Probabilistic concepts were developed around 50 years back, based on probability theory. Probability of failure is tried to quantify and later on extended to **Reliability-based methods,** but it is not fully popular because of its complicated and intractable nature. Probabilistic (reliability-based) approach had to be reduced to a deterministic format involving multiple factors.

 DOI: 10.1201/9781003208204-4

Probabilistic Analysis and Design

Uncertainties in Design

The safety margins are assigned in terms of permissible stresses in WSM. The science of reliability-based design evolved with a rational solution in terms of adequate safety. The main variables in design calculations that are subject to varying degrees of uncertainty and randomness are the loads, material properties, and dimensions. Further, there are simplifying assumptions in structural analysis and design. Several other variables and unforeseen factors are also influencing the prediction of strength and serviceability - such as workmanship and quality control, possible future change of use, frequency of loading, etc. (ref. to Figures 4.1 and 4.2).

Classical Reliability Models

A simple reliability-based design is introduced. Two simple 'classical' models are considered, one for **strength design** and the other for **serviceability design**.

Strength Design Model

The lifetime maximum load effect S on a structure and the ultimate resistance R of the structure, expressed in terms of stress resultant at a critical section) are treated as random variables whose respective probability density functions $f_S(S)$ and $f_R(R)$ are shown in Figure 4.3. It is also assumed that S and R are statistically independent, which is approximately true for cases of normal static loading.

If $S<R$, the structure is expected to be safe, and if $S<R$, the structure is expected to fail. It is evident from Figure 4.3 that there is always a probability,

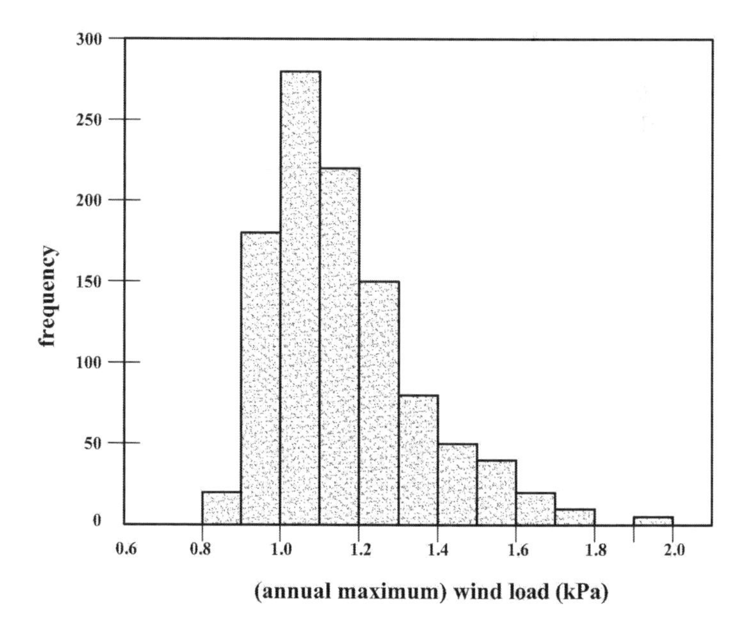

FIGURE 4.1 Typical example of frequency distribution of wind loads on a structure.

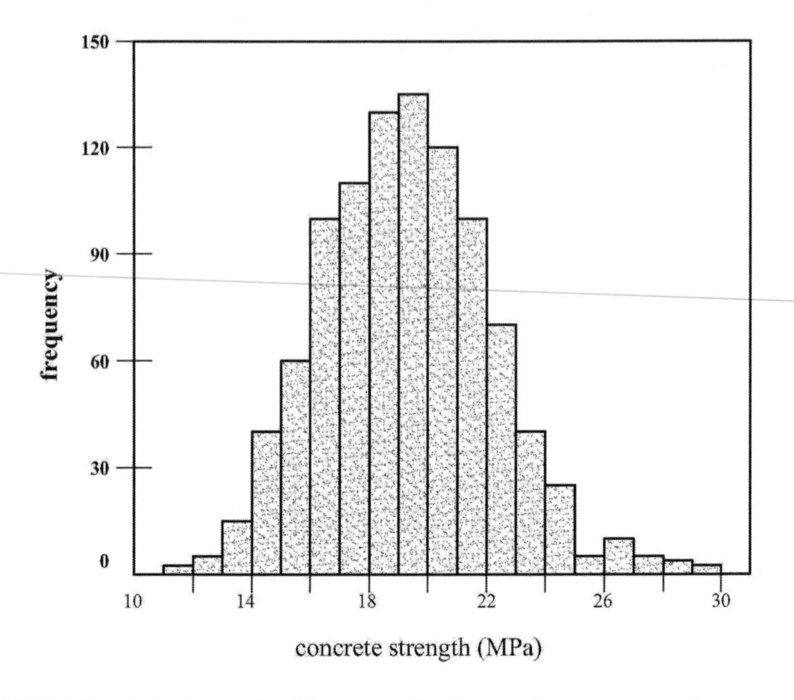

FIGURE 4.2 Typical example of frequency distribution of concrete strength.

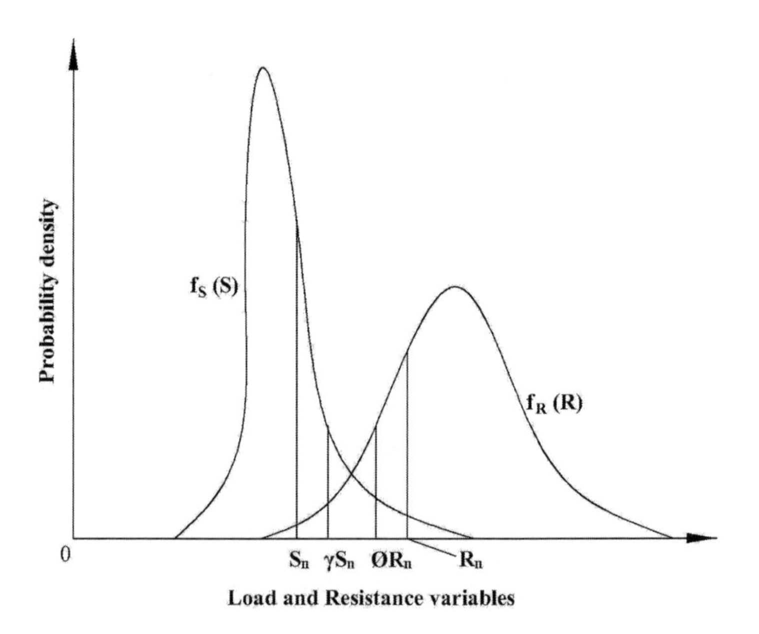

FIGURE 4.3 Classical reliability model for strength design.

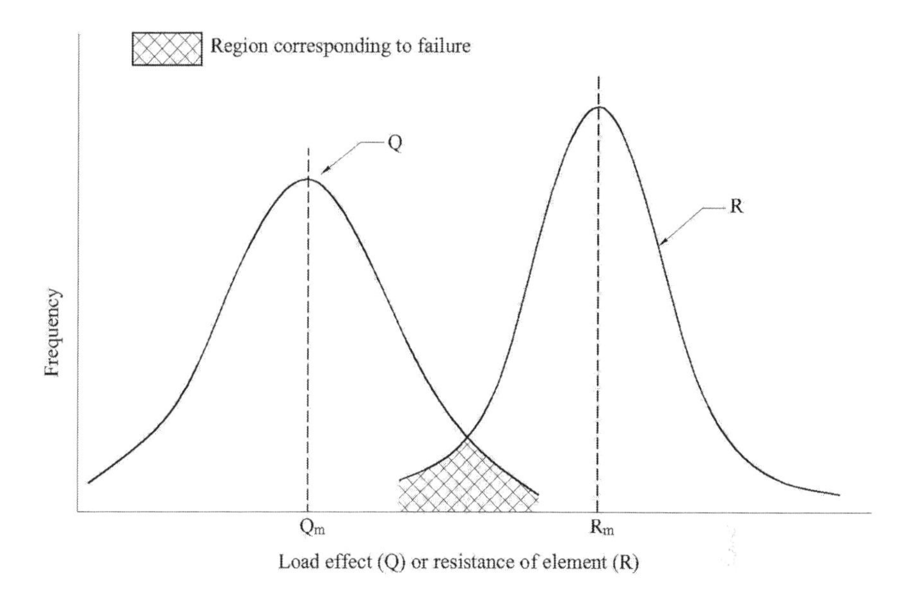

FIGURE 4.4 Frequency distribution curves.

however small, that failure may occur due to the exceeding of the load-bearing capacity of the structure (or structural element under consideration).

The probability of failure P_f, may be calculated as follows:

$$P_f = \text{Prob}\left[\{R < S\} \cap \{0 < S < \infty\}\right]$$

$$P_f = \int_0 f_s\,(S)\left[\int_0^s f_R\,(R)dR\right]dS$$

Reliability is expressed as the complement of the probability of failure i.e., equal to $(1 - P_f)$.

Evaluating the probability of failure P_f (or the reliability) underlying a given structure is termed reliability analysis, whereas designing a structure to meet the target reliability is termed reliability design. The problem becomes complicated when load and resistance basic variables are involved as well as the joint probability distribution of the multiple variables. Target reliabilities are hard to define since losses associated with failures are influenced by economic considerations, which are difficult to quantify (Figure 4.4).

Levels of Reliability Methods

Levels of reliability analysis are differentiated by the extent of probabilistic information that is available. A full-scale probabilistic analysis is generally described as a Level III reliability method. It is highly advanced mathematically difficult and still in research level. Level II reliability method is based on the basic variables to their second moment statistics in terms of mean and variance. It evaluates the risk underlying a structural design in terms of a reliability index β (in lieu of the 'probability of failure' P_f used in Level I method). However, it requires the application of optimization techniques for the determination of β. For codal use, the method must be as simple as possible-using deterministic rather than probabilistic data. Such a method is called a Level I reliability method. The multiple safety factor format of limit states design comes under this category. It provides deterministic measures of safety such as **permissible stress** and **load factors** (ref. to Figures 4.5 and 4.6).

Limit States Method (LSM)

Limit States

Limit state is a state of impending failure, beyond which a structure ceases to perform its intended function satisfactorily, in terms of either safety or serviceability. i.e. it either collapses or becomes unserviceable. There are two types of limit states:

Ultimate limit states (or 'limit states of collapse') deal with strength, overturning, sliding, buckling, fatigue fracture, etc.

Serviceability limit states which deal with discomfort to occupancy and/or malfunction, caused by excessive deflection, crack width, vibration, etc., and also loss of durability, etc.

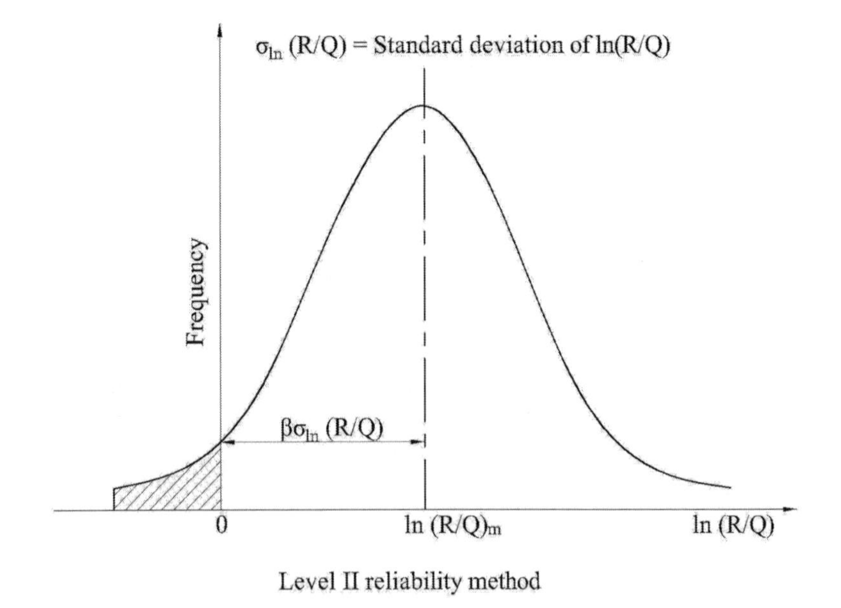

FIGURE 4.5 Level II frequency curve.

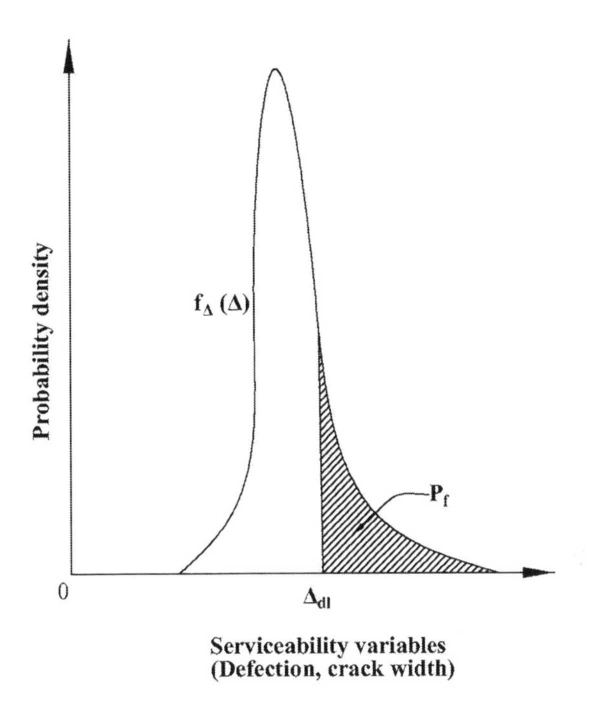

FIGURE 4.6 Probability density curve.

There are two main limit states: (i) limit state of collapse and (ii) limit state of serviceability

i. Limit state of collapse deals with the strength and stability of structures subjected to the maximum design loads out of the possible combinations of several types of loads. Therefore, this limit state ensures that neither any part nor the whole structure should collapse or become unstable under any combination of expected overloads.
ii. Limit state of serviceability deals with deflection and cracking of structures under service loads, durability under working environment during their anticipated exposure conditions during service, stability of structures as a whole, fire resistance, etc. All relevant limit states have to be considered in the design to ensure adequate degree of safety and serviceability. The structure shall be designed on the basis of the most critical limit state and shall be checked for other limit states.

Load and Resistance Factor Design Format

Simplest to understand is the Load and Resistance Factor Design (LRFD) format, which is adopted by the ACI Code. Applying the LRFD concept to the classical reliability model adequate safety requires the following conditions to be satisfied:

Design Resistance (Φ R_n) $\geq$ Design Load effect (γ S_n), where R_n and S_n denote the nominal or characteristic values of resistance R and load effect S respectively. Φ and γ denote the **resistance factor** and **load factors** respectively. The resistance factor Φ accounts for understrength', i.e. possible shortfall in the computed 'nominal' resistance, owing to uncertainties related to material strengths, dimensions, theoretical assumptions, etc., and accordingly it is less than unity. On the contrary, the load factor γ, which accounts for 'overloading' and the uncertainties associated with S_n is generally greater than unity.

PARTIAL SAFETY FACTOR FORMAT

The multiple safety factor format recommended by CEB-FIP, and adopted by the Code, is the so-called partial safety factor format, which may be expressed as follows:

$$R_d \geq S_d$$

where R_d is the design resistance computed using the reduced material strengths $0.67f_{ck} / \gamma_c$ and f_y / γ_s, involving two separate partial (material) safety factors γ_c (for concrete) and γ_s (for steel), in lieu of a single overall resistance factor Φ and S_d is the design load effect computed for the enhanced loads (γ_D. DL, γ_L. LL, γ_Q. QL…..) involving separate partial load factors γ_D (for dead load), γ_L (for live load), γ_Q (for wind or earthquake load). The other terms involved are the nominal compressive strength of concrete $0.67\, f_{ck}$ and the nominal yield strength of steel f_y, on the side of the resistance, and the nominal load effects DL, LL, and QL representing dead loads, live loads, and wind/earthquake loads, respectively. It may be noted that, whereas the multiplication factor Φ is generally less than unity, the dividing factors γ_C, and γ_S, are greater than unity - giving the same effect. All the load factors are generally greater than unity because over-estimation usually results in improved safety. However, one notable exception to this rule is the dead load factor γ_D which is taken as 0.8 or 0.9 while considering stability against overturning or sliding, or while considering reversal of stresses when dead loads are combined with wind/earthquake loads; in such cases, under-estimating the counteracting effects of dead load results in greater safety. One other effect to be considered in the selection of load factors is the reduced probability of different types of loads (DL, LL, and QL) acting simultaneously at their peak values. Thus, it is usual to reduce the load factors when three or more types of loads are considered acting concurrently; this is referred to sometimes as the 'load combination effect'.

PARTIAL SAFETY FACTORS AND CHARACTERISTIC STRENGTHS AND LOADS

The general definition of the **characteristic strength** of a material (concrete or steel) was given in IS 456. It corresponds to the 5% strength value. In the case of reinforcing steel, it refers to the 'yield / proof stress as mentioned in IS 456.

The **characteristic load** is defined as the load that "has a 95% probability of not being exceeded during the life of the structure" (Cl. 36.2 of the Code).

However, in the absence of statistical data regarding loads, the nominal values specified for dead, live, and wind loads are to be taken from IS 875, and the values for 'seismic loads' (earthquake loads) from IS 1893

The characteristic values of loads are based on statistical data. It is assumed that in 95% of cases the characteristic loads will not be exceeded during the life of the structures. However, structures are subjected to overloading also. Hence, structures should be designed with loads obtained by multiplying the characteristic loads with suitable factors of safety depending on the nature of loads or their combinations, and the limit state being considered. These factors of safety for loads are termed as partial safety factors (γ_f) for loads. Thus, the design loads are calculated as (Design load F_d) = (Characteristic load F) (Partial safety factor for load γ_f) Respective values of γ_f for loads in the two limit states are given in Table 18 of IS 456 for different combinations of loads.

The code recommends the following weighted combinations for estimating the ultimate load effect (UL) and the serviceability load effect (SL):

Ultimate Limit States

- UL = 1.5 (DL + LL)
- UL = 1.5 (DL + QL) or (0.9DL + 1.5 QL)
- UL = 1.2 (DL + LL + QL)

The reduced load factor of 1.2 in the third combination above recognizes the reduced probability of all three loads acting together at their possible peak values.

For the purpose of structural design, the design resistance (using the material partial safety factors) should be greater than or equal to the maximum load effect that arises from the above load combinations. However, when live loads and wind loads are combined, it is improbable that both will reach their characteristic values simultaneously; hence a lower load factor is assigned to LL and QL in the third combination, to account for the reduced probability of joint occurrence.

Partial Safety Factors for Materials

Similarly, the characteristic strength of a material as obtained from the statistical approach is the strength of that material below which not more than 5% of the test results are expected to fall. However, such characteristic strengths may differ from sample to sample also. Accordingly, the design strength is calculated by dividing the characteristic strength further by the partial safety factor for the material (γ_f), where γ_f depends on the material and the limit state being considered.

Design strength of the material (f_d) = [Characteristic strength of the material (f_{ck})]/Partial safety factor of the material (γ_f). Clause 36.4.2 of IS 456 states that γ_f for concrete and steel should be taken as 1.5 and 1.15, respectively when assessing the strength of the structures or structural members employing a limit state of collapse. However, when assessing the deflection, the material properties such as modulus of elasticity should be taken as those associated with the characteristic strength of the material. It is worth mentioning that the partial safety factor for steel (1.15) is comparatively lower than that of concrete (1.5) because the steel for reinforcement is produced in steel plants and commercially available in specific diameters with expected better quality control than that of concrete. Further, in case of concrete, the characteristic strength is calculated based on test results on 150 mm standard cubes. However, the concrete in the structure has different sizes. To take the size effect into account, it is assumed that the concrete in the structure develops a strength of 0.67 times the characteristic strength of cubes. Accordingly, in the calculation of strength employing the limit state of collapse, the characteristic strength (f_{ck}) is first multiplied with 0.67 (size effect) and then divided by 1.5 (γ_m for concrete) to have 0.446 f_{ck} as the maximum strength of concrete in the stress block (Figures 4.7 and 4.8).

DESIGN UNDER FLEXURE

Assumptions

The behavior of reinforced concrete beam sections at ultimate loads has been explained in detail in IS 456. The basic assumptions involved in the analysis at the ultimate limit state of flexure (Cl. 38.1 of the Code) are listed here. (Most of these assumptions have already been explained earlier.) (a) Plane sections normal to the beam axis remain plane after bending, i.e., in an initially straight beam, strain varies linearly over the depth of the section. (b) The maximum compressive strain in concrete (at the outermost fiber) ε_{cu} shall be taken as 0.0035. This is so because regardless of whether the beam is under-reinforced or over-reinforced, collapse invariably occurs by the crushing of concrete (Figures 4.9 and 4.10).

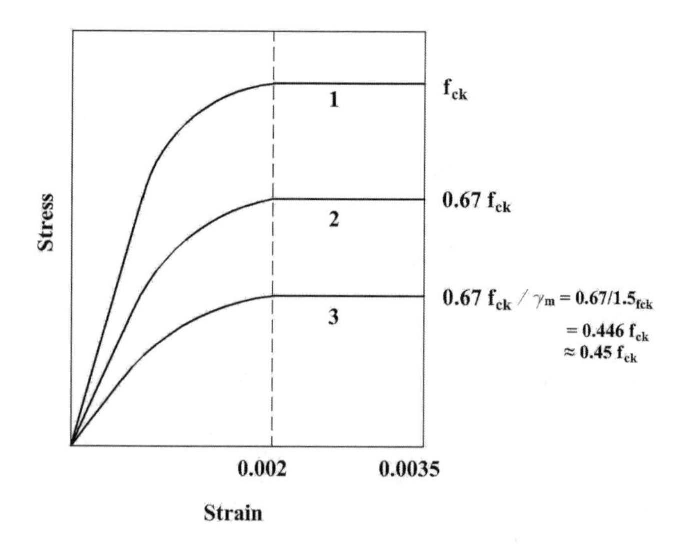

FIGURE 4.7 Stress-strain curve of concrete.

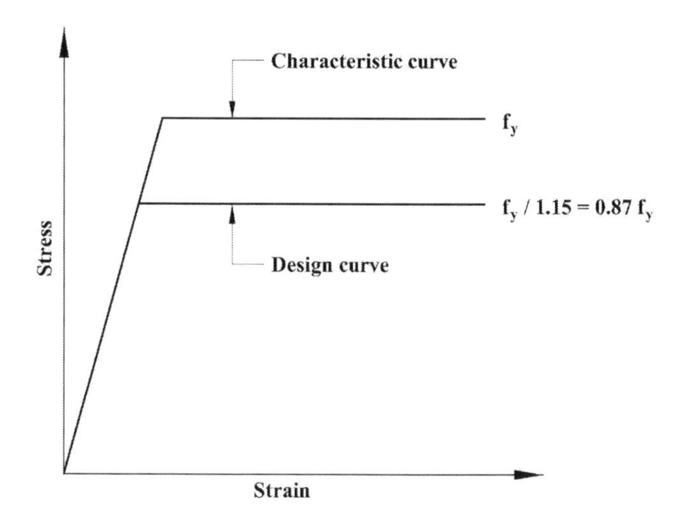

FIGURE 4.8 Stress-strain curve of steel.

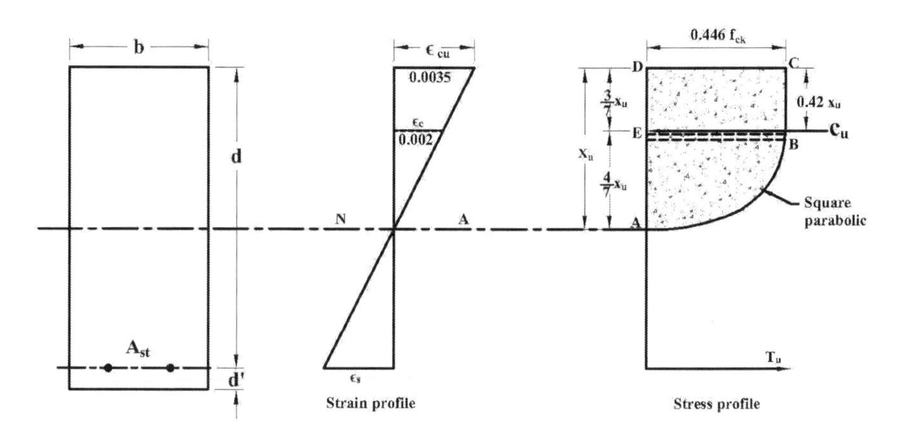

FIGURE 4.9 Compression stress block (LSD).

Singly Reinforced Rectangular and Flanged Sections

An 'under-reinforced section' is one in which the area of tension steel is such that as the ultimate limit state is approached, the yield strain ε_y is reached in the steel before the ultimate compressive strain is reached in the extreme fiber of the concrete (Figures 4.11–4.13).

Effective depth of a beam is defined as 'the distance between the centroid of the area of tension reinforcement and the maximum compression fibre' (Cl. 23.0 of the Code). Reinforcing bars are usually provided in multiple numbers, and sometimes in multiple layers, due to size and spacing constraints. In flexural computations, it is generally assumed that the entire steel area resisting tension

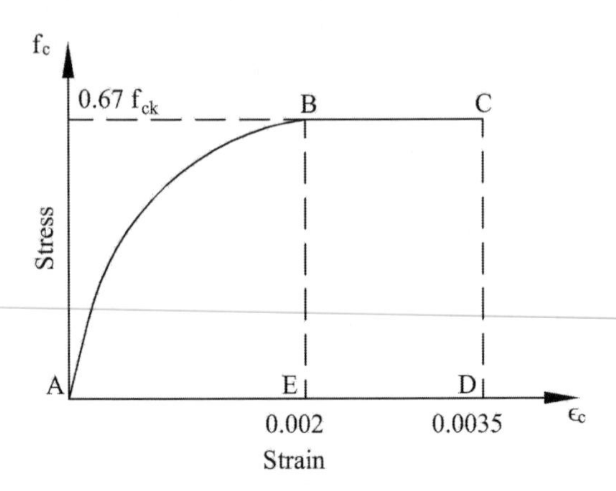

FIGURE 4.10 Idealized design stress-strain curve of concrete (LSD).

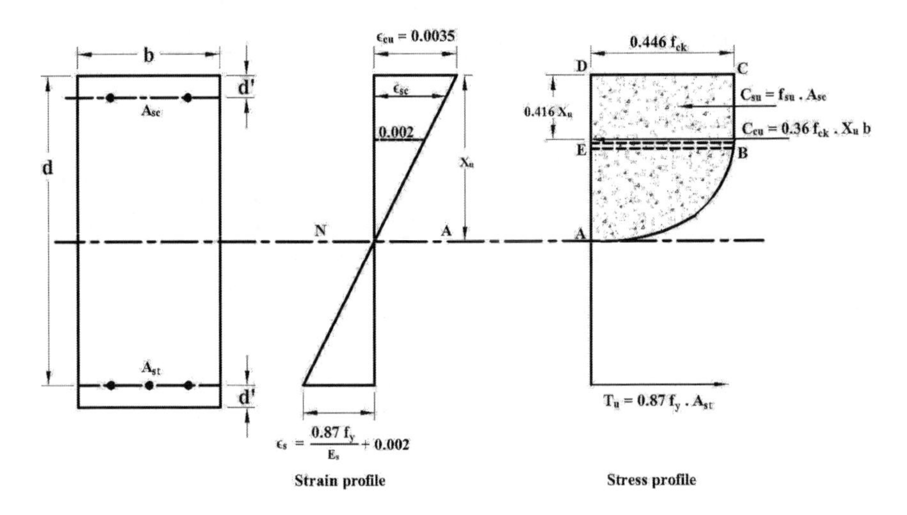

FIGURE 4.11 Compression stress block of rectangular section (LSD).

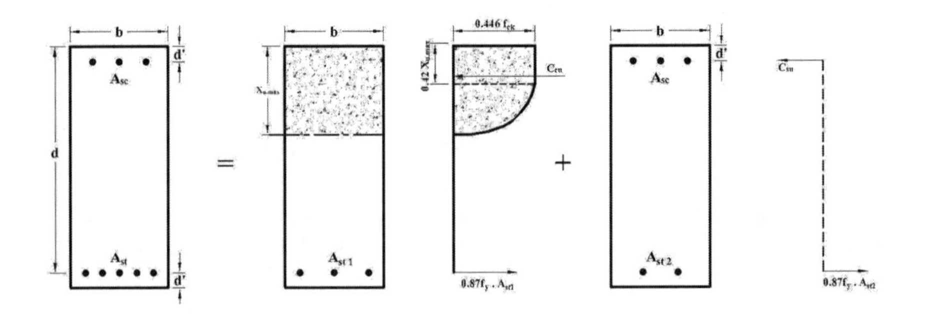

FIGURE 4.12 Force diagram of doubly reinforced section (LSD).

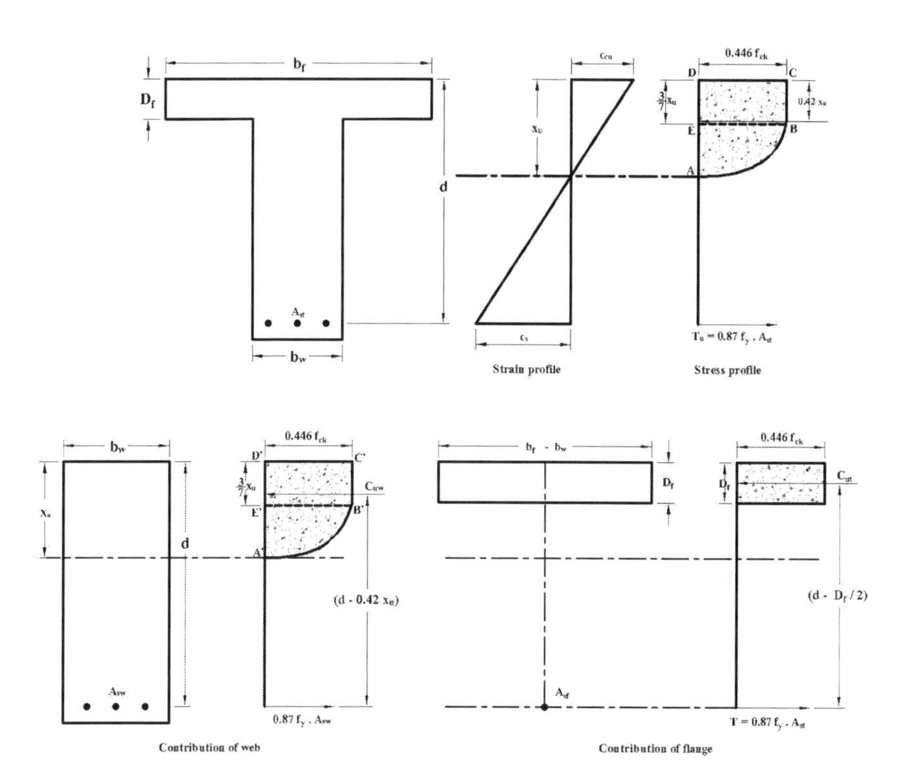

FIGURE 4.13 Compression stress block of flanged section (LSD).

is located at the centroid of the bar group, and that all the bars carry the same stress — corresponding to the centroid level (i.e., at the effective depth).

The failure of an under-reinforced beam is termed as tension failure — so called because the primary cause of failure is the yielding in tension of the steel. The onset of failure is gradual, giving ample prior warning of the impending collapse by way of increased curvatures, deflections, and cracking. Hence, such a mode of failure is highly preferred in design practice. The actual collapse, although triggered by the yielding of steel, occurs by means of the eventual crushing of concrete in compression ('secondary compression failure'). A sketch of the moment-curvature relation for an under-reinforced beam. The large increase in curvature (rotation per unit length), prior to collapse, is indicative of a typical ductile mode of failure.

An 'over-reinforced' section is one in which the area of tension steel is such that at the ultimate limit state, the ultimate compressive strain in concrete is reached, however the tensile strain in the reinforcing steel is less than the yield strain ε_y. this type of failure is termed compression failure. The failure occurs (often, explosively) without warning. In this case, the tension steel remains in the elastic range up to collapse. As the limit state of collapse is approached, the tensile stress in steel increases proportionately with the tensile strain, whereas the compressive stress in concrete does not increase proportionately with the compressive strain (Figures 4.11– 4.13).

Singly Reinforced Non-rectangular Sections

Flange Width as per IS 456

$$b_f = l_o/6 + b_w + 6D_f, \text{ for T-beam}$$

$$= l_o/12 + b_w + 3D_f, \text{ for L-beam}$$

$$= l_o/(l_o/b + 4) + b_w, \text{ for Isolated T-beam}$$

$$= 0.5l_o/(l_o/b + 4) + b_w, \text{ for Isolated L-Beam}$$

Basic Equations of Flanged Beam

If kd > D_f

$$(b_f - b_w)D_f(kd - D_f/2) + b_w(kd)^2/2 = mA_{st}(d - kd)\frac{1}{2}\sigma_{cbc}(kd)$$

$$-\frac{1}{2}\sigma_{cbc1}(b_f - b_w)(kd - D_f) = A_{st}\sigma_{st}$$

$$\sigma_{cbc1} = \sigma_{cbc}\frac{(kd - D_f)}{kd}$$

$$M = \frac{1}{2}\sigma_{cbc}b_f(kd)\left(d - \frac{kd}{3}\right) - \frac{1}{2}\sigma_{cbc1}(b_f - b_w)(kd - D_f)\left(d - D_f - \frac{(kd - D_f)}{3}\right)$$

If kd < D_f

$$(b_f - b_w)D_f(kd - D_f/2) + b_w(kd)^2/2 + (1.5m - 1)A_{sc}(kd - d') = mA_{st}(d - kd)$$

$$C_c = 0.5\sigma_{cbc}\left[b_f(kd) - (b_f - b_w)\frac{(kd - D_f)^2}{kd}\right]$$

$$M = \frac{1}{2}\sigma_{cbc}\left[b_f(kd)\left(d - \frac{kd}{3}\right) - (b_f - b_w)\frac{(kd - D_f)^2}{kd}\left(d - D_f - \frac{(kd - D_f)}{3}\right)\right]$$

$$+ (1.5m - 1)A_{sc}\sigma_{cbc}\frac{(kd - d')}{kd}(d - d')$$

Example 1

Determine the ultimate moment of resistance of a simply supported T-beam having a span of 6 m and cross-sectional dimensions $D = 600$ mm, $d = 600 - 50 = 550$ mm $b_f = 1000$ mm, $D_f = 100$ mm, $b_w = 250$ mm. Assuming M20 concrete and Fe 415 steel and reinforced with 6 nos. 25Φ.

$$A_{st} = 6 \times (\pi / 4) \times 25^2 = 2946 \text{ mm}^2$$

For M20 concrete and Fe 415 steel,

$$x_{u,\max} / d = 0.48 \rightarrow x_{u,\max} = 0.48 \times 550 = 264 \text{ mm}$$

- First assuming, $x_u \leq D_f$ and $x_u \leq x_{u,\max}$
  Considering force equilibrium,

$$\begin{aligned} & C_u = T_u \\ & \rightarrow 0.36 \times f_{ck} \times b_f \times x_u = 0.87 \times f_d \times A_{st} \\ & \rightarrow x_u = (0.87 \times 415 \times 2946) / (0.36 \times 20 \times 1000) \\ & \rightarrow x_u = 147.73 \text{ mm} > D_f = 100 \text{ mm} \end{aligned}$$

  Hence, this value of x_u is not correct as $x_u > D_f$.
- As $x_u > D_f$, the neutral axis falls in the web and the compression in the web is given by

$$\begin{aligned} C_{uw} &= 0.36(f_{ck})(b_x)(x_u) \\ &= (0.36)(20)(250)(x_u) \\ &= (1800 \times x_u)\text{N} \end{aligned}$$

  Assuming $x_u \geq 7/3 \times D_f = 7/3 \times 100 = 233.33$ mm, the compression in the flange is given by

$$\begin{aligned} C_{uf} &= 0.447 \times f_{ck} (b_f - b_w) D_f \\ &= 0.447 \times 20 \times (1000 - 250) \times 100 \\ &= 670{,}500 \text{ N} \end{aligned}$$

Also assuming $x_u \leq x_{u,max} = 264$ mm

$$\begin{aligned} T_u &= 0.87 \times (f_y)(A_{st}) \\ &= 0.87 \times 415 \times 2946 \\ &= 1{,}063{,}653.3 \text{ N} \end{aligned}$$

Applying force equilibrium,

$$1800 \times x_u + 670{,}500 = 1{,}063{,}653.3$$
$$\rightarrow x_u = 218.42 \text{ mm} < 7/3 \times D_f = 7/3 \times 100 = 233.33 \text{ mm}$$

Hence, this value of x_u is also not correct.

- As $D_f < x_u < 7/3 \times D_f$, the depth γ_f of the equivalent concrete stress block is obtained as:

$$\begin{aligned}\gamma_f &= 0.15 \times x_u + 0.65 \times D_f \\ &= 0.15 \times x_u + 0.65 \times 100 \\ &= (0.15 \times x_u + 65)\text{mm}\end{aligned}$$

Hence,

$$\begin{aligned}C_{\text{uf}} &= 0.447(f_{\text{ck}})(b_f - b_w)(\gamma_f) \\ &= 0.447(20)(1000 - 250) \times (0.15(x_u) + 65) \\ &= (1005.75(x_u) + 435{,}825)\text{N}\end{aligned}$$

$$C_{\text{uw}} + C_{\text{uf}} = T_u$$
$$\rightarrow 1800 \times x_u + 1005.75 \times x_u + 435{,}825 = 1{,}063{,}653.3$$
$$\rightarrow x_u = 223.76 \text{ mm} < x_{u,\max} = 264 \text{ mm}$$

So, $\gamma_f = (0.15 \times x_u + 65)$ mm $= (0.15 \times 223.76 + 65) = 98.56$ mm

Therefore,

$$\text{Ultimate moment of resistance, } M_{\text{uR}} = C_{\text{uw}} \times (d - 0.416 \times x_u) + \text{C}_{\text{uf}} \times (\text{d} - \text{y}_\text{f}/2)$$

$$= 1800(x_u)(d - 0.416 \times x_u) + 1005.75(x_u) + 435{,}825) \times (550 - 98.56/2)$$

$$= 1800 \times 223.76 \times (550 - 0.416 \times 223.76) + (1005.75 \times 223.76 + 435{,}825)$$
$$\times (550 - 98.56/2)$$

$$= 514{,}942{,}716.6 \text{ N-mm} = 514.94 \text{ kN-m}$$

Example 2

Determine the ultimate moment of resistance of a simply supported T-beam having a span of 6m and cross-sectional dimensions $D = 600$ mm, $d = 600-50 = 550$ mm $b_f = 1000$ mm, $D_f = 100$ mm, $b_w = 250$ mm. Assuming M20 concrete and Fe 415 steel and reinforced with 5 nos. 28Φ.

$$A_{st} = 5\times(\pi/4)\times 28^2 = 3079\ \text{mm}^2$$

For M20 concrete and Fe 415 steel,

$$x_{u,\max}/d = 0.48 \rightarrow x_{u,\max} = 0.48\times 550 = 264\ \text{mm}$$

- First assuming, $x_u \le D_f$ and $x_u \le x_{u,max}$
  Considering force equilibrium,

$$\begin{aligned}
&C_u = T_u\\
&\rightarrow 0.36(f_{ck})(b_f)(x_u) = 0.87(f_y)(A_{st})\\
&\rightarrow x_u = ((0.87)(415)(3079))/((0.36)(20)(1000))\\
&\rightarrow x_u = 154.4\ \text{mm} > D_f = 100\ \text{mm}
\end{aligned}$$

  Hence this value of x_u is not correct as $x_u > D_f$.
- As $x_u > D_f$, the neutral axis falls in the web and the compression in the web is given by

$$\begin{aligned}
C_{uw} &= 0.36(f_{ck})(b_w)(x_u)\\
&= 0.36\times 20\times 250\times x_u\\
&= (1800\times x_u)\text{N}
\end{aligned}$$

  Assuming $x_u \ge 7/3 \times D_f = 7/3 \times 100 = 233.33$ mm, the compression in the flange is given by

$$\begin{aligned}
C_{uf} &= 0.447(f_{ck})(b_f - b_w)(D_f)\\
&= 0.447(20)(1000-250)\times 100\\
&= 670{,}500\ \text{N}
\end{aligned}$$

  Also assuming $x_u \le x_{u,max} = 264$ mm

$$\begin{aligned}
T_u &= 0.87\times f_y\times A_{st}\\
&= 0.87\times 415\times 3079\\
&= 1{,}111{,}672.95\ \text{N}
\end{aligned}$$

Applying force equilibrium,

$$1800 \times x_u + 670{,}100 = 1{,}111{,}672.95$$
$$\rightarrow x_u = 245.1 \text{ mm} > 7/3 \times D_f = 7/3 \times 100 = 233.33 \text{ mm}$$

- Therefore,

Ultimate moment of resistance, $M_{uR} = C_{uw} \times (d - 0.416 \times x_u)$

$$+ C_{uff} \times (d - D_f/2)$$
$$= 1800 \times x_u \times (d - 0.416 \times x_u) + 670{,}500 \times (550 - 100/2)$$
$$= 1800 \times 245.1 \times (550 - 0.416 \times 245.1) + 670{,}500 \times (550 - 100/2)$$
$$= 532{,}915{,}581.3 \text{ N-mm}$$
$$= 532.92 \text{ kN-m}$$

DOUBLY REINFORCED SECTIONS

If there is some restriction of depth or width of the headroom consideration or some other reasons and demand of the same considering singly reinforced section or economic considerations, doubly reinforced concept is proposed. It is a balanced section having higher moment of resistance than singly reinforced balanced section. By providing additional steel in tension face as well providing steel in compression face, the moment capacity is increased as required.

Example 3

A simply supported T-beam a cross-sectional dimensions $D = 600$ mm, Effective cover = 50 mm, $d = 600 - 50 = 550$ mm $b_f = 1000$ mm, $D_f = 100$ mm, $b_w = 250$ mm, is subjected to a bending moment of 100 kNm. Compute the stresses in concrete and steel. M20 concrete and Fe 415 steel and the beam is reinforced with 2 nos. 25Φ. Find also moment of resistance (Figure 4.14).

For M20 concrete, m = 13, $A_{st} = 2 \times (\pi/4) \times 25^2 = 982 \text{ mm}^2$

- First assuming, neutral axis depth $kd \le D_f$ and equating moments of compression area and tension area about the neutral axis,

$$b_f \times (kd)^2/2 = m(A_{st})(d - kd)$$
$$\rightarrow 1000 \times (kd)^2/2 = 13 \times 982 \times (550 - kd) \rightarrow kd = 106.42 \text{ mm}$$

$kd > D_f = 100$ mm, so the initial assumption $kd \le D_f$ is incorrect. Considering $kd > D_f$, ie. the neutral axis is located in the web

$$(b_f - b_w) \times D_f \times (kd - D_f/2) + b_w \times (kd)^2/2 = m \times A_{st} \times (d - kd)$$
$$\rightarrow (1000 - 250) \times 100 \times (kd - 100/2) + 250 \times (kd)^2/2 = 13 \times 982 \times (550 - kd)$$
$$\rightarrow kd = 106.56 \text{ mm}$$

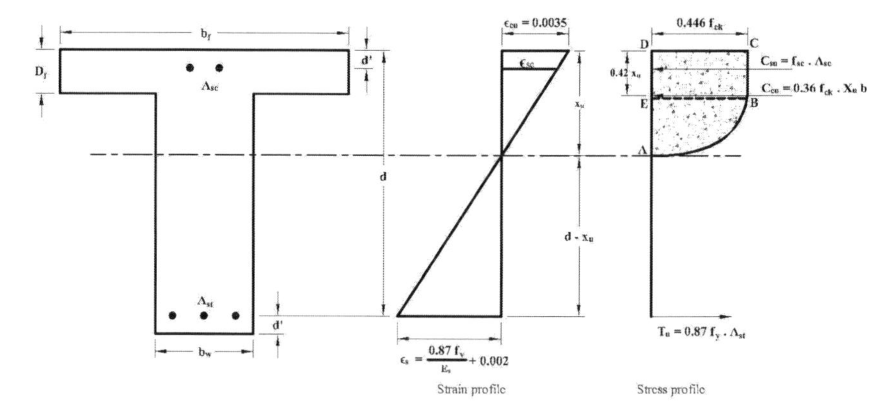

FIGURE 4.14 Stress distribution diagram for doubly reinforced flanged beam.

We know,

Compressive force $C = 0.5 \times f_c \times [b_f \times (kd) - (b_f - b_w) \times (kd - D_f)^2 / (kd)]$

Taking moments of forces about tension steel,

$$M = 0.5(f_c)\Big[b_f(kd)(d - kd/3) - (b_f - b_w)(kd - D_f)^2/(kd)\big(d - D_f - (kd - D_f)/3\big)\Big]$$

$$\rightarrow 100 \times 10^6 = 0.5 \times f_c \times [1000 \times 106.56 \times (550 - 106.56/3) - (1000 - 250)$$

$$\times (106.56 - 100)^2/(106.56) \times (550 - 100 - (106.56 - 100)/3)] \rightarrow f_c = 3.66 \text{ N/mm}^2$$

We know $C = T$

$$0.5 \times f_c \times \Big[b_f \times (kd) - (b_f - b_{wx}) \times (kd - D_f)^2 / (kd)\Big] = A_{st} \times f_{stt}$$

$$\rightarrow 0.5 \times 3.66 \times \big[1000 \times (106.56) - (1000 - 250) \times (106.56 - 100)^2 / (106.56)\big] = 982 \times f_{st}$$

$$\rightarrow f_{st} = 198 \text{ N / mm}^2$$

If steel stress is equal to allowable stress, $\sigma_{st} = 230$ N/mm^2

Stress in concrete, $f_c = kd/(d - kd) \times (1/m) = 106.56/(550 - 106.56) \times (1/13) = 4.25$ N/mm^2

Therefore, it is an under-reinforced section

Hence, Moment of resistance

$$M = 0.5(f_c)\Big[(b_f)(kd)(d - kd / 3) - (b_f - b_w)(kd - D_f)^2 / (kd)\big(d - D_f - (kd - D_f) / 3\big)\Big]$$

$$= (0.5)(4.25)\big[(1000)(106.56)(550 - 106.56 / 3) - (1000 - 250)(106.56 - 100)^2 /$$

$$(106.56)(550 - 100 - (106.56 - 100) / 3)\big]$$

$$= 116.21 \text{ kN-m}$$

Example 4

Determine the ultimate moment of resistance of a simply supported T-beam having a span of 6 m and cross-sectional dimensions $D = 600\,\text{mm}$, $d = 600 - 50 = 550\,\text{mm}$ $b_f = 1000\,\text{mm}$, $D_f = 100\,\text{mm}$, $b_w = 250\,\text{mm}$. Assuming M20 concrete and Fe 415 steel & reinforced with 6 nos. 25Φ.

$$A_{st} = 6 \times (\pi / 4) \times 25^2 = 2946\ \text{mm}^2$$

For M20 concrete and Fe 415 steel,

$$x_{u,\max} / d = 0.48 \rightarrow x_{u,\max} = 0.48 \times 550 = 264\ \text{mm}$$

- First assuming, $x_u \le D_f$ and $x_u \le x_{u,max}$
  Considering force equilibrium,

$$C_u = T_u$$
$$\rightarrow 0.36 \times f_{ck} \times b_f \times x_u = 0.87 \times f_y \times A_{st}$$
$$\rightarrow x_u = (0.87 \times 415 \times 2946) / (0.36 \times 20 \times 1000)$$
$$\rightarrow x_u = 147.73\ \text{mm} > D_f = 100\ \text{mm}$$

  Hence, this value of x_u is not correct as $x_u > D_f$.
- As $x_u > D_f$, the neutral axis falls in the web and the compression in the web is given by

$$C_{uw} = 0.36 \times f_{ck} \times b_w \times x_u$$
$$= 0.36 \times 20 \times 250 \times x_u$$
$$= (1800 \times x_u)\,\text{N}$$

  Assuming $x_u \ge 7/3 \times D_f = 7/3 \times 100 = 233.33\,\text{mm}$, the compression in the flange is given by

$$C_{uf} = 0.447 \times f_{ck} \times \left(b_f - b_{wv}\right) \times D_f$$
$$= 0.447 \times 20 \times (1000 - 250) \times 100$$
$$= 670{,}500\ \text{N}$$

  Also assuming $x_u \le x_{u,max} = 264\,\text{mm}$

$$T_u = 0.87 \times f_y \times A_{st}$$
$$= 0.87 \times 1415 \times 2946 = 1{,}063{,}653.3\ \text{N}$$

Applying force equilibrium,

$$1800 \times x_u + 670{,}500 = 1{,}063{,}653.3$$
$$\rightarrow x_u = 218.42 \text{ mm} < 7/3 \times D_f = 7/3 \times 100 = 233.33 \text{ mm}$$

Hence, this value of x_u is also not correct.

- As $D_f < x_u < 7/3 \times D_f$, the depth y_f of the equivalent concrete stress block is obtained as:

$$\gamma_f = 0.15 \times x_u + 0.65 \times D_f$$
$$= 0.15 \times x_u + 0.65 \times 100$$
$$= (0.15 \times x_u + 65)\text{mm}$$

Hence,

$$C_{uf} = 0.447 \times f_{ck} \times (b_f - b_w) \times \gamma_f$$
$$= 0.447 \times 20 \times (1000 - 250) \times (0.15 \times x_u + 65)$$
$$= (1005.75 \times x_u + 435{,}825)\text{N}$$

- $C_{uw} + C_{uf} = T_u$

$$\rightarrow 1800 \times x_u + 1005.75 \times x_u + 435{,}825 = 1{,}063{,}653.3$$

$$\rightarrow x_u = 223.76 \text{ mm} < x_{u,\max} = 264 \text{ mm}$$

$$\text{So, } \gamma_f = (0.15 \times x_u + 65)\text{mm}$$
$$= (0.15 \times 223.76 + 65) = 98.56 \text{ mm}$$

- Therefore,

$$\text{Ultimate moment of resistance, } M_{uR} = C_{uw} \times (d - 0.416 \times x_u) + C_{uf} \times (d - \gamma_f / 2)$$

$$= 1800 \times x_u \times (d - 0.416 \times x_u) + (1005.75 \times x_u + 435{,}825) \times (550 - 98.56/2)$$
$$= 1800 \times 223.76 \times (550 - 0.416 \times 223.76)$$
$$+ (1005.75 \times 223.76 + 435{,}825) \times (550 - 98.56/2)$$
$$= 514{,}942{,}716.6 \text{ N-mm} = 514.94 \text{ kN-m}$$

Example 5

Determine the ultimate moment of resistance of a simply supported T-beam having a span of 6m and cross-sectional dimensions $D = 600\,\text{mm}$, $d = 600 - 50 = 550\,\text{mm}$ $b_f = 1000\,\text{mm}$, $D_f = 100\,\text{mm}$, $b_w = 250\,\text{mm}$. Assuming M20 concrete and Fe 415 steel & reinforced with 5 nos. 28Φ.

$$A_{st} = 5 \times (\pi / 4) \times 28^2 = 3079 \text{ mm}^2$$

For M20 concrete and Fe 415 steel,

$$x_{u,\max} / d = 0.48 \rightarrow x_{u,\max} = 0.48 \times 550 = 264 \text{ mm}$$

- First assuming, $x_u \leq D_f$ and $x_u \leq x_{u,max}$
  Considering force equilibrium,

$$C_u = T_u$$
$$\rightarrow 0.36 \times f_{ck} \times b_f \times x_u = 0.87 \times f_y \times A_{st}$$
$$\rightarrow x_u = (0.87 \times 415 \times 3079) / (0.36 \times 20 \times 1000)$$
$$\rightarrow x_u = 154.4 \text{ mm} > D_f = 100 \text{ mm}$$

  Hence this value of x_u is not correct as $x_u > D_f$.
- As $x_u > D_f$, the neutral axis falls in the web and the compression in the web is given by

$$C_{uw} = 0.36 \times f_{ck} \times b_w \times x_u$$
$$= 0.36 \times 20 \times 250 \times x_u$$
$$= (1800 \times x_u)\text{N}$$

  Assuming $x_u \geq 7/3 \times D_f = 7/3 \times 100 = 233.33\,\text{mm}$, the compression in the flange is given by

$$C_{uf} = 0.447 \times f_{ck} \times \left(b_f - b_w\right) \times D_f$$
$$= 0.447 \times 20 \times (1000 - 250) \times 100$$
$$= 670{,}500 \text{ N}$$

  Also assuming $x_u \leq x_{u,max} = 264\,\text{mm}$

$$T_u = 0.871 \times f_y \times A_{stt} = 0.87 \times 415 \times 3079 = 1{,}111{,}672.95 \text{ N}$$

  Applying force equilibrium,

$$1800 \times x_u + 670{,}500 = 1{,}111{,}672.95$$
$$\rightarrow x_u = 245.1 \text{ mm} > 7 / 3 \times D_f = 7 / 3 \times 100 = 233.33 \text{ mm}$$

- Therefore,

Ultimate moment of resistance, $M_{uR} = C_{uw} \times (d - 0.416 \times x_u) + C_{uf} \times (d - D_f / 2)$

$= 1800 \times x_u \times (d - 0.416 \times x_u) + 670{,}500 \times (550 - 100 / 2)$

$= 1800 \times 245.1 \times (550 - 0.416 \times 245.1) + 670{,}500 \times (550 - 100 / 2)$

$= 532915581.3$ N-mm

$= 532.92$ kN-m

DESIGN UNDER SHEAR

The shear-resisting mechanisms are related to the bond between concrete and the embedded reinforcement as well as anchorage at the end. Shear capacity depends on the tensile and compression strength of the concrete. Shear failure is non-ductile. For earthquake-resistant RCC structures insufficient attention was paid on ductility through IS13920 and IS 4326. The designer must ensure that a shear failure does not occur as well as ductility is ensured. The analogy between the shear resistance of a parallel chord truss and a web-reinforced concrete beam is one of the concepts. The analogy, postulated implies that the web of the equivalent truss consists of stirrups acting as tension members and concrete struts running parallel to diagonal. The flexural concrete compression zone and the flexural reinforcement form the top and bottom chords of this analogous pin-jointed truss. The forces in the truss can be determined from considerations of equilibrium only.

The slope of the compression diagonals has been traditionally assumed to be 45° to the beam axis. It has been observed, however, that the slope of the diagonal cracks at the boundaries of the struts varies along the beam. This is often the case, and design equations based on compression struts at 45° are conservative. On the other hand, the struts are steeper in the vicinity of point loads. However, in these areas, local arch action boosts the capacity of the other shear-carrying mechanisms. Generally a beam has high concrete strength and low web steel content, representing a less rigid tension system. The compression struts arc at an angle less than 45°, hence the stirrups arc is more effective than in a 45° truss.

The transverse (external) shear force is denoted as V (and has a maximum value near the support, equal to the support reaction). It is resisted by various mechanisms, the major ones being:

1. Shear resistance V_{cz} of the uncracked portion of concrete;
2. Vertical component of the 'interface shear' (*aggregate interlock*) force V_a;
3. Dowel force V_d in the tension reinforcement (due to dowel action); and
4. Shear resistance V_s carried by the shear (transverse) reinforcement, if any.

The interface shear V_a is a tangential force transmitted along the inclined plane of the crack, resulting from the friction against relative slip between the interlocking surfaces of the crack. Its contribution can be significant, if the crack width is limited. The dowel force V_d comes from 'dowel action' of longitudinal reinforcement.

Examples 6

A reinforced concrete beam of 250 mm and 450 mm effective depth is subjected to shear force of 110 kN at the support. The tensile reinforcement at the support is 0.8%. Find the spacing of 12 mm diameter 2-legged stirrup to resist the shearing stress at supports, for M25 concrete, use Fe415 HYSD bars. Also design the minimum shear reinforcement at the mid-span of the beam.

Solution:

$$P = \frac{100A_{st}}{bd} = 0.8\%$$

Hence from table 19 for M25 concrete, we get $\tau_c = 0.58$ N/mm^2

From Table 20, maximum shear stress $\tau_{c\ max} = 3.1$ N/mm^2

The nominal shear stress in the beam is $\tau_v = \frac{V}{bd} = \frac{110 \times 1.5 \times 1000}{250 \times 450} = 1.47 \text{ N/mm}^2$

This is greater than the permissible shear stress t_c but less than the maximum shear stress $t_{c\ max}$.

The shear reinforcement is provided to carry a shear V_s equal to $V - \tau_c \cdot bd$.

$$V_s = V - \tau_c bd = \left(1.5 \times 110 \times 10^3\right) - (0.58 \times 250 \times 450) = 99{,}750 \text{ N}$$

$$S_v = \frac{0.87 f_y A_{sv} d}{V_{us}} = \frac{0.87 \times 415 \times A_{sv} \times 450}{99{,}750} A_{xv} = 1.63 A_{sv} \text{ mm.}$$

Using 12 mm φ bars two-legged stirrups,

$$A_{sv} = 2 \times \frac{\pi}{4} \times (12)^2 = 226.2 \text{ mm}^2$$

$$S_v = 1.63 \times 226.2 = 369 \text{ mm.}$$

Subject to a maximum of 0.75 $d = 0.75 \times 450 = 337.5$ mm.

Hence provide the stirrups @ 337.5 mm c/c.

For nominal (minimum) shear reinforcement, let us use 8 mm dia bars.

$$A_{xv} = 2 \times \frac{\pi}{4} \times (8)^2 = 100.53 \text{ mm}^2$$

$$S_v = \frac{0.87 \times 415 \times A_{sv} \times 2.5}{250} = 363 \text{ mm.}$$

Subject to a maximum spacing equal to 0.75 $d = 0.75 \times 450 = 337.5$ mm

Hence provide 8 mm dia. 2-legged stirrups at 337.5 mm spacing at the location where nominal reinforcement is required.

Example 7

A simply supported beam, 300 mm wide and 500 mm effective depth carries a uniformly distributed load of 50 kN/m including its own weight, over an effective span of 4 meters. Design the shear stirrups in the form of vertical stirrups. Use M15 concrete. Take $\sigma_{xv} = \sigma_{xv} = 140$ N/mm² and $f_y = 250$ N/mm². Assume that the beam contains 0.75% reinforcement throughout the length. using HYSD bars, taking $\sigma_{xt} = \sigma_{xv} = 230$ N/mm² and $f_y = 415$ N/mm².

$$V = \frac{wL}{2} = \frac{1.5 \times 50{,}000 \times 6}{2} = 225{,}000 \text{ N}$$

$$\tau_v = \frac{V_u}{bd} = \frac{225{,}000}{300 \times 550} = 1.364 \text{ N/mm}^2$$

For $\frac{100As}{bd} = 0.8\%$, we get $\tau_c = 0.58$ N/mm² from Table 19.

Hence shear reinforcement is necessary at the ends.

Using 8 mm dia 2- lgd vertical stirrups, $A_{sv} = 2\frac{\pi}{4}(8)^2 = 100.53 \text{ mm}^2$

Now $V_s = V - \tau_c\, bd = 225{,}000 - 0.58 \times 300 \times 550 = 129{,}300$ N

$$\text{S}_\text{v} = \frac{0.87 \times 415 \times \text{A}_\text{SV} \times 550}{129{,}300} = 154.4 \text{ mm say } 150 \text{ mm}$$

Maximum spacing of stirrups is given by

$$S_v = \frac{2.5\,\text{A}_\text{XV} \cdot f_y}{V_s} = \frac{0.87 \times 415 \times 100.53 \times 2.5}{300} = 300 \text{ mm} \tag{4.1}$$

In no case should the spacing exceed $0.75d = 0.75 \times 550 = 412.5$ mm

Hence the spacing is to be varied from 150 mm at the end section to 300 mm at a section distant x m from the mid-span where minimum shear reinforcement, given by Eq. 4.1 is to be provided.

$$\text{At that section}, V_s = \frac{V}{2}(x) = \frac{225{,}000}{2}x = 112{,}500x$$

$$V = V_x - \tau_c bd = 112{,}500x - 0.58 \times 300 \times 550 = 112{,}500x - 95{,}700$$

$$S_v = \frac{\sigma_{\text{XV}} \cdot A_{\text{XV}} \cdot d}{V_s} = \frac{0.87 \times 415 \times 100.53 \times 550}{112{,}500x - 97{,}500} \tag{4.2}$$

Equating 4.1 and 4.2, we get $\frac{0.87 \times 415 \times 100.53 \times 550}{112{,}500x - 97{,}500} = 300$

From which we get $x = 1.44$ m

Thus vary the spacing from 150 mm c/c at supports to 300 mm c/c at 1.44 m from mid-span. For the remaining length, provide the stirrups at 300 mm c/c.

Example 8

A singly reinforced concrete beam has an effective depth of 650 mm and width of 300 mm and is reinforced with 6 bars of 20 mm diameter at the center of the span. The maximum S.F. at the ends is 180 kN. Design the stirrup reinforcement required. Take $\sigma_{xv} = 230$ N/mm Use M25 concrete.

For M 15 concrete J = 0.867

Lever arm $a = jd = 0.867 \times 600 = 520$ mm

Let us bend two bars at a distance = 1.414 $a = 1.414 \times 520 = 735$ mm from the support.

Shear resistance of the two bars bent at 45° is

$$V_{sl} = A_{xv} \cdot \sigma_{xv} \sin 45° = 0.707 \cdot A_{xv} \cdot \sigma_{xv} = 0.707\left\{2 \times \frac{\pi}{4} \times (20)^2\right\} \times 140 = 62{,}190 \text{ N.}$$

However, maximum contribution toward shear resistance shall not be more than half the total shear reinforcement.

$$\text{Also, } \frac{100As}{bd} = \frac{100 \times 4\left(\frac{\pi}{4}\right)(20)2}{300 \times 600} = 0.7\%$$

Hence from Table 3.1, $\tau_c = 0.33$ N/mm²

$$V_c = \tau_c \cdot bd = 0.33 \times 300 \times 600 = 59{,}400 \text{ N}$$

$$V_s = V - V_c = 160{,}000 - 59{,}400 \text{ N} = 100{,}600 \text{ N}$$

Max. contribution V_{sl} due to bent up bars $= \frac{1}{2} V_s = \frac{1}{2} \times 100{,}600 = 50{,}300$ N.

$$V_{s2} \text{ from stirrups } = 50{,}300 \text{ N.}$$

Using 8 mm φ2-legged stirrups $A_{sv} = 2\frac{\pi}{4}(8)^2 = 100.5 \text{ mm}^2$

$$S_v = \frac{\sigma_{xv} \cdot A_{xv} \cdot d}{V_{s2}} = \frac{140 \times 100.5 \times 600}{50{,}300} = 167.8 \text{ mm.}$$

This spacing should not be more than S_v corresponding to minimum reinforcement requirement.

$$\text{i.e. } S_v \leq \frac{2.5 \cdot A_{xv} \cdot f_y}{b} = \frac{2.5 \times 100.5 \times 250}{300} = 209 \text{ mm}$$

Also, spacing should not be more than 0.75 $d = 0.75 \times 600 = 450$ mm

Nor more than 450 mm.

Hence provide 2- lgd 8 mm dia. Stirrups @ 160 mm c/c.

DESIGN UNDER TORSION

Torsion usually occurs in combination with flexure and transverse shear. The interactive behavior of torsion with bending moment and flexural shear in reinforced concrete beams is a bit complex, considering the non-homogeneous, nonlinear and composite nature of concrete in the presence of cracks. Codes prescribe highly simplified design procedures, which provide a proper place of theoretical considerations and experimental results. In reinforced concrete design, the terms **equilibrium torsion** and **compatibility torsion** generally refer to two different torsion-inducing situations. In **equilibrium torsion**, the torsion is induced by an eccentric loading (w r t shear center), and equilibrium conditions alone suffice in determining the twisting moments. In **compatibility torsion**, the torsion is induced due to the member undergoing an angle of twist to maintain deformation compatibility, and the resulting twisting moment depends on the torsional stiffness of the member. This is associated with twisting moments that are developed in a structural member to maintain static equilibrium with the external loads and are independent of the torsional stiffness of the member. Such torsion must be necessarily considered in design (clause 41.1). The magnitude of the twisting moment does not depend on the torsional stiffness of the member and is entirely determinable from statics alone. The member has to be designed for the full torsion, which is transmitted by the member to the supports.These moments are generally statically indeterminate and their analysis necessarily involves (rotational) compatibility conditions, so it is termed as compatibility **torsion**. In statically indeterminate structures, the torsional restraints are 'redundant', and releasing such redundant restraints will eliminate the compatibility torsion. As per clause 41.1-IS 456, where torsion can be eliminated by releasing redundant restraints, no specific design for torsion is necessary, provided torsional stiffness is neglected in the calculation of internal forces.

ESTIMATION OF TORSIONAL STIFFNESS

The observed behavior of reinforced concrete members under torsion [Section 7.3- IS456] shows that the torsional stiffness is little influenced by the amount of torsional reinforcement in the linear elastic phase, and may be taken as that of the plain concrete section. However, once torsional cracking occurs, there is a drastic reduction in the torsional stiffness. The post-cracking torsional stiffness is only a small fraction (less than 10 percent) of the pre-cracking stiffness and depends on the amount of torsional reinforcement, provided in the form of closed stirrups and longitudinal bars. Heavy torsional reinforcement can, no doubt, increase the torsional resistance (strength) to a large extent, but this can be realized only at very large angles of twist (accompanied by very large cracks). Hence, even with torsional reinforcement provided, in most practical situations, the maximum twisting moment in a reinforced concrete member under compatibility torsion is the value corresponding to the torsional cracking of the member. This 'cracking torque' is very nearly the same as the failure strength obtained for an identical plain concrete section.

Numerical Example 9

A fixed ended beam AB as shown on the plan in Figure 4.15 is subjected to a uniformly distributed live load of 30 kN/m and a concentrated point live load *P*. The distance of the point of application of the concentrated load *P*, $a = 1$ m. Design the beam against the torsional moment, bending moment, and shear force.

Considering, width of the beam = 250 mm and overall depth = 600 mm

$$\text{Self-Weight} = (25 \times 0.25 \times 0.6)\ \text{kN/m} = 3.75\ \text{kN/m}$$

Bending Moment at end support (M_1), due to the uniformly distributed load of 30 kN/m

And self-weight of the beam = (3.75 + 30) $(6)^2/12$ = 101.25 kNm

Bending Moment at end support (M_2), due to the concentrated load $P = 15$ kN:

$$= PL/8 = 15(6)/8\ \text{kNm} = 11.25\ \text{kNm}$$

Torsional Moment at end (T) = $P_a/2$ = 15 × ½ = 7.5 kNm

Design bending moment (M) = $M_1 + M_2$ = 101.25 + 11.25 = 112.5 kNm

Design torsional moment (T) = 7.5 kNm

$$\text{Factored design bending moment}\left(M_u\right) = 1.5\left(M_1 + M_2\right)$$

$$= 1.5(101.25 + 11.25) = 168.75\ \text{kNm}$$

Factored design torsional moment (T_u) = (1.5 × T) = 11.25 kNm

According to IS 456, the longitudinal reinforcement shall be designed to resist an

$$\text{Equivalent bending moment}\left(M_{e1}\right) = M_u + M_t$$

Where,

M_u = Factored design bending moment and

M_t = Factored equivalent bending moment of the torsional moment (T_u)

$$= T_u\left(1 + D/b\right)/1.7 = 11.25\left[\left(1 + 600/250\right)/1.7\right] = 22.5\ \text{kNm}$$

Equivalent bending moment M_{e1} = (168.75 + 22.5) kNm = 191.25 kNm

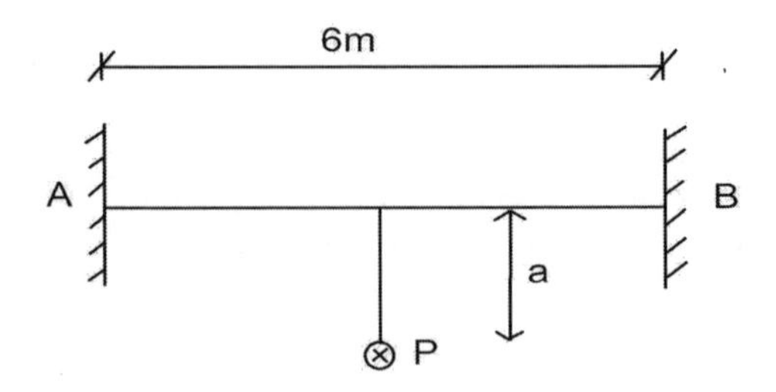

FIGURE 4.15 Plan of a fixed beam under eccentric vertical load.

Considering, M25 grade concrete, Fe500 grade HYSD bars
$Q_b = 3.34$ N/mm² (from SP-16, page-10)
Let us consider 20Φ bars then, $d = (600 - 20/2 - 25)$ mm = 565 mm

$$Q = M_{e1} / bd^2 = \left(191.25\times10^6\right)/\left(250\times565^2\right)$$

$$= 2.40 \text{ N/mm}^2$$

From SP 16, Page-49, $p_t = 0.632\%$

$$A_{st} = (0.632/100)(250\times565)\text{mm}^2 = 892.7 \text{ mm}^2$$

According to IS 456, Equivalent Shear, $V_e = V_u + 1.6\ (T_u/b)$

$$V_u = \text{Factored design shear force } = 1.5[(33.75(6/2)) + 15/2]\text{kN} = 163.125 \text{ kN}$$

$$V_e = 163.125 + 1.6(11.25/0.250) = 235.125 \text{ kN}$$

Equivalent Nominal Shear Stress (ζ_{ve}), as per clause 40.1(IS 456:2000)

$$V_e / bd = \left(235.125\times10^3 / (250\times(600-25-20/2-8))\right) = 1.69 \text{ N/mm}^2$$

$\zeta_{ve} < \zeta_{c\,max} = 3.1$ N/mm² from Table 20 of IS456
For $p_t = 0.632\%$, ζ_c from Table 19, IS456

$$\zeta_c = 0.49 + (0.57-0.49)(0.632-0.49)/(0.75-0.50) = 0.535 \text{ N/mm}^2$$

$\zeta_{ve} > \zeta_c$, Therefore, transverse reinforcement is to be designed.

TRANSVERSE REINFORCEMENT

2-legged closed hoops/stirrups, enclosing the corner longitudinal bars, shall have an area of cross section A_{sv} given as;

$$A_{sv} = \left[T_u S_v / (0.87 f_y\ bd\ )\right] + \left[(V_u S_v)/2.5 d_1 (0.87 f_y)\right]$$

Considering $S_v = 200$ mm c/c, $b_1 = 250 - 25 - 25 - 20/2 - 20/2 = 180$ mm and

$$d_1 = 600 - 25 - 25 - 20/2 - 20/2 = 530 \text{ mm}$$

Therefore,

$$A_{sv} = \left[\left(11.25\times10^6(250)\right)/(180\times530\times0.87 f_y)\right]$$

$$+\left[\left(163.125\times10^3\times200\right)/(2.5\times530\times0.87\times500)\right]$$

$$= 67.773 + 56.604 \text{ mm}^2 = 124.377 \text{ mm}^2$$

TABLE 4.1
Comparative Result of Different Torsion Related Design Problems

Cases	b = Breadth (mm)	D = Depth (mm)	P (kN)	A_{st} Required (mm²)	A_{sv} Required (mm²)
1	250	600	15	892	132
2	250	600	20	963	154
3	250	600	25	1038	174
4	300	600	25	1000	154
5	350	600	25	979	138

However, minimum transverse reinforcement shall not be less than

$$A_{sv,\min} = \left[\left(\zeta_v - \zeta_c\right) \times bS_v\right] / (0.87 f_y)$$
$$= [(1.69 - 0.535) \times 250 \times 200] / (0.87 \times 500)\text{mm}^2 = 132.759 \text{ mm}^2$$

A few cases are designed, and some useful information is observed and furnished in Table 4.1

It is quite evident that when the concentrated load (P) increases, both shear force and torsional moment increase. If the beam section is kept unchanged, then the requirement for longitudinal reinforcement and transverse reinforcements increases with the increase of torsional moment due to increase of concentrated load P. However, if the width is increased, keeping the depth unchanged, then the requirement for longitudinal and transverse reinforcements decreases. Torsional moment plays a major role in the design output and width plays a major role.

Numerical Examples 10

Calculate the flexural longitudinal reinforcement required for a simply supported beam of span

$l = 6$ m subjected to an imposed uniformly distributed load $w_L = 40$ kN/m (Dl +LL) over its entire span, using M25 concrete and Fe 415 steel. Use SP16. Apply LSD. Check against deflection and crack width.

$$M_u = 1.5 \times M = 1.5 \times 211.5 \text{ kNm} = 317.25 \text{ kN-m}$$

$$M_u / bd^2 = 317.25 \times 10^6 / \left(350 \times 750^2\right) = 1.61 \text{ N/mm}^2$$

$$P_t = 0.4854 \left(\text{SP-16, table} - 3\right)$$

$$A_{st} = p_t / 100 \times b \times d = 0.4854/100 \times 350 \times 750 = 1274.2 \text{mm}^2$$

Check against Deflection

Short-Term Deflection

As per IS456,

$$I_{\text{eff}} = \frac{I_r}{1.2 - \frac{M_r}{M}\frac{z}{d}\left(1 - \frac{x}{d}\right)\frac{b_W}{b}}$$

where

I_r = moment of inertia of the cracked section

$M_r = f_{cr}\, I_{gr}/y_t$

f_{cr} = Modulus of rupture of concrete = $0.7\sqrt{f_{ck}} = 0.7\sqrt{25} = 3.5$ N/mm^2

I_{gr} = Moment of inertia of the gross section about the centroidal axis, neglecting the reinforcement

$= bD^3/12 = 350 \times 800^3/12 = 1.49 \times 10^{10}$ mm^4

y_t = distance from the centroidal axis of the gross section to the extreme fibre in tension

$= D/2 = 800/2 = 400$ mm

Therefore, $M_r = f_{cr}\, I_{gr}/y_t = 3.5 \times 1.49 \times 10^{10} / 400 = 13.067 \times 10^7$ N-mm

M = Maximum moment under service loads = 211.5×10^6 Nmm

d = effective depth = 750 mm

x_u / d = depth of neutral axis =

$$\frac{0.87 f_y A_{st}}{0.36 f_{ck}\, bd} = 0.87 \times 415 \times 1274.2 / (0.36 \times 25 \times 350 \times 750) = 0.195$$

z = lever arm = $d - x_u/3 = d\,(1 - x_u/3d)$

b_w = breadth of beam

$$I_{\text{eff}} = \frac{I_r}{1.2 - \frac{M_r}{M}\left(1 - \frac{x_u}{3d}\right)\left(1 - \frac{x_u}{d}\right)\frac{b_W}{b}} = \frac{I_r}{1.2 - \frac{13.067 \times 10^7}{211.5 \times 10^6}\left(1 - \frac{0.195}{3}\right)(1 - 0.195)}$$

$$\frac{I_{\text{eff}}}{I_r} = 1.36$$

$$p_t = 0.4854,\ \ p_c = 0,\ \ E_c = 5000\sqrt{f_{ck}} = 5000\sqrt{25} = 25{,}000 \text{ N/mm}^2,$$

$$E_s = 2 \times 10^5 \text{ N/mm}^2$$

$$m = E_s / E_c = 2 \times 10^5 / 25{,}000 = 8$$

$$\frac{p_c(m-1)}{p_t m} = 0,\ p_t m = 0.4854 \times 8 = 3.88, \frac{d'}{d} = \frac{50}{750} = 0.067$$

Referring to table 88, SP 16

$$\frac{I_r}{\left(\frac{bd^3}{12}\right)} = 0.324$$

$$I_r = 0.324 \times \frac{bd^3}{12} = 0.324 \times 350 \times 750^3/12 = 3.987 \times 10^9 \text{ mm}^4$$

So, $I_{eff} = 1.36 \times I_r = 1.36 \times 3.987 \times 10^9 \text{mm}^4 = 5.42 \times 10^9 \text{mm}^4$

Elastic deflection of a simply supported beam with UDL = $\frac{5ML^2}{48EI_{eff}}$

$$E_c = 5000\sqrt{f_{ck}} = 5000\sqrt{25} = 25{,}000 \text{ N/mm}^2$$

Short-term deflection = $5 \times 211.5 \times 10^6 \times 6000^2 \;/\; 48 \times 25000 \times 5.42 \times 10^9 =$ 5.85 mm

LONG-TERM DEFLECTION

Deflection due to Shrinkage

$$a_{cs} = k_3 \times \Psi_{cs} \times l^2$$

where $k_3 = 0.125$ for simply supported members

$$\Psi_{cs} = k_4 \times \frac{\varepsilon_{cs}}{D}$$

$$p_t - p_c = 0.4854 - 0 = 0.4854$$

$$k_4 = 0.72 \times \frac{(p_t - p_c)}{\sqrt{p_t}} \leq 1 \text{ for } 0.25 \leq p_t - p_c < 1$$

$$= 0.72 \times \frac{(0.4854 - 0)}{\sqrt{0.4854}} = 0.5$$

$$\varepsilon_{cs} = 0.0003 \text{ (Cl.6.2.4 IS - 456)}$$

$$\Psi_{cs} = 0.5 \times 0.0003/800 = 1.88 \times 10^{-7}$$

$$a_{cs} = 0.125 \times 1.88 \times 10^{-7} \times (6000)^2 = 0.846 \text{ mm}$$

Deflection due to Creep

As per IS456

$$a_{cc(perm)} = a_{i,cc(perm)} - a_{i,(perm)}$$

In the absence of data, the age at loading is assumed to be 28 days and the value of creep coefficient, θ is taken as 1.6 from cl 6.2.5 IS 456

$$E_{ce} = E_c/(1+\theta)$$
$$= 25{,}000/(1+1.6) = 9.62\times10^3 \text{ N/mm}^2$$

$$m = E_s/E_{ce} = 2\times10^5/9.62\times10^3 = 20.79$$

$$p_t = 0.4854;\ p_c = 0$$

$$\frac{p_c(m-1)}{p_t\, m} = 0$$

$$p_t m = 0.4854\times20.79 = 10.1$$

Referring to table 88, SP – 16

$$\frac{I_r}{\left(\frac{bd^3}{12}\right)} = 0.683$$

$$I_r = 0.683\times\frac{bd^3}{12} = 0.683\times350\times750^3/12 = 8.4\times10^9 \text{ mm}^4$$

So, $I_{eff} = 1.36\times I_r = 1.36\times8.4\times10^9\text{mm}^4 = 11.424\times10^9\text{mm}^4 < I_{gr} = 1.49\times10^{10}\text{mm}^4$

$$a_{i,cc(perm)} = \frac{5ML^2}{48E_{ce}I_{eff}} = 5\times211.5\times10^6\times6000^2/48\times9.62$$
$$\times10^3\times11.424\times10^9 = 7.22 \text{ mm}$$

$$a_{i,(perm)} = \frac{5ML^2}{48E_C I_{eff}} = 5\times211.5\times10^6\times6000^2/48\times25{,}000\times5.42\times10^9$$
$$= 5.85 \text{ mm}$$

$$a_{cc(perm)} = 7.22 - 5.85 = 1.37 \text{ mm}$$

Total deflection including deflection due to shrinkage and creep

$$= 5.85 + 0.846 + 1.37 = 8.1 \text{ mm}$$

Permissible total deflection = Span/250 = 6000/250 = 24 mm. > 8.1 mm Ok.

Long-term deflection = 0.846 + 1.37 = 2.216 mm

Permissible long-term deflection = Span/350 = 6000/350 = 17.143 mm > 2.216 mm Ok.

Crack Width Analysis

Numerical Examples 11

Calculate the flexural longitudinal reinforcement required for a simply supported beam of span

$l = 6\,\text{m}$ subjected to an imposed uniformly distributed load $w_L = 40$ kN/m (Dl + LL) over its entire span, using M25 concrete and Fe 415 steel. Use SP16. Apply LSD. Check against deflection and crack width.

$$M_u = 1.5 \times M = 1.5 \times 211.5 \text{ kNm} = 317.25 \text{ kN-m}$$

$$M_u / bd^2 = 317.25 \times 10^6 / \left(350 \times 750^2\right) = 1.61 \text{ N/mm}^2$$

$$P_t = 0.4854 (\text{SP-16, table } -3)$$

$$A_{st} = p_t / 100 \times b \times d = 0.4854 \times 350 \times 750 / 100 = 1274.2 \text{ mm}^2$$

$$W_{cr} = \frac{3 a_{cr} \varepsilon_m}{1 + \dfrac{2(a_{cr} - c_{min})}{(h - x)}}$$

$$a_{cr} = 30 \text{ mm}, C_{min} = 25 \text{ mm}$$

$$a = 800 - 50 - 30 \text{ mm} = 720 \text{ mm}$$

$$h = 800 \text{ mm}$$

$$x = 0.195 \times 750 = 146.25 \text{ mm}$$

$$\varepsilon_m = \varepsilon - b(h - x)(a - x) / 3E_s A_s (d - x)$$

$$= 0.87 \times 415 / \left(2 \times 10^5\right) - \frac{350 \times (800 - 146.25) \times (720 - 146.25)}{3 \times 2 \times 10^5 \times 1274.2 \times (750 - 146.25)}$$

$$= 1.81 \times 10^{-3} - 2.84 \times 10^{-4}$$

$$= 1.53 \times 10^{-3}$$

$$W_{cr} = \frac{3 \times 30 \times 1.53 \times 10^{-3}}{1 + \dfrac{2(30 - 25)}{(800 - 146.25)}}$$

$$= 0.136 \text{ mm} < 0.3 \text{ mm Ok.}$$

DUCTILITY OF RCC STRUCTURE

It may be said that the ductility of a building is its capacity to accommodate large lateral deformations, quantified primarily with a ratio of maximum deformation sustained just prior to collapse or significant loss of strength to the deformation at yield. Large inelastic deformation capacity without significant loss of strength capacity, is the most important observation. Buildings are designed and detailed to develop favorable failure mechanisms which involve specified lateral strength and reasonable stiffness. Ductility helps in dissipating input earthquake energy through hysteretic behavior. Ductile chain design concept in building as per capacity design, is extremely popular. When a brittle chain alone is pulled on either side, it would break suddenly. Among the chain, the weakest link would break first. If we make the weakest link as the ductile one, it provides more elongation though failure load may be more or less the same i.e., gains ductile. Similar concept is applied in case building design. Inertia forces due to seismic activity are transmitted from the floors to the beams and then to the columns. The stability of the building is affected mostly failure of the column rather than the failure of beams. Therefore, it is better to make beams as weak ductile links than columns. This concept is known as **"Strong column-Weak beam concept"**. The structure gains a larger ductility other than the contributions of material ductility due to special joint detailing or making stirrups closer as recommended in different design codes of practices. The structure are going through the plastic stage where stiffness decreases appreciably and deformation is drastically increasing. It should be ensured that the structure sustains these loads undergoing smaller damage (repairable) but no collapse. Ductility plays an important role to achieve this target. A ductile building can withstand inelastic actions without losing stability, avoiding collapse and undue loss of strength at deformation levels beyond the elastic limit. Ductility of a building is assessed through different concepts of ductility such as global ductility, member ductility, sectional ductility, and material ductility. These ductility concepts are interrelated. The stirrups are primarily designed for resisting shear force but provide confinement to core concrete also. Similarly, lateral ties are provided in columns to avoid buckling of longitudinal reinforcement as well as carry shear force, but provide confinement to core concrete also. The core concrete is prevented from dilating in the transverse direction thereby enhancing its peak strength and ultimate strain capacities. The use of closer spacing of stirrups in beams or closer spacing of lateral ties makes confining pressure more uniform and effective. The concrete, a brittle material gains ductility when provided with closer stirrups/lateral ties to confine core concrete. Steel by nature is far more ductile than even confined concrete. However, it must be ensured that such steels have at least the prescribed minimum elongation specified in different seismic design codes; for instance, as per IS: 13920, steels used in earthquake-resistant constructions should have at least 14.5% elongation at fracture. Material ductility is reflected through moment-curvature relationship. Sectional ductility is assessed from moment-curvature relationship. If a building behaves elastically during earthquake shaking, no damage is occurred. If there is a requirement imposed on the structure to undergo inelastic action i.e., demand for the ductility, then it will undergo some inelastic deformation beyond yield deformation but no collapse, allowing small pre-decided damages. The global ductility of the structure can be obtained from the ratio of the maximum displacement of the building to its yield displacement.

Different types of failure occur in RCC structures i.e., shear failure, bond slip failure, flexural over-reinforced failure, flexural under-reinforced failure, torsional failure, etc. In general, flexural under-reinforced failure is preferred because where the RCC member stretches in flexure on the tension side without any failure of compression concrete and exploits the ductility of the steel bars. This plastic action spreads over a small length of the member forming plastic hinge. The effect of enhanced ultimate strain of concrete is reflected in the moment-curvature relation of the section. The curvature ductility is significantly increased. However, the moment carrying capacity of the section is governed by tensile capacity of the steel and not the strength of the concrete. The maximum strain capacity of concrete is increased for confined concrete and as a result, curvature ductility is also enhanced and ultimately reflected in the global load-deformation response of structures. Pushover responses of the building with and without transverse confining reinforcement also change appreciably. The global drift capacity of the building is significantly increased with additional confining reinforcement. RCC columns must not be designed to carry axial compression load above the balanced condition as this leads to brittle compression failure. Curvature ductility is significantly enhanced due to imposed confinement below the balanced level and will prevent brittle collapse of columns in case of extreme shaking due to earthquake. Quantification of ductility is still not common. Design for ductility is normally done through prescriptive advices, like recommending certain specific reinforcement detailing, on structural configurations, imposing material specifications, and demanding an acceptable sequencing of possible failure modes. Finally, the level of ductility of the building structure or a structural element is obtained by applying the **"Pushover Analysis".**

It is a common understanding to achieve economy and safety for building structures under earthquake shaking, let the structure undergo small damages due to plasticity, fracture, cracking, etc., ensuring its strength to carry the vertical load to prevent collapse. Lateral ties are provided to confine concrete and to avoid buckling of longitudinal reinforcement. It would enable the column to continue taking vertical loads even if it is subjected to cracking/ concrete crushing/yielding of steel reinforcement, but the stiffness of the structural elements/ structure would fall drastically. It may be noted here that, if the structure had remained elastic there would be higher internal forces and total base shear i.e., the sum of internal shear forces in all the columns. Therefore, the incorporation of ductility in the material, allowed the structure to get damaged, thus reducing internal forces. Provision in this regard is provided in building codes worldwide. The ductility factor depends on the lateral structural system provided. Another aspect is, determining the failure mechanism of members. The idea is to force a member/structure to undergo failure in a ductile manner.

5 Design of Simple Beams

PREAMBLE

In this chapter, different types of simple beams having different support conditions, loading, etc., are designed as per IS456 and detail of reinforcements are also presented.

DESIGN OF A SIMPLY SUPPORTED BEAM

Design Example 1

Calculate the flexural longitudinal reinforcement required for a simply supported beam of span l = 6 m subjected to an imposed uniformly distributed load $w_L = 40$ kN/m (*D*l + LL) over its entire span, using M25 concrete and Fe 415 steel. Use SP16.

Apply
From Table 22, IS 456, allowable $\sigma_{st} = 230$ N/mm^2 and from Table 21, IS 456, $\sigma_{cbc} = 8.5$ N/mm^2 (for M25 concrete). Apply both WSD and LSD and compare the design output.

Applying WSD

Therefore, $m = 280 / (3)(8.5) = 10.98$, $k_b = 0.289$ (calculated by using methods described earlier), $J_b = (1 - k_b/3) = (1 - (0.289/3)) = 0.904$

$$Q_b = \tfrac{1}{2}\,\sigma Q_b = \tfrac{1}{2}\,{}_{\text{cbc}}k_b j_b = \tfrac{1}{2}(8.5)(0.289)(0.904) = 1.11\ \text{N/mm}^2$$

Considering a beam section 350 mm × 800 mm, ie. $D = 800$ mm, $d' = 50$ mm and $d = 750$ mm

$$w_d = \text{Self weight of beam} = bD\gamma_{\text{conc}} = (0.35 \times 0.8 \times 25) = 7\ \text{kN/m}$$

$$w = \text{Design load} = w_d + w_L = 7 + 40 = 47\ \text{kN/m}$$

$$M = \text{Design bending moment} = wl^2/8 = (47 \times 6)^2 / 8 = 211.5\ \text{kNm}$$

For a balance section, $M_b = Q_b\, bd^2 \rightarrow 211.5 \times 10^6 = (1.11)(350)(d_b)^2 \rightarrow d_b =$ 737.84 mm

$$d = 750\ \text{mm} > d_b = 737.84\ \text{mm}$$

DOI: 10.1201/9781003208204-5

$$M/bd^2 = 211.5 \times 10^6/(350 \times 750^2) = 1.074 \text{ N/mm}^2$$

$$P_t = 0.516$$

$$A_{st} = p_t/100 \times b \times d = 0.516/100 \times 350 \times 750 = 1354.5 \text{ mm}^2$$

Check Against Shear

$$V = wl/2 = (47)(6)/2 = 141 \text{ kN}$$

$$\tau_v = V/bd = 141 \times 10^3/(350 \times 750) = 0.54 \text{ N/mm}^2$$

$$\tau_s = 0.3132 \text{ N/mm}^2 \text{ (Table 23- IS 456; for M 25 and } P_t = 0.516) < \tau_v$$

$$V_{us} = V - \tau_c bd = 141 - 0.3132 \times 350 \times 750 \times 10^{-3} = 58.785 \text{ kN}$$

$$V_{us}/d = 58.785/(750/10) \text{ kN/cm} = 0.7838 \text{ kN/cm}$$

Providing 8 mm 2-legged vertical stirrups,
Spacing, $S_v = 29.584$ cm (Table – 81: SP 16)

Check Against Deflection

Short Term Deflection

$$I_{eff} = \frac{I_r}{1.2 - \frac{M_r}{M}\frac{z}{d}\left(1 - \frac{x}{d}\right)\frac{b_W}{b}}$$

where
I_r = moment of inertia of the cracked section

$$M_r = f_{cr} I_{gr}/y_t$$

f_{cr} = Modulus of rupture of concrete = $0.7\sqrt{f_{ck}} = 0.7\sqrt{25} = 3.5$ N/mm^2
I_{gr} = Moment of inertia of the gross section about the centroidal axis, neglecting the reinforcement

$$= bD^3/12 = 350 \times 800^3/12 = 1.49 \times 10^{10} \text{mm}^4$$

y_t = distance from the centroidal axis of the gross section to the extreme fiber in tension
= $D/2 = 800/2 = 400$ mm

Therefore, $M_r = f_{cr} I_{gr}/y_t = 3.5 \times 1.49 \times 10^{10} / 400 = 13.067 \times 10^7$ N-mm

$$M = \text{maximum moment under service loads}$$

$$= 211.5 \times 10^6 \text{ Nmm}$$

$$d = \text{effective depth } = 750 \text{ mm}$$

$$p_t = 0.516;\ p_c = 0$$

$$\frac{p_c(m-1)}{p_t m} = 0$$

$$p_t m = 0.516 \times 10.98 = 5.67$$

$$\frac{d'}{d} = \frac{50}{750} = 0.067$$

Referring to Table – 92 SP-16

$$x/d = 0.285 \text{ mm}$$

$$z = \text{lever arm}$$

$$= d - x/3$$

$$= d(1 - x/3d)$$

b_w = breadth of web
b = breadth of compression face

$$I_{\text{eff}} = \frac{I_r}{1.2 - \frac{M_r}{M}\left(1 - \frac{x}{3d}\right)\left(1 - \frac{x}{d}\right)\frac{b_W}{b}}$$

$$= \frac{I_r}{1.2 - \frac{13.067 \times 10^7}{211.5 \times 10^6}\left(1 - \frac{0.285}{3}\right)(1 - 0.285)}$$

$$\frac{I_{\text{eff}}}{I_r} = 1.25$$

Referring to Table 88, SP – 16

$$\frac{I_r}{\left(\frac{bd^3}{12}\right)} = 0.44$$

$$I_r = 0.44 \times \frac{bd^3}{12} = 0.44 \times 350 \times 750^3/12 = 5.414 \times 10^9 \text{ mm}^4$$

So, $I_{\text{eff}} = 1.25 \times I_r = 1.25 \times 5.414 \times 10^9 \text{mm}^4 = 6.77 \times 10^9 \text{mm}^4$

$$\text{Elastic deflection of a simply supported beam with UDL} = \frac{5\ ML^2}{48EI_{\text{eff}}}$$

$$E_c = 5000\sqrt{f_{\text{ck}}} = 5000\sqrt{25} = 25{,}000 \text{ N/mm}^2$$

$$\text{Elastic deflection } = 5 \times 211.5 \times 10^6 \times 6000^2/48 \times 25{,}000 \times 6.77 \times 10^9 = 4.686 \text{ mm}$$

Long Term Deflection

Deflection due to Shrinkage

$$a_{\text{cs}} = k_3 \times \Psi_{\text{cs}} \times l^2$$

where $k_3 = 0.125$ for simply supported members

$$\Psi_{\text{cs}} = k_4 \times \frac{\varepsilon_{\text{cs}}}{D}$$

$$p_t - p_c = 0.516 - 0 = 0.516$$

$$k_4 = 0.72 \times \frac{(p_t - p_c)}{\sqrt{p_t}} \le 1 \text{ for } 0.25 \le p_t - p_c < 1$$

$$= 0.72 \times \frac{(0.516 - 0)}{\sqrt{0.516}} = 0.5172$$

$$\varepsilon_{\text{cs}} = 0.0003 \text{ (Cl.6.2.4IS-456)}$$

$$\Psi_{\text{cs}} = 0.5172 \times 0.0003/800 = 1.94 \times 10^{-7}$$

$$a_{\text{cs}} = 0.125 \times 1.94 \times 10^{-7} \times 6000^2 = 0.873 \text{ mm}$$

Deflection due to Creep

$$a_{\text{cc(perm)}} = a_{i,\text{ cc(perm)}} - a_{i\text{(perm)}}$$

In the absence of data, the age at loading is assumed to be 28 days and the value of creep coefficient, θ is taken as 1.6 from cl 6.2.5 IS 456

$$E_{\text{ce}} = E_c/(1+\theta) = 25{,}000/(1+1.6) = 9.62 \times 10^3 \text{ N/mm}^2$$

$$m = E_s / E_{ce} = 2 \times 10^5 / 9.62 \times 10^3 = 20.79$$

$$p_t = 0.516;\ p_c = 0$$

$$\frac{p_c(m-1)}{p_t m} = 0$$

$$p_t m = 0.516 \times 20.79 = 10.72$$

Referring to Table 88, SP – 16

$$\frac{I_r}{\left(\frac{bd^3}{12}\right)} = 0.72$$

$$I_L = 0.72 \times \frac{bd^3}{12} = 0.72 \times 350 \times 750^3 / 12 = 8.86 \times 10^9 \text{ mm}^4$$

So, $I_{\text{eff}} = 1.25 \times I_r = 1.25 \times 8.86 \times 10^9 \text{mm}^4 = 11.075 \times 10^9 \text{mm}^4 < I_{\text{gr}} = 1.49 \times 10^{10} \text{mm}^4$

$$a_{i,\text{cc(perm)}} = \frac{5M^2}{48 E_{\text{ce}} I_{\text{eff}}} = 5 \times 211.5 \times 10^6 \times 6000^2 / 48 \times 9.62 \times 10^3 \times 11.075 \times 10^9$$

$$= 7.44 \text{ mm}$$

$$a_{i,\text{(perm)}} = \frac{5ML^2}{48 E_c I_{\text{eff}}} = 5 \times 211.5 \times 10^6 \times 6000^2 / 48 \times 25{,}000 \times 6.77 \times 10^9$$

$$= 4.686 \text{ mm}$$

$$a_{\text{cc(perm)}} = 7.44 - 4.686 = 2.754 \text{ mm}$$

Total deflection (long term) due to initial load, shrinkage and creep
$= 4.686 + 0.873 + 2.754 = 8.313$ mm

The final deflection should not exceed span/250.

Permissible deflection $= 6000/250 = 24$ mm.

The calculated deflection is within the limit.

Check Against Crack Width

$$w_{cr} = \frac{3a_{cr}\varepsilon_m}{1+\dfrac{2(a_{cr}-C_{min})}{(h-x)}}$$

$$a_{cr} = 30 \text{ mm}$$

$$a = 800 - 50 - 30 \text{ mm} = 720 \text{ mm}$$

$$C_{min} = 25 \text{ mm}$$

$$h = 800 \text{ mm}$$

$$x = 0.285 \times 750 = 213.75 \text{ mm}$$

$$\varepsilon_m = \varepsilon - b(h-x)(a-x)/3E_s A_s(d-x)$$

$$= 230/(2\times 10^5) - \frac{350\times(800-213.75)\times(720-213.75)}{3\times 2\times 10^5\times 1354.5\times(750-213.75)}$$

$$= 1.15\times 10^{-3} - 2.38\times 10^{-4} = 9.12\times 10^{-4}$$

$$w_{cr} = \frac{3\times 30\times 9.12\times 10^{-4}}{1+\dfrac{2(30-25)}{(800-213.75)}} = 0.081 \text{ mm} < 0.3 \text{ mm}$$

Apply LSD

$$f_{ck} = 25 \text{ N/mm}^2; f_y = 415 \text{ N/mm}^2$$

$$M_u = 1.5\times M = 1.5\times 211.5 \text{ kNm} = 317.25 \text{ kN-m}$$

$$M_u/bd^2 = 317.25\times 10^6/(350\times 750^2) = 1.61 \text{ N/mm}^2$$

$$P_t = 0.4854 \text{ (SP} - 16, \text{ table } -3)$$

$$A_{st} = p_t/100\times b\times d = 0.4854/100\times 350\times 750 = 1274.2 \text{ mm}^2$$

Check Against Shear

$$V = wl/2 = (47 \times 6)/2 = 141 \text{ kN}$$

$$\tau_v = 1.5V/bd = 1.5 \times 141 \times 10^3/(350 \times 750) = 0.806 \text{ N/mm}^2$$

$$\tau_c = 0.482 \text{ N/mm}^2 \left(\text{Table 19- S456; for M25 and } P_t = 0.4854\right) < \tau_v$$

$$V_{us} = 1.5V - \tau_c bd = 1.5 \times 141 - 0.482 \times 350 \times 750 \times 10^{-3} = 84.975 \text{ kN}$$

$$V_{us}/d = 84.975/(750/10) \text{ kN/cm} = 1.133 \text{ kN/cm}$$

Providing 8 mm 2-legged vertical stirrups,
Spacing, $S_v = 32.23$ cm (Table – 62: SP 16)

Check Against Deflection

Short Term Deflection

$$I_{eff} = \frac{I_r}{1.2 - \dfrac{M_r}{M}\dfrac{z}{d}\left(1 - \dfrac{x}{d}\right)\dfrac{b_w}{b}}$$

where
I_r = moment of inertia of the cracked section

$$M_r = f_{cr} I_{gr}/y_t$$

f_{cr} = Modulus of rupture of concrete = $0.7\sqrt{f_{ck}} = 0.7\sqrt{25} = 3.5$ N/mm^2
I_{gr} = Moment of inertia of the gross section about the centroidal axis, neglecting the reinforcement

$= bD^3/12 = 350 \times 800^3/12 = 1.49 \times 10^{10}$ mm^4

y_t = distance from the centroidal axis of the gross section to the extreme fiber in tension
$= D/2 = 800/2 = 400$ mm

Therefore, $M_r = f_{cr} I_{gr}/y_t = 3.5 \times 1.49 \times 10^{10}/400 = 13.067 \times 10^7$ N-mm

$$M = \text{ maximum moment under service loads } = 211.5 \times 10^6 \text{ Nmm}$$

$$d = \text{ effective depth } = 750 \text{ mm}$$

$$x_u/d = \text{ depth of neutral axis } = \frac{0.87\, f_y A_{st}}{0.36\, f_{ck} bd}$$

$$= 0.87 \times 415 \times 1274.2/(0.36 \times 25 \times 350 \times 750) = 0.19$$

$$z = \text{ lever arm } = d - x_u/3 = d(1 - x_u/3d)$$

b_w = breadth of web, b = breadth of compression face

$$I_{\text{eff}} = \frac{I_r}{1.2 - \frac{M_r}{M}\left(1 - \frac{x_u}{3d}\right)\left(1 - \frac{x_u}{d}\right)\frac{b_w}{b}}$$

$$= \frac{I_r}{1.2 - \frac{13.067 \times 10^7}{211.5 \times 10^6}\left(1 \quad \frac{0.195}{3}\right)(1 - 0.195)}$$

$$\frac{I_{\text{eff}}}{I_r} = 1.36$$

$$p_t = 0.4854 \text{ and } p_c = 0$$

$$E_c = 5000\sqrt{f_{\text{ck}}} = 5000\sqrt{25} = 25{,}000 \text{ N/mm}^2, E_s = 2 \times 10^5 \text{ N/mm}^2$$

$$m = E_S / E_S = 2 \times 10^5 / 25{,}000 = 8$$

$$\frac{p_c(m-1)}{p_t m} = 0, \quad p_m = 0.4854 \times 8 = 3.88, \quad \frac{d'}{d} = \frac{50}{750} = 0.067$$

Referring to Table 88, SP – 16, $\frac{I_r}{\left(\frac{bd^3}{12}\right)} = 0.324$

$$I_K = 0.324 \times \frac{bd^3}{12} = 0.324 \times 350 \times 750^3 / 12 = 3.987 \times 10^9 \text{ mm}^4$$

So, $I_{\text{eff}} = 1.36 \times I_r = 1.36 \times 3.987 \times 10^9 \text{ mm}^4 = 5.42 \times 10^9 \text{ mm}^4$

Elastic deflection $= 5 \times 211.5 \times 10^6 \times 60002 / 48 \times 25{,}000 \times 5.42 \times 10^9 = 5.85$

Long Term Deflection

Deflection due to Shrinkage

$$a_{\text{cs}} = k_3 \times \Psi_{\text{cs}} \times l^2$$

where $k_3 = 0.125$ for simply supported members

$$\Psi_{\text{cs}} = k_4 \times \frac{\varepsilon_{c5}}{D}$$

$$p_t - p_c = 0.4854 - 0 = 0.4854$$

$$k_4 = 0.72 \times \frac{(p_t - p_c)}{\sqrt{p_t}} \leq 1 \text{ for } 0.25 \leq p_t - p_c < 1$$

$$= 0.72 \times \frac{(0.4854 - 0)}{\sqrt{0.4854}} = 0.5$$

$$\varepsilon_{cs} = 0.0003 \text{ (Cl.6.2.4 IS-456)}$$

$$\Psi_{cs} = 0.5 \times 0.0003/800 = 1.88 \times 10^{-7}$$

$$a_{cs} = 0.125 \times 1.88 \times 10^{-7} \times 6000^2 = 0.846 \text{ mm}$$

Deflection due to Creep

$$a_{cc(perm)} = a_{i,cc(perm)} - a_{i,(perm)}$$

In the absence of data, the age at loading is assumed to be 28 days, and the value of creep coefficient, θ, is taken as 1.6 from cl 6.2.5 IS 456

$$E_{ce} = E_d/(1+\theta)$$

$$= 25{,}000/(1+1.6) = 9.62 \times 10^3 \text{ N/mm}^2$$

$$m = E_s / E_{ce} = 2 \times 10^5/9.62 \times 10^3 = 20.79$$

$$p_t = 0.4854, \quad p_c = 0$$

$$\frac{p_c(m-1)}{p_t m} = 0, \ p_t m = 0.4854 \times 20.79 = 10.1,$$

Referring to Table 88, SP – 16, $\dfrac{I_r}{\left(\dfrac{bd^3}{12}\right)} = 0.683$

$$I_s = 0.683 \times \frac{bd^3}{12} = 0.683 \times 350 \times 750^3/12 = 8.4 \times 10^9 \text{ mm}^4$$

So, $I_{eff} = 1.36 \times I_r = 1.36 \times 8.4 \times 10^9 \text{mm}^4 = 11.424 \times 10^9 \text{mm}^4 < I_{gr} = 1.49 \times 10^{10} \text{mm}^4$

$$a_{i,cc(perm)} = \frac{5\,ML^2}{48\,E_{ce} I_{eff}} = 5 \times 211.5 \times 10^6 \times 6000^2/48 \times 9.62 \times 10^3$$

$$\times 11.424 \times 10^9 = 7.22 \text{ mm}$$

$$a_{i,(\text{perm})} = \frac{5\,M^2}{48\,E_c I_{\text{eff}}} = 5 \times 211.5 \times 10^6 \times 6000^2 / 48 \times 25{,}000 \times 5.42 \times 10^9 = 5.85 \text{ mm}$$

$$a_{\text{cc(perm)}} = 7.22 - 5.85 = 1.37 \text{ mm}$$

Total deflection (long term) due to initial load, shrinkage and creep $= 5.85 + 0.846 + 1.37 = 8.1$ mm

The final deflection should not exceed span/250.

Permissible deflection = 6000/250 = 24 mm.

The calculated deflection is within the limit.

Check Against Crack Width

$$W_{\text{cr}} = \frac{3a_{\text{cr}}\varepsilon_m}{1 + \dfrac{2(a_{\text{cr}} - C_{\min})}{(h - x)}}$$

$$a_{\text{cr}} = 30 \text{ mm}$$

$$C_{\min} = 25 \text{ mm}$$

$$a = 800 - 50 - 30 \text{ mm} = 720 \text{ mm}$$

$$h = 800 \text{ mm}, x = 0.195 \times 750 = 146.25 \text{ mm}$$

$$\varepsilon_m = \varepsilon - b(h - x)(a - x)/3E_s A_s (d - x)$$

$$= 0.87 \times 415/\left(2 \times 10^5\right) - \frac{350 \times (800 - 146.25) \times (720 - 146.25)}{3 \times 2 \times 10^5 \times 1274.2 \times (750 - 146.25)}$$

$$= 1.81 \times 10^{-3} - 2.84 \times 10^{-4} = 1.53 \times 10^{-3}$$

$$W_{\text{cr}} = \frac{3 \times 30 \times 1.53 \times 10^{-3}}{1 + \dfrac{2(30 - 25)}{(800 - 146.25)}} = 0.136 \text{ mm} < 0.3 \text{ mm}$$

Design Example 2

Design the beam of span 6 m and having a cross-sectional dimension $250 \times 500\,\text{mm}^2$, Effective cover = 40 mm, Use M25 grade concrete and Fe 415 HYSD bars. Apply LSD.

Floor to floor height = 3.3 m, Slab thickness = 110 mm, Live load on floor slab = 5 kN/m^2, 250 mm thk. (Finished to 280 mm) brick wall on beam to be considered. Design the beam and show the detail of reinforcements (Figure 5.1).

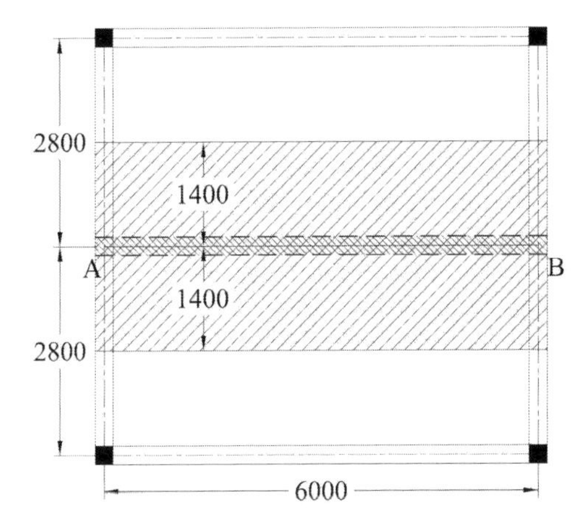

FIGURE 5.1 Plan of RCC beam.

Load Calculation

Load due to brick work on beam = 0.28 × (3.3 – 0.5) × 20 = 15.68 kN/m

Self-weight of beam rib (250 × (500 – 110), ie. 250 × 390) = 0.25 × 0.39 × 25 = 2.44 kN/m

Self-weight of slab (110 thk.)	= 0.11 × 25	= 2.75 kN/m^2
Floor finish & celling plaster		= 1.0 kN/m^2
Live load		= 5.0 kN/m^2
		Total = 8.75 kN/m^2

$\frac{l_y}{l_x} = \frac{6}{2.8} > 2$ One-way slab, i.e. almost 95% load is transferred in shorter direction

Considering 100% load is transferred in shorter direction

$$\text{Load from slab to beam} = 8.75 \times 2.8 = 24.5 \text{ kN/m}$$

$$\mathbf{w = 15.68 + 2.44 + 24.5 = 42.62\ kN/m \approx 43\ kN/m}$$

Considering T-action

$$D_f = 110 \text{ mm}, b_w = 250 \text{ mm}, D = 500 \text{ mm}, d = 500 - 40 = 460 \text{ mm}$$

$$b_f = \frac{l_0}{6} + b_w + 6\,D_f = \frac{6}{2.8} + 250 + 16 \times 100 = 1000 + 250 + 600 = 1850 \text{ mm}$$

$$< 1_{x1}/2 + l_{x2}/2 = 2800 \text{ mm}$$

Therefore, $b_f = 1850\,\text{mm}$

$$\text{Design moment} = \frac{43 \times 6^2}{8} = 193.5 \text{ kN-m}$$

$$M_u = 1.5 \times 193.5 = 290 \text{ kN-m} = 290 \times 10^6 \text{ N-mm.}$$

Assuming $X_u = D_f$, we get

$$M_{u1} = 0.36\ f_{ck} b_f\ D_f \left(d - 0.416\ D_f\right)$$

$$= 0.36 \times 25 \times 1850 \times 110(460 - 0.416 \times 110) = 758 \times 10^6 \text{ N-mm} < M_{uD}$$

Since $M_u < M_{u1}$, Therefore, $x_u < D_f$

NA is within flange.

Assuming lever arm = $[d - 0.5\ D_f]$ and under-reinforced section, it can be written

$$M_u = 0.87 f_y A_{st}\left[d - 0.5 D_f\right]$$

$$290 \times 10^6 \text{ N-mm} = 0.87 \times 415 \times \text{A}_{st} \times [460 - 0.5 \times 110]$$

$A_{st} = 1984\,\text{mm}^2$, providing 3–25 ϕ + 3–16 ϕ, in two layers ie. A_{st} provided = $2076\,\text{mm}^2$

$$P_t = 100 A_{st}\ /bd = 100 \times 2076/250 \times 460 = 1.8\%$$

Design for Shear

$$V = \text{ Design Shear force } = 43 \times 6/2 = 129 \text{ kN}$$

$$V_u = 1.5 \times 129 = 193.5 \text{ kN}$$

$$\tau_v = V_u /bd = 193.5 \times 10^3 /(250 \times 460) = 1.68 \text{ N/mm}^2$$

$$\tau_c = 0.79 \text{ N/mm}^2 \text{ (Table 19- IS 456; for M 25 and } p_t = 1.8\%\big) < \tau_v$$

$$V_{us} = V_u - \tau_c bd = 193.5 \times 10^3 - 0.79 \times 250 \times 460 = 102.65 \text{ kN}$$

$$V_{us}\ /d = 102.65/46 \text{ kN/cm} = 2.23 \text{ kN/cm}$$

From Table 62, SP16, Fe415

Providing 2L-8ϕ vertical stirrups @ 150 c/c,

Check for Deflection

Short Term Deflection

From IS 456, we have

$$I_{\text{eff}} = \frac{I_r}{1.2 - \dfrac{M_r}{M}\dfrac{z}{d}\left(1 - \dfrac{x}{d}\right)\dfrac{b_w}{b}}$$

where

I_r = moment of inertia of the cracked section

$M_r = f_{\text{cr}} I_{\text{gr}} / y_t$

f_{cr} = Modulus of rupture of concrete $= 0.7\sqrt{f_{\text{ck}}} = 0.7\sqrt{25} = 3.5\ \text{N/mm}^2$

I_g = Moment of inertia of the gross section about the centroidal axis, neglecting the reinforcement

$$= 1850 \times 110^3/12 + 1850 \times 110 \times (650 - 438.5 - 110/2)^2$$
$$+ 350 \times (650 - 110)^3/12 + 350 \times (650 - 110) \times (438.5 - (650 - 110)/2)^2$$
$$= 1.515 \times 10^{10}\ \text{mm}^4$$

y_t = distance from the centroidal axis of the gross section to the extreme fibre in tension

$$= \frac{250 \times \dfrac{(500 - 110)^2}{2} + 1850 \times 110 \times \left(500 - \dfrac{110}{2}\right)}{250 \times (500 - 110) + 1850 \times 110}$$

$$= 438.5\ \text{mm}$$

Therefore, $M_r = f_{\text{cr}}\, I_{\text{gr}}/y_t = 3.5 \times 1.515 \times 10^{10} / 438.5 = 12.1 \times 10^7$ N-mm

M = maximum moment under service loads $= 129 \times 10^6$ Nmm

d = effective depth $= 460$ mm

Assuming $x = D_f/2 = 110/2 = 55\text{mm}$, $x/d = 55/460 = 0.19\,\text{mm}$
z = lever arm $= d - x/3 = 460 - 55/3 = 441.67\,\text{mm}$

$$b_w = \text{breadth of web} = 250\text{ mm}$$

$$b = \text{breadth of compression face} = b_f$$

$$I_{\text{eff}} = \frac{I_r}{1.2 - \frac{M_r}{M}\left(1 - \frac{x}{3d}\right)\left(1 - \frac{x}{d}\right)\frac{b_w}{b}}$$

$$= \frac{I_r}{1.2 - \frac{12.1 \times 10^7}{680 \times 10^6}\left(1 - \frac{0.19}{3}\right)(1 - 0.19)\left(\frac{350}{1850}\right)}$$

$$\frac{I_{\text{eff}}}{I_r} = 0.85$$

$$\text{But } \frac{I_{\text{eff}}}{I_r} \geq 1$$

$$p_t = 2.42;\ p_c = 0$$

$$E_c = 5000\sqrt{f_{\text{ck}}} = 5000\sqrt{25} = 25{,}000\text{ N/mm}^2$$

$$E_s = 2 \times 10^5\,\text{N/mm}^2$$

$$m = E_s / E_c = 2 \times 10^5 / 25{,}000 = 8$$

$$\frac{p_c(m-1)}{p_t m} = 0$$

$$p_t m = 2.42 \times 8 = 19.36$$

$$\frac{d'}{d} = \frac{50}{600} = 0.1$$

Referring to Table 88, SP – 16

$$\frac{I_r}{\left(\frac{bd^3}{12}\right)} = 1.067$$

$$I_r = 1.067 \times \frac{bd^3}{12} = 1.067 \times 1850 \times 600^3/12 = 3.55 \times 10^{10}\ \text{mm}^4$$

So, $I_{eff} = 1 \times I_r = 1 \times 3.55 \times 10^{10} mm^4 = 3.55 \times 10^{10} mm^4 > I_{gr} = 1.515 \times 10^{10} mm^4$
Hence, $I_{eff} = 1.515 \times 10^{10} mm^4$

$$\text{Elastic deflection of a simply supported beam with UDL} = \frac{5\ ML^2}{48\ EI_{eff}}$$

$$E_c = 5000\sqrt{f_{ck}} = 5000\sqrt{25} = 25{,}000 \text{ N/mm}^2$$

$$\begin{aligned}\text{Elastic deflection } &= 5 \times 680 \times 10^6 \times 6000^2 / 48 \times 25{,}000 \times 1.515 \times 10^{10} \\ &= 6.73 \text{ mm}\end{aligned}$$

Long Term Deflection
Deflection due to Shrinkage

$$a_{css} = k_3 \times \Psi_{cs} \times 1^2$$

where $k_3 = 0.125$ for simply supported beams

$$\Psi_{cs} = k_4 \times \frac{\varepsilon_{cs}}{D}$$

$$p_t - p_c = 2.42 - 0 = 2.42$$

$$k_4 = 0.65 \times \frac{(p_t - p_c)}{\sqrt{p_t}} \le 1 \text{ for } p_t - p_c \ge 1$$

$$= 0.65 \times \frac{(2.42 - 0)}{\sqrt{2.42}} = 1.011$$

$$\varepsilon_{cs} = 0.0003 \text{ (Cl.6.2.4IS-456)}$$

$$\Psi_{cs} = 1 \times 0.0003/650 = 4.62 \times 10^{-7}$$

$$a_{cs} = 0.125 \times 4.62 \times 10^{-7} \times 6000^2 = 2.1 \text{ mm}$$

Deflection due to Creep

$$a_{cc(perm)} = a_{i,cc(perm)} - a_{i(perm)}$$

In the absence of data, the age at loading is assumed to be 28 days and the value of creep coefficient, θ is taken as 1.6 from cl 6.2.5 IS 456

$$E_{ce} = E_c / (1+\theta) = 25{,}000 / (1+1.6) = 9.62 \times 10^3 \text{ N/mm}^2$$

$$m = E_s / E_{ce} = 2 \times 10^5 / 9.62 \times 10^3 = 20.79$$

$$p_t = 2.42,\ p_c = 0$$

$$\frac{p_c(m-1)}{p_t m} = 0$$

$$p_{tm} = 2.42 \times 20.79 = 50.3$$

Referring to Table 88, SP – 16

$$\frac{I_r}{\left(\frac{bd^3}{12}\right)} = 1.391$$

$$I_r = 1.391 \times \frac{bd^3}{12} = 1.391 \times 1850 \times 600^3 / 12 = 4.63 \times 10^{10} \text{ mm}^4$$

So, $I_{eff} = 1 \times I_r = 1 \times 4.63 \times 10^{10}\text{mm}^4 = 4.63 \times 10^{10}\text{mm}^4 > I_{gr} = 1.515 \times 10^{10}\text{mm}^4$
Hence, $I_{eff} = 1.515 \times 10^{10}\text{mm}^4$

$$a_{i,cc(perm)} = \frac{5M^2}{48E_{ce}I_{eff}} = 5 \times 680 \times 10^6 \times 6000^2 / 48 \times 9.62 \times 10^3 \times 1.515 \times 10^{10}$$

$$= 17.5 \text{ mm}$$

$$a_{i,(perm)} = \frac{5ML^2}{48E_cI_{eff}} = 5 \times 680 \times 10^6 \times 6000^2 / 48 \times 25{,}000 \times 1.515 \times 10^{10} = 6.73 \text{ mm}$$

$$a_{cc(perm)} = 17.5 - 6.73 = 10.77 \text{ mm}$$

Total deflection (long term) due to initial load, shrinkage and creep
$= 6.73 + 2.1 + 10.77 = 19.6$ mm

The final deflection should not exceed span/250.
Permissible deflection = 6000/250 = 24 mm.
The calculated deflection is within the limit.

Check Against Crack Width

$$W_{cr} = \frac{3\, a_{cr}\, \varepsilon_m}{1 + \dfrac{2(a_{cr} - C_{min})}{(h - x)}}$$

$$a_{cr} = 30 \text{ mm}$$

$$C_{min} = 25 \text{ mm}$$

$$a = 650 - 50 - 30 \text{ mm} = 570 \text{ mm}$$

$$h = 650 \text{ mm}$$

$$x = 113.34$$

$$\varepsilon_m = \varepsilon - b(h - x)(a - x)/3E_s A_s (d - x)$$

$$= 0.87 \times 415/\left(2 \times 10^5\right) - \frac{350 \times (650 - 113.34) \times (570 - 113.34)}{3 \times 2 \times 10^5 \times 5088.45 \times (600 - 113.34)}$$

$$= 1.81 \times 10^{-3} - 5.77 \times 10^{-5}$$

$$= 1.75 \times 10^{-3}$$

$$W_{cr} = \frac{3 \times 30 \times 1.75 \times 10^{-3}}{1 + \dfrac{2(30 - 25)}{(650 - 113.34)}}$$

$$= 0.155 \text{ mm} < 0.3 \text{ mm}$$

Design Example 3

Design a cantilever beam 3 m, $w = 40$ kN/m, as uniform/tapered both (Figure 5.2).

Apply limit state method of design as per IS456 and use SP16.

Provide uniform cross section of the beam.

$$f_{ck} = 25 \text{ N/mm}^2; f_y = 415 \text{ N/mm}^2; Q_b = 3.45 \text{ N/mm}^2$$

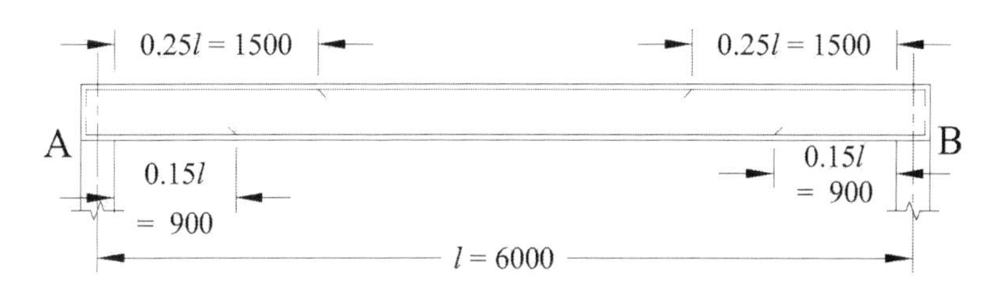

FIGURE 5.2 Section and RCC details of beam.

Considering a beam section 350 mm × 550 mm, ie. $D = 550$ mm, $d' = 50$ mm and $d = 500$ mm

$$w_d = \text{ Self weight of beam} = bD\gamma_{\text{conc}} = (0.35 \times 0.55 \times 25) = 4.8 \text{ kN/m}$$

$$w = \text{Design load} = w_d + w_L = 4.8 + 40 = 45 \text{ kN/m}$$

$$M = \text{Design bending moment} = wl^2/2 = (45)(3)^2/2 = 202.5 \text{ kNm}$$

$$M_u = 1.5 \times M = 1.5 \times 202.5 \text{ kNm} = 304 \text{ kN-m}$$

For a balance section, $M_{u,\,\text{lim}} = Q_b\ bd^2 \rightarrow 304 \times 10^6 = (3.45)\ (350)\ (d_b)^2 \rightarrow d_b =$ 502 mm

$$D = 575 \text{ mm}, d' = 50 \text{ mm}; d = D - d' = 575 - 50 = 525 \text{ mm}$$

$$d = 525 \text{ mm} > d_b = 502 \text{ mm}$$

$$M_u/bd^2 = 304 \times 10^6/\left(350 \times 525^2\right) = 3.15 \text{ N/mm}^2$$

$$P_t = 1.061$$

$$A_{\text{st}} = p_t/100 \times b \times d = 1.061/100 \times 350 \times 525 = 1950 \text{ mm}^2$$

Check Against Shear

$$V_u = 1.5\ wl = 1.5 \times 45 \times 3 = 202.5 \text{ kN}$$

$$\tau_v = V_u/bd = 202.5 \times 10^3/(350 \times 525) = 1.1 \text{ N/mm}^2$$

$$\tau_c = 0.65 \text{ N/mm}^2\left(\text{ Table 19- IS 456; for M 25 and } P_t = 1.061\right) < \tau_v$$

$$V_{\text{us}} = 202.5 - \tau_c bd = 202.5 - 0.65 \times 350 \times 525 \times 10^{-3} = 83.1 \text{ kN}$$

$$V_{\text{us}}/d = 83.1/(525/10) \text{ kN/cm} = 1.6 \text{ kN/cm}$$

Providing 8 mm 2-legged vertical stirrups,
Spacing, $S_v = 23$ cm (Table – 62: SP 16)

Check Against Deflection

Short Term Deflection

$$I_{\text{eff}} = \frac{I_r}{1.2 - \frac{M_r}{M}\frac{z}{d}\left(1 - \frac{x}{d}\right)\frac{b_w}{b}}$$

where

I_r = moment of inertia of the cracked section

$M_r = f_{\text{cr}} I_{\text{gr}} / \gamma_t$

f_{cr} = Modulus of rupture of concrete $= 0.7\sqrt{f_{\text{ck}}} = 0.7\sqrt{25} = 3.5 \text{ N/mm}^2$

I_{gr} = Moment of inertia of the gross section about the centroidal axis, neglecting the reinforcement

$= bD^3/12 = 350 \times 575^3/12 = 0.55 \times 10^{10} \text{ mm}^4$

γ_t = distance from the centroidal axis of the gross section to the extreme fibre in tension

$= D/2 = 575/2 = 287.5$ mm

Therefore, $M_r = f_{\text{cr}}\, I_{\text{gr}}/y_t = 3.5 \times 0.55 \times 10^{10} / 287.5 = 6.7 \times 10^7$ N-mm

M = maximum moment under service loads
$= 202.5 \times 10^6$ Nmm

d = effective depth = 525 mm

x_u/d = depth of neutral axis $= \frac{0.87\, f_y A_{\text{st}}}{0.36\, f_{\text{ck}} bd} = 0.87 \times 415 \times 1950/$

$(0.36 \times 25 \times 350 \times 525) = 0.426$

z = lever arm $= d - x_u/3 = \text{d}(1 - x_u/3d)$

b_w = breadth of web

b = breadth of compression face

$$I_{\text{eff}} = \frac{I_r}{1.2 - \frac{M_r}{M}\left(1 - \frac{x_u}{3d}\right)\left(1 - \frac{x_u}{d}\right)\frac{b_w}{b}} = \frac{I_r}{1.2 - \frac{6.7 \times 10^7}{202.5 \times 10^6}\left(1 - \frac{0.426}{3}\right)(1 - 0.426)}$$

$$\frac{I_{\text{eff}}}{I_r} = 0.96$$

$$\text{But } \frac{I_{\text{eff}}}{I_r} \geq 1$$

$$p_t = 1.061;\ p_c = 0$$

$$E_c = 5000\sqrt{f_{\text{ck}}} = 5000\sqrt{25} = 25{,}000 \text{ N/mm}^2$$

$$E_c = 2 \times 10^5 \text{ N/mm}^2$$

$$m = E_s / E_c = 2 \times 10^5 / 25{,}000 = 8$$

$$\frac{p_c(m-1)}{p_t m} = 0$$

$$p_t m = 1.061 \times 8 = 8.5$$

$$\frac{d'}{d} = \frac{50}{525} = 0.1$$

Referring to Table 88, SP – 16

$$\frac{I_r}{\left(\frac{bd^3}{12}\right)} = 0.601$$

$$I_r = 0.601 \times \frac{bd^3}{12} = 0.601 \times 350 \times 525^3 / 12 = 2.54 \times 10^9 \text{ mm}^4$$

So, $I_{\text{eff}} = 1 \times I_r = 1 \times 2.54 \times 10^9 \text{mm}^4 = 2.54 \times 10^9 \text{mm}^4$

$$\text{Elastic deflection of a simply supported beam with UDL} = \frac{ML^2}{4\, EI_{\text{eff}}}$$

$$E_c = 5000\sqrt{f_{\text{ck}}} = 5000\sqrt{25} = 25{,}000 \text{ N/mm}^2$$

$$\text{Elastic deflection } = 202.5 \times 10^6 \times 3000^2 / 4 \times 25{,}000 \times 2.54 \times 10^9 = 7.2 \text{ mm}$$

Long Term Deflection

Deflection due to Shrinkage

$$a_{cs} = k_3 \times \Psi_{cs} \times 1^2$$

where $k_3 = 0.5$ for cantilevers

$$\Psi_{cs} = k_4 \times \frac{\varepsilon_{cs}}{D}$$

$$p_t - p_c = 1.061 - 0 = 1.061$$

$$k_4 = 0.65 \times \frac{(p_t - p_c)}{\sqrt{p_t}} \leq 1 \text{ for } p_t - p_c \geq 1$$

$$= 0.65 \times \frac{(1.061 - 0)}{\sqrt{1.061}} = 0.67$$

$$\varepsilon_{cs} = 0.0003 \text{ (Cl.6.2.4IS-456)}$$

$$\Psi_{cs} = 0.67 \times 0.0003 / 575 = 3.5 \times 10^{-7}$$

$$a_{cs} = 0.5 \times 3.5 \times 10^{-7} \times 3000^2 = 1.58 \text{ mm}$$

Deflection due to Creep

$$a_{cc(perm)} = a_{i,cc(perm)} - a_{i(perm)}$$

In the absence of data, the age at loading is assumed to be 28 days and the value of creep coefficient, θ, is taken as 1.6 from cl 6.2.5 IS 456

$$E_{ce} = E_c / (1 + \theta) = 25{,}000 / (1 + 1.6) = 9.62 \times 10^3 \text{ N/mm}^2$$

$$m = E_s / E_{ce} = 2 \times 10^5 / 9.62 \times 10^3 = 20.79$$

$$p_t = 1.061;\ p_c = 0$$

$$\frac{p_c\ (m - 1)}{p_t\ m} = 0$$

$$p_t m = 1.061 \times 20.79 = 22.1$$

Referring to Table 88, SP – 16

$$\frac{I_r}{\left(\frac{bd^3}{12}\right)} = 1.16$$

$$I_r = 1.16 \times \frac{bd^3}{12} = 1.16 \times 350 \times 525^3 / 12 = 4.9 \times 10^9 \text{ mm}^4$$

So, $I_{\text{eff}} = 1 \times I_r = 1 \times 4.9 \times 10^9 \text{mm}^4 = 4.9 \times 10^9 \text{mm}^4 < I_{\text{gr}} = 0.55 \times 10^{10} \text{mm}^4$

Hence, $I_{\text{eff}} = 4.9 \times 10^9 \text{mm}^4$

$$a_{i,\text{cc(perm)}} = \frac{ML^2}{4E_{\text{ce}}I_{\text{eff}}} = 202.5 \times 10^6 \times 3000^2 / 4 \times 9.62 \times 10^3 \times 4.9 \times 10^9 = 9.7 \text{ mm}$$

$$a_{i,\text{(perm)}} = \frac{ML^2}{4E_c I_{\text{eff}}} = 202.5 \times 10^6 \times 3000^2 / 4 \times 25{,}000 \times 2.54 \times 10^9 = 7.2 \text{ mm}$$

$$a_{\text{cc(perm)}} = 9.7 - 7.2 = 2.5 \text{ mm}$$

Total deflection (long term) due to initial load, shrinkage and creep
$= 7.2 + 1.58 + 2.5 = 11.28$ mm

The final deflection should not exceed span/250.

Permissible deflection = 3000/250 = 12 mm.

The calculated deflection is within the limit.

Check Against Crack Width

$$W_{\text{cr}} = \frac{3a_{\text{cr}}\varepsilon_m}{1 + \frac{2(a_{\text{cr}} - C_{\text{min}})}{(h - x)}}$$

$$a_{\text{cr}} = 30 \text{ mm}$$

$$C_{\text{min}} = 25 \text{ mm}$$

$$a = 575 - 50 - 30 \text{ mm} = 495 \text{ mm}$$

$$h = 575 \text{ mm}$$

$$x = 0.426 \times 525 = 223.65 \text{ mm}$$

$$\varepsilon_m = \varepsilon - b(h - x)(a - x) / 3E_s A_s (d - x)$$

$$= 0.87 \times 415 / \left(2 \times 10^5\right) - \frac{350 \times (575 - 223.65)\text{x}(495 - 223.65)}{3 \times 2 \times 10^5 \times 1950 \times (525 - 223.65)}$$

$$= 1.81 \times 10^{-3} - 9.5 \times 10^{-5} = 1.71 \times 10^{-3}$$

$$W_{\text{cr}} = \frac{3 \times 30 \times 1.71 \times 10^{-3}}{1 + \frac{2(30 - 25)}{(575 - 223.65)}} = 0.15 \text{ mm} < 0.3 \text{ mm}$$

Now the effective depth (d) at fixed end is kept equal to 525 mm, which is gradually decreased to 325 mm at the free end.

Check Against Shear

$$V_u = 202.5 \text{ kN}$$

$$\tau_v = \left(V_u - \frac{M_u}{d} \tan\beta \right) / bd = \left(202.5 \times 10^3 - 304 \times 10^6 / 525 \times (200 / 3000)\right) /$$

$$(350 \times 525) = 0.89 \text{ N/mm}^2$$

$$\tau_c = 0.65 \text{ N/mm}^2 \left(\text{Table 19- IS 456; for M 25 and } P_t = 1.061\right) < \tau_y$$

$$V_{\text{us}} = 202.5 - \tau_c bd = 202.5 - 0.65 \times 350 \times 525 \times 10^{-3} = 83.1 \text{ kN}$$

$$V_{\text{us}} / d = 83.1 / (525 / 10) \text{ kN/cm} = 1.6 \text{ kN/cm}$$

Providing 8 mm 2-legged vertical stirrups,
Spacing, $S_v = 23$ cm (Table – 62: SP 16)

6 Design Deep Beams

PREAMBLE

Such structures are found in transfer girders and in shear wall structures that resist lateral forces in buildings. It is also found in some of the industrial buildings.

RC deep beams, which fail with shear compression, are the structural members having a shear span to effective depth ratio, *a/d*, not exceeding 1.RC deep beams have many useful applications in building structures such as transfer girders, wall footings, foundation pile caps, floor diaphragms, and shear walls. Particularly, the use of deep beams at the lower levels in tall buildings for both residential and commercial purposes has increased rapidly because of their convenience and economic efficiency. Deep beams are structural elements loaded as simple beams in which a significant amount of the load is carried to the supports by a compression force combining the load and the reaction. As a result, the strain distribution is no longer considered linear, and the shear deformations become significant when compared to pure flexure. Floor slabs under horizontal load, short span beams carrying heavy loads, and transfer girders are examples of deep beams. Deep beam is a beam having large depth/thickness ratio and shear span depth ratio of less than 2.5 for concentrated load and less than 5.0 for distributed load. Because of the geometry of deep beams, their behavior is different with slender beam or intermediate beam. Two-dimensional action, because of the dimension of deep beam, they behave as two-dimensional action rather than one-dimensional action. Plane section does not remain plane, the assumption of plane section remains plane and cannot be used in the deep beam design. The strain distribution is no longer linear. Shear deformation cannot be neglected as in the ordinary beam. The stress distribution is not linear even in the elastic stage. At the ultimate limit state, the shape of concrete compressive stress block is not parabolic shape again. Deep beams are structural elements loaded as beams but having a large depth/thickness ratio and a shear span/depth ratio not exceeding 2 for simple span and 2.5 for continuous span, where the shear span is the clear span (l_n) of the beam for distributed load and the distance between the point of application of the load and the face of the support for concentrated load. This definition is somewhat arbitrary. Better definition is a deep beam in which a significant amount of the load is carried to the supports by a compression thrust joining the load and the reactions. This occurs if the above proportions are maintained. Deep beams are usually found in transfer girders (girders support one or more columns transferring it laterally to other columns) used in multistoried buildings to provide column offsets, in foundation walls, pile caps, walls of rectangular tanks and bins, floor diaphragms, and shear walls.

The traditional principles of stress analysis are neither suitable nor adequate to determine the strength of reinforced concrete deep beams. In deep beams, the bending stress distribution across any transverse section deviates appreciably from

DOI: 10.1201/9781003208204-6

straight line distribution assumed in the elementary beam theory. The behavior of a deep beam depends also on how they are loaded and special considerations should be given to this aspect in design. Here cracking will occur at one-third to one-half of the ultimate load. In the single-span beam supporting a concentrated load at mid-span, the compressive stresses act roughly parallel to the lines joining the load and the supports and the tensile stresses act parallel to the bottom of the beam. The flexural stresses at the bottom are constant over much of the span. A single-span beam supporting a uniform load acting on the top has the stress trajectories, crack pattern, and simplified truss. Single-span beam supporting a uniform load acting on the lower face of the beam has the stress trajectories, crack pattern, and simplified truss. The compression trajectories form an arch with the loads hanging from it. The crack pattern shows that the load is transferred upward by reinforcement until it acts on the compression arch, which then transfers the load down to the supports. The force in the longitudinal tension ties will be constant along the length of the deep beam. This is the reason that the steel must be anchored at the joints over the reaction, failure of which is a major cause of distress.

Deep Beams Design as per IS 456

A deep beam is a structural member whose span to depth ratio is relatively small so that shear deformation dominates the behavior. According to Clause 29.1 of IS 456, a beam is considered a deep beam when the effective span to overall depth ratio (*L/D* ratio) is less than (i) 2.0 for simply supported beams and (ii) 2.5 for continuous beams.

The assumptions of linear-elastic flexural theory and plane sections remaining plane even after bending are not valid for deep beams. Hence, these beams have to be designed taking into account nonlinear stress distribution along the depth and lateral buckling. Arch action is more predominant than bending in deep beams. Hence, these beams require special considerations for their design and detailing. RC deep beams are often found as single-span or continuous transfer girders, pile-supported foundations, foundation walls supporting strip footings or raft slabs, walls of silos and bunkers, bridge bents, or shear wall structures, and in offshore structures. One such example of a deep beam is shown in Figure 6.1.

A brief introduction to the IS code provisions is provided here, and strut-and-tie modeling is discussed. It should be noted that deep beams are sensitive to loading at the boundaries, and the length of bearing may affect the stress distribution in the vicinity of the supports. Similarly, stiffening ribs, cross walls, or extended columns at supports will also influence the stress distribution (Park and Paulay, 1975). The concrete compression stresses are seldom critical. However, the considerable increase of diagonal compression stresses near the support after the onset of cracking and anchorage of tensile steel are important considerations in the design and detailing (Park and Paulay, 1975).

The 'Simple Rules' provided in IS 456, based on the CIRIA Guide 2, are intended primarily for uniformly loaded (from the top) deep beams and are intended to control the crack width rather than the ultimate strength. In addition,

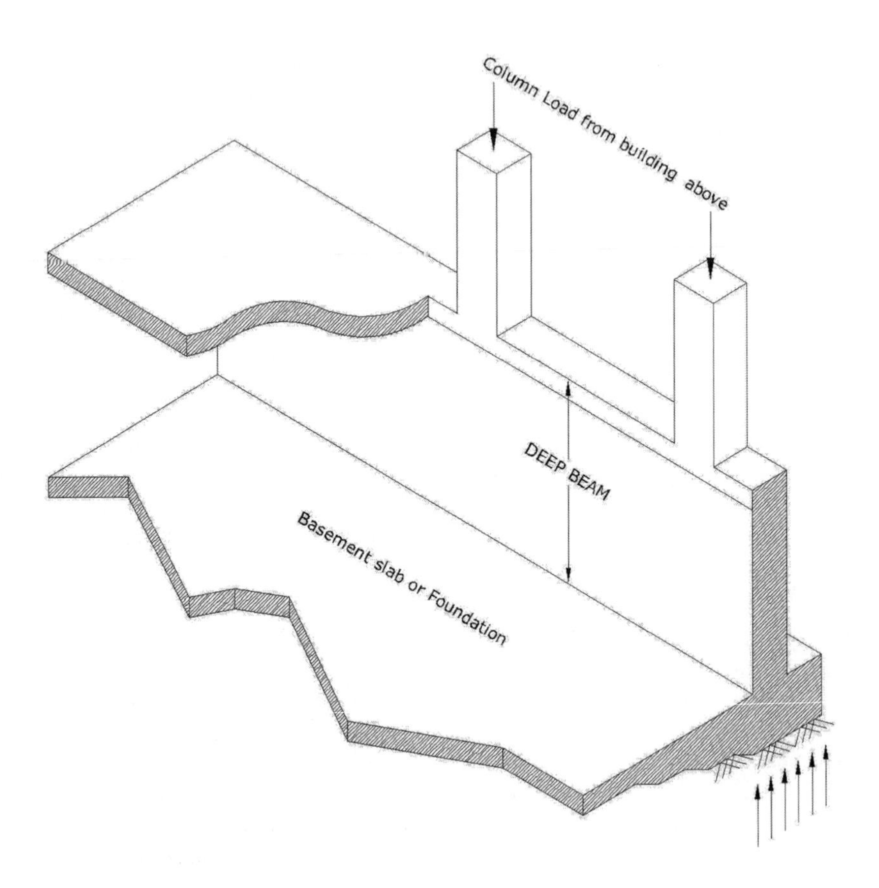

FIGURE 6.1 A typical deep beam supporting columns.

the active height of a deep beam is limited to a depth equal to the span; the part of the beam above this height is merely taken as a load-bearing wall between the supports. These rules are provided for single-span and continuous beams. The steps required for the flexural design of deep beams as per IS 456 are as follows.

If $L/D \leq 1.5$, go to step 4. If $L/D > 1.5$, check whether the applied moment M_u does not exceed Mn of Eq. (5.89), where L is the effective span taken as the c/c between the supports or 1.15 times the clear span, whichever is lesser (see Clause 29.2 of IS 456), and D is the overall depth.

Calculate the area A_{st} of the main longitudinal reinforcement:

$$A_{st} \geq M_u / \left(0.87 f_y \times Z\right)$$

where M_u is the factored applied moment, f_y the characteristic yield strength of steel, and Z the lever arm, which is to be taken as follows (Clause 29.2 of IS 456):

For single-span beams:

$$Z = 0.2L + 0.4D \text{ if } (1 \le L/D \le 2) \text{ and } 0.6L \text{ if } (L/D < 1)$$

For continuous beams:

$$Z = 0.2L + 0.3D \text{ if } (1 \le L/D \le 2.5) \text{ and } 0.5L \text{ if } (L/D < 1)$$

It should be noted that in deep beams the requirement of flexural reinforcement is not large, and hence the approximate lever arms, as determined from experiments and given here, are sufficient to arrive at them. It is also important to detail the reinforcement properly as the deep beam behavior is different from that of normally sized beams. The recommendations given in IS 456 are as follows:

Reinforcement for Positive Moment

In a simply supported beam, due to the arching action, the tension steel serves as a tie connecting the concrete compression struts. The cracking will occur at one-third to one-half of the ultimate load. The flexural stress at the bottom is constant over much of the span. The non-uniform stress distribution due to uniformly distributed load is considered. The tests on deep beams have shown that the tension zone in the bottom is relatively small. Accordingly, Clause 29.3.1 of IS 456 suggests that the tensile reinforcement for a positive moment should be placed within a tension zone of depth equal to $0.25D - 0.05L$ from the extreme tension fiber at the mid-span.

The force in the longitudinal tension ties will be constant along the length of the deep beam. This implies that the force must be well anchored at the supports; else, it will result in major cause of distress. Hence, Clause 29.3.1(b) of IS 456 suggests that the bottom reinforcement should be extended into the supports without curtailment and embedded beyond the face of each support to a length of $0.8L_d$ where L_d is the development length for the design stress in the reinforcement. If sufficient embedment length is not available, the longitudinal reinforcement may be adequately anchored by hooks or welding to special mechanical anchorage devices. Bent-up bars are not recommended. The beam should be proportioned in such a way that the strength of the steel tension ties governs the design.

Reinforcement for Negative Moment

In the case of continuous deep beams, the tensile reinforcement for negative moment should satisfy the following requirements (Clause 29.3.2 of IS 456):

Termination of reinforcement: Negative reinforcement can be curtailed only in deep beams with $l/D > 1$ Not more than 50% of the reinforcement may be terminated at a distance of $0.5D$ from the face of the support and the remaining should be extended over the full span (it has to be noted that/denotes the clear span and not the effective span of deep beam).

Distribution of reinforcement: When the l/D ratio is less than 1.0, the negative reinforcement should be evenly distributed over a depth of $0.8D$ measured from the

TABLE 6.1
Side Face Reinforcement (Clause 32.5 of IS 456)

Type of Reinforcement	Side Face Reinforcement of Gross Area of Concrete (Percentage)	
	Vertical	Horizontal
Bars of diameter ≤ 16 mm and $f_y \geq 415$ MPa	0.12[a]	0.20
Bars of Fe 250 grade steel	0.15	0.25
Welded wire fabric made with bars of diameter ≤ 16 mm	0.12	0.20

[a] It should be noted that as per the 2011 version of the ACI code, both vertical and horizontal reinforcement should not be less than $0.0025b_w\, s$ and the spacing 's' should not exceed the smaller of d/5 and 300 mm. As per Clause 32.5(b) and (d) of IS 456, both vertical and horizontal spacing of reinforcement should not exceed three times the thickness of the beam or 450 mm.

top tension fiber at the support. However, when the l/D ratio is in the range 1.0–2.5, the negative reinforcement should be provided in two zones and described as follows (Leonhardt and Mönnig 1977):

a. A zone of depth $0.2D$ from the tension fiber should be provided with $(0.5l/D-0.25)$ times the reinforcement calculated for negative moment, where l is clear span of beam.
b. A zone of 0.6D from this zone should contain the remaining reinforcement for negative moment and shall be evenly distributed.

When the depth of the beam is much larger than the span, the portion above a depth equal to 0.8 times the span can be merely considered a load-bearing wall. Beam action should be considered only in the lower portion. Continuous deep beams are very sensitive to differential settlement of their supports.

Vertical Reinforcement

The loads applied at the bottom of the beam, induce hanging action (e.g., as in the face of bunker walls). Hence, Clause 29.3.3 of IS 456 suggests that suspension stirrups should be provided to carry the concerned loads (see also SP 24:1983 for detailing of suspended stirrups). Tests have shown that vertical shear reinforcement (perpendicular to the longitudinal axis of the member) is more effective for member strength than horizontal shear reinforcement (parallel to the longitudinal axis of the member) in deep beams. However, equal minimum reinforcement in both directions is specified in the ACI code to control the growth and width of diagonal cracks, as shown in Table 5.10. The maximum spacing of bars has also been reduced in the ACT code from 450 to 300 mm. Hence, there is an urgent need to revise the IS code clauses. IS 456 stipulates that the spacing of vertical reinforcement should not exceed three times the thickness of the beam or 450 mm (see Table 5.10). Suspender stirrups should completely surround the bottom reinforcement and extend into the compression zone of the beam (see SP 24).

Side Face or Web Reinforcement

IS 456 suggests that the side face reinforcements should be provided as per the minimum requirements of walls. The requirement of minimum vertical and horizontal side face reinforcements, as per the code, is given in Table 5.10. For deep beams of thickness more than 200 mm, the vertical and horizontal reinforcements should be provided in two grids, one near each face of the beam (Clause 32.5.1). The horizontal and vertical steel placed on both the faces of the deep beam serve not only as shrinkage and temperature-reinforcement but also as shear reinforcement (Table 6.1).

Shear Reinforcement

A deep beam provided with the reinforcements is deemed to satisfy the provision for shear, that is, the main tension and the web steels together with concrete will carry the applied shear, and hence, a separate check for shear is not required. However, Clause 11.7.3 of ACI 318:2011 code stipulates that the shear in deep beams should not exceed $0.74\varphi\sqrt{f_{ck}}.b_w.d$ (where φ is the strength reduction factor = 0.75).

Bearing Strength

In addition, the local failure of deep beams due to bearing stresses at the supports as well as loading points should be checked. To estimate the bearing stress at the support, the reaction may be considered uniformly distributed over the area equal to the beam width $b_w \times$ effective support length. The permissible ultimate stress is limited to $0.45f_{ck}$ as per Clause 34.4 of IS 456. The support areas may be strengthened by vertical steel and spiral reinforcement to prevent brittle failure at support.

Lateral Buckling Check

To prevent the lateral buckling of simply supported deep beams, the breadth, b, should be such that the following conditions are satisfied (Clause 23.3 of IS 456):

$$L/b \leq 60 \text{ and } L.d/b^2 \leq 250$$

where L is the clear distance between the lateral restraints and d is the effective depth of the beam.

Example 1

A simply supported deep beam girder 6 m length, supports two columns located at 2 m apart. Transfer column loads = 4000 kN, Total depth of the beam = 4 m and width of support = 500 mm. Concrete grade = M30, Fe 415 steel. Design and detail the girder.

$$l_{eff} = \text{(i) C/C distance between supports}$$
$$\text{(ii) } 1.15\times \text{clear span whichever is less}$$

Step 1:

Check for bearing capacity at support

Let B = Beam width

Allowable stress = $0.45 \times 30 = 13.5$ MPa [As per Cl. 34.4 of IS 456]

Support width = 500 mm

Effective width of support = $0.2 \times$ clear span = $0.2 \times (6000 - 2 \times 500)$ = 1000 mm

Adopt 500 mm

$13.5 = 1.5 \times 4000 \times 10^3/(500 \times B)$; $B = 888.88$ mm ≈ 900 mm

$l_{eff} = 6000 - 500 = 5500$ mm or $1.15 \times (6000 - 2 \times 500) = 5750$ mm ;

$l_{eff} = 5500$ mm [As per Cl. 29.2 of IS 456]

$D/b < 25$: $4000/900 = 4.44$;

or $L/b < 50$: $5500/900 = 6.11$;

$L/D = 5500/4000 = 1.375 > 1$ and < 2; Deep beam category as per CL 29.1 of IS 456

Step 2:

Center to center distance between the column = 2 m;

$a = (5500 - 2000)/2 = 1750$ mm = 1.75 m;

∴ Factored Moments, $M_u = 1.5 \times (4000 \times 1.75) = 7000$ kNm

Lever arm $Z = 0.2\ (l_{eff} + 2D) = 0.2 \times (5500 + 2 \times 4000) = 2700$ mm [As per CL 29.2 (a) of IS 456]

$A_{st} = M_u/(0.87 f_y\ Z) = 7000 \times 10^6/(0.87 \times 415 \times 2700) = 7180.7\ \text{mm}^2$

$A_{st\ min} = 0.85 \times b \times D/415 = 0.85 \times 900 \times 4000/415 = 7373.49\ \text{mm}^2$ [As per CL 26.5.1.1 of IS 456]

Let's provide 12 nos. 25Ø + 6 nos. 20Ø bars, $A_{st} = 7776\ \text{mm}^2 > 7373.49\ \text{mm}^2$ Ok.

Step 3:

Detailing of reinforcement

Tension Zone depth = $0.25\ D - 0.05\ l_{eff} = 0.25 \times 4000 - 0.05 \times 5500$ = 725 mm; [As per CL 29.3.1 c) of IS 456]

Assume clear bottom and side cover = 40 mm

Arrange bars in 6 rows in a depth = 725 mm

Step 4:

Detailing of vertical reinforcement

$A_{st\ min}$/m length = $0.0012 \times 900 \times 1000 = 1080\ \text{mm}^2/\text{m}$ [As per CL 32.5 a) 1) of IS 456]

Provide on each face = $1080/2 = 540\ \text{mm}^2/\text{m}$

Use 16Ø bars. ∴ $A_{Ø} = \pi \times 16^2/4 = 201\ \text{mm}^2$.

Spacing of 16Ø bars = $1000 \times 201/540 = 372$ mm < 450 mm

Let's provide 16 T bars @ 300 c/c

Step 5:

Detailing of horizontal reinforcement

Side face reinforcement shall comply with requirements of minimum reinforcement of walls.

$A_{st\ min}$/m length = $0.002 \times 900 \times 1000 = 1800\ \text{mm}^2/\text{m}$ [As per CL 29.3.4 of IS 456]

Provide on each face = 1800/2 = 900 mm^2/m
Use 20Ø bars. ∴ $A_{Ø} = \pi \times 20^2/4 = 314\,mm^2$.
Spacing of 20Ø bars = 1000 × 314/900 = 349 mm < 450 mm
Let's provide 20 T bars @ 300 c/c

Example 4

Fixed ends and continuous deep beam

A reinforced girder 4.0 m deep is continuous over two spans 8 m c/c, resting on column supports 800 mm width is to be designed to support a total load of 400 kN/m including its own weight. M30 and Fe415

Step 1:
Check for bearing capacity at support
Concrete Grade = M30; Allowable stress = 0.45 × 30 = 13.5 MPa [As per Cl. 34.4 of IS 456]
Support width = 800 mm;
Effective width of support = 0.2 × Clear Span = 0.2 × (8000 − 2 × 800/2) = 0.2 × 7200 = 1440 mm
Adopt 800 mm
Total Load $W = 400 \times (2 \times 8.0 + 2 \times 0.8/2) = 400 \times 16.8 = 6720$ kN
Reaction at interior support = W = 6720/2 = 3360 kN
Now, $1.5 \times 3360 \times 10^3/(800 \times B) = 13.5$; $B = 466.67$ mm
$L_{eff} = 8000$ mm or 1.15 × 7200 = 8280 mm
Adopt $L_{eff} = 8000$ mm [As per Cl. 29.2 of IS 456]
$D/b = 4000/800 = 5 < 25$;
or $L/b = 8000/800 = 10 < 50$;
$L/D = 8000/4000 = 2 > 1$ and < 2.5
Deep beam category as per CL 29.1 of IS 456

Step 2:
Factored Moments, A_{st} Span Moment
$M_u = 1.5 \times 400 \times 9^2/24 = 2050$ kNm.
Lever arm $Z = 0.2\ (L_{eff} + 1.5D) = 0.2 \times (9000 + 1.5 \times 4500) = 3150$ mm CL 29.2 (b)
$A_{st} = M_u/(0.87 f_y Z) = 1012.5 \times 10^6/(0.87 \times 415 \times 3150) = 890\,mm^2$
$A_{st\,min} = 0.85 \times b \times D/415 = 0.85 \times 350 \times 4500/415 = 3226\,mm^2$ CL 26.5.1.1
Adopt 8 nos. - 25T in 4 rows Support Moment
$M_u = 1.5 \times 200 \times 9^2/12 = 2025$ kNm
$A_{st} = M_u/(0.87 fy\ Z) = 2025 \times 10^6/(0.87 \times 415 \times 3150) = 1780\,mm^2$
$A_{st\,min} = 3226\,mm^2$ CL 26.5.1.1
Adopt $A_{st} = 3226\,mm^2$

Step 3:
Detailing of rebars in span region CL 29.3.1
Tension Zone Depth = $0.25\ D - 0.05 L_{eff} = 0.25 \times 4500 - 0.05 \times 9000 =$ 675 mm

Assume Clear bottom and side cover = 40 mm
Arrange bars in 4 rows in a depth = 675 mm 675 45 210 210 210 350

Step 4:
Detailing of rebars in support region CL 29.3.2
Clear Span/D = 8100/4500 = 1.8 > 1 and < 2.5
Rebars are placed in two zones CL 29.3.2 (b)
$A_{st} = 3226\ \text{mm}^2$
Zone1 • Depth = 0.2D = 0.2 × 4500 = 900 mm
$A_{st1} = 3226 \times 0.5 \times (1.8 - 0.5) = 2097\ \text{mm}^2$
Adopt 8 - 20 – 2512 mm^2 in four rows
Zone2 • Depth = 0.3D = 0.3 × 4500 = 1350 mm on both sides of mid-depth
$A_{st1} = (3226 - 2097) = 1129\ \text{mm}^2$
Adopt 6 nos. – 16T = 1206 mm^2 in three rows

Step 5:
Detailing of vertical rebars: CL 32.5
$A_{st\ min}$/m length = 0.0012 × 350 × 1000 = 420 mm^2/m
Provide on each face: 210 mm^2/m
Spacing of 10T rebars = 1000 × 78.5/210 = 373 mm < 450 mm
Adopt 10T @300 mm c/c (stirrups)

Step 6:
Detailing of horizontal rebars. Side face reinforcement shall comply with requirements of minimum reinforcement of walls
A_{st} min/m length = 0.002 × 350 × 1000 = 700 mm^2/m
Provide on each face: 350 mm^2/m
Spacing of 12T rebars = 1000 × 113/350 = 322 mm < 450 mm
Adopt 12T @300 mm c/c

7 Design of Continuous Beams

DESIGN CONCEPT

Continuous beam is a multi-span beam. It may have two or more spans part of a frame. Location of live load provides the maximum value of bending moment and shear force at critical sections. However, the effect of support yielding (settlement, rotation, etc.) as well as temperature play a major role in the behavior of continuous beam out of a frame. Redundancy of continuous beam has a direct effect on its behavior. Hence, to make exact analysis to some extent, computer-aided analysis is essential using software. IS 456 recommended bending moment and shear force coefficients, to obtain design moments. However, there are some restrictions in using these recommendations. Some case studies are included in this chapter to appreciate these restrictions. After sectional design, reinforcement detailing needs to be done as per IS 456, SP34, IS 13920, etc. Some working drawings have been furnished regarding reinforcement detailing.

APPROXIMATE METHOD OF ANALYSIS USING MOMENT AND SHEAR COEFFICIENT UNDER DEAD LOAD AND LIVE LOAD AS PER IS 456

Design bending moment and shear force at critical sections can be obtained using moment coefficients under dead and live load instead of analyzing different load cases, provided the following conditions are fulfilled as per clause no. 22.4 of IS 456. The simplifying assumptions as given in 22.4.1 to 22.4.3 as per IS 456, may be used in the analysis of frames.

Arrangement of imposed load (live loads)

Consideration may be limited to combinations of:

- Design dead load on all spans with full design imposed load on two adjacent spans; Design dead load on all spans with full design imposed load on alternate spans.
- When design imposed load does not exceed three-fourths of the design dead toad, the load arrangement may be design dead load and design imposed load on all the spans.
- For beams and slabs continuous over support clause 22.4.1 may be considered.

DOI: 10.1201/9781003208204-7

Moment and Shear Co-Efficient for Continuous Beams

As recommended in clause 22.5.1 of IS 456, unless more exact estimates are made, for beams of uniform cross-section that support substantially uniformly distributed loads over three or more spans, which do not differ by more than 15% of the longest, the bending moments and shear forces used in design may be obtained using the coefficients given in Table 12 and Table 13 of IS 456, respectively. For moments at supports where two unequal spans meet or in case where the spans are not equally loaded, the average of the two values for the negative moment at the support may be taken for design. However, where the coefficient given in Table 12 is used for the calculation of bending moments, redistribution referred to in clause 22.7 shall not be permitted.

As per clause 22.5.2 of IS 456, for beams and slabs over free end supports where a member is built into a masonry wall that develops partial fixity, the member shall be designed to resist a negative moment at the face of the support of W1/24 where W is the total design load and 1 is the effective span, or such other restraining moment as may be shown to be applicable, For such a condition shear coefficient given in Table 13 at the end support may be increased by 0.05 (Figures 7.1–7.4).

Conversion from Trapezoidal Load (w kN/m) to Equivalent UDL (w_E)

We know from handbook, $w_e = \frac{wL_x}{2}\left(1+\frac{1}{3\beta^2}\right)$, where $\beta = \frac{L_y}{L_x}$ and w_e = Equivalent UDL

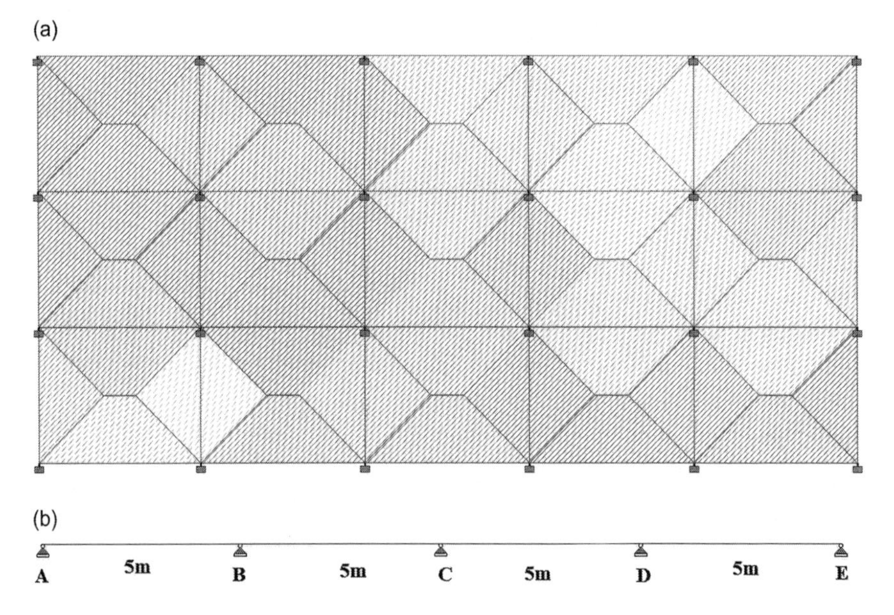

FIGURE 7.1 Load transfer from slab continuous beam A1 – F1 (dead load = 4.75 kN/m^2 and live load = 4 kN/m^2) – fach slab panel –4 m × 5 m.

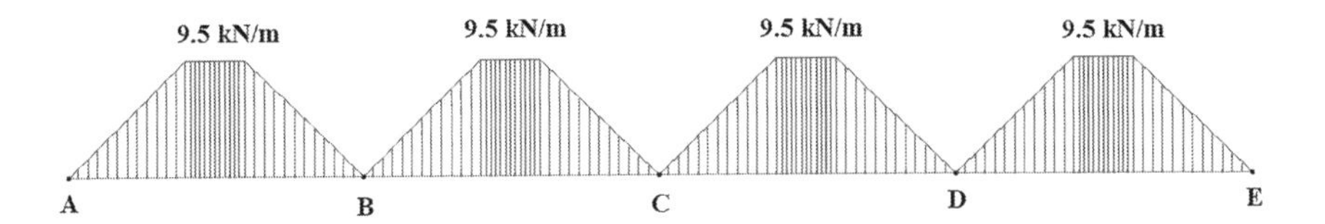

FIGURE 7.2 Trapezoidal load transferred from floor (dead load).

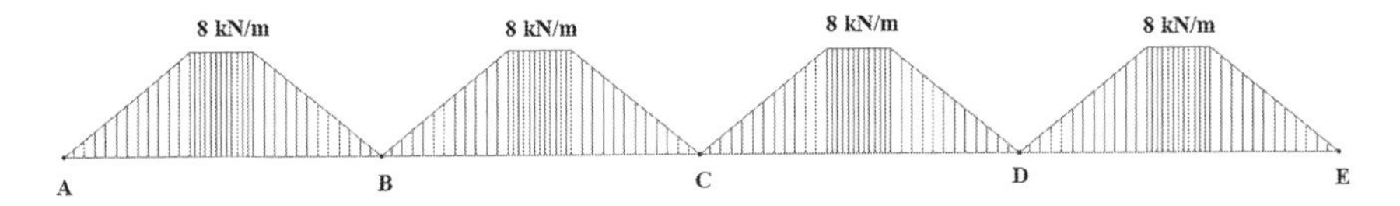

FIGURE 7.3 Trapezoidal load transferred from floor (live load).

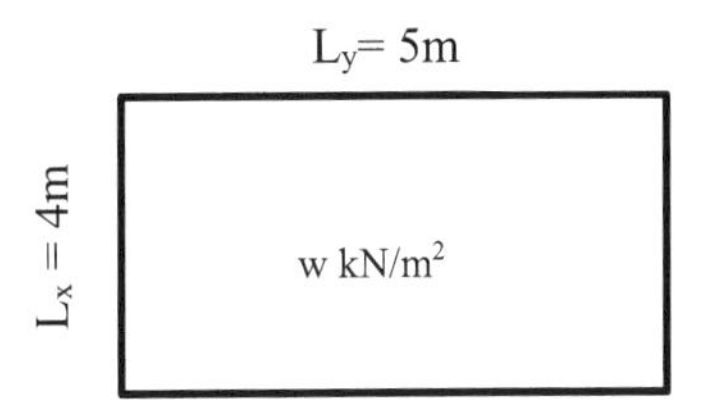

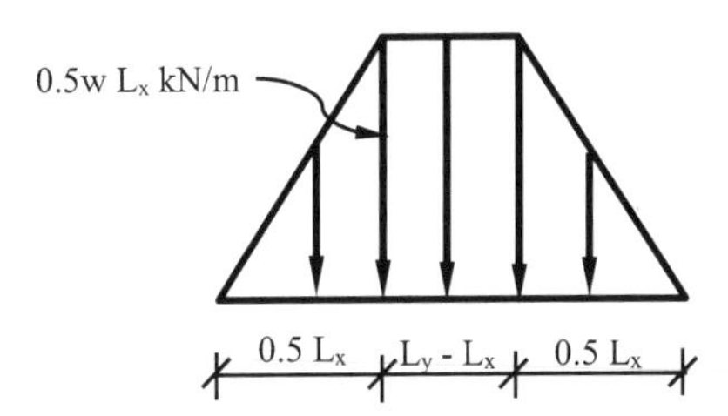

FIGURE 7.4 Typical trapezoidal load transferred from slab.

$$L_x = \text{ Shorter span} = 4 \text{ m and } L_y = \text{Shorter span} = 5 \text{ m}$$

$$w_d = \text{Equivalent UDL (Dead Load)} = 9.5 * 4/2\left(1 + 1/\left(3 * 1.25^2\right)\right) = 23 \text{ kN/m}$$

$$w_L = \text{Equivalent UDL(Live Load)} = 8 * 4/2\left(1 + 1/\left(3 * 1.25^2\right)\right) = 19.4 \text{ kN/m}$$

Let us use Tables 12 and 13 of IS456 (Tables 7.1 and 7.2),

SUBSTITUTE FRAME ANALYSIS

For determining the design moments and shear forces at any floor or roof level due to gravity loads, the beams at that level together with columns above and below with their far ends fixed may be considered to constitute the frame, namely **Substitute frame**.

TABLE 7.1
Design Bending Moment

	Span Moments		Support Moments	
Type of Load	**Near Middle of End Span**	**At Middle of Interior Span**	**At Support Next to End Support**	**At Other Interior Support**
Dead load (Fixed type)	$+w_d L^2/12$	$+w_d L^2/16$	$-w_d L^2/10$	$-w_d L^2/12$
	$= 23(5)^2/12$	$= 23(5)^2/16$	$= -23(5)^2/10$	$= -23(5)^2/12$
	= 47.92 kN-m	= 35.94 kN-m	= −57.5 kN-m	= −47.92 kN-m
Live load (Not Fixed type)	$+w_L L^2/10$	$+w_L L^2/12$	$-w_L L^2/9$	$-w_L L^2/9$
	$= 19.4(5)^2/10$	$= 19.4(5)^2/12$	$= -19.4(5)^2/9$	$=19.4(5)^2/9$
	= 48.5 kN-m	= 40.42 kN-m	= −53.9 kN-m	= −53.9 kN-m
Total load	96.42 kN-m	76.36 kN-m	−111.4 kN-m	−101.82 kN-m

TABLE 7.2
Design Shear Force

		At Support Next to End Support		At all Other Interior Support
Type of Load	**At End Support**	**Out Side**	**Inner Side**	
Dead load (Fixed type)	$0.4\ w_d L$	$0.6\ w_d L$	$0.55\ w_d L$	$0.5\ w_d L$
	= 0.4(23)(5)	= 0.6(23)(5)	= 0.55(23)(5)	= 0.5(23)(5)
	= 46 kN	= 69 kN	= 63.25 kN	= 57.5 kN
Live load (Not Fixed type)	$0.45\ w_L L$	$0.6\ w_L L$	$0.6\ w_L L$	$0.6\ w_L L$
	= 0.45(19.4)(5)	= 0.6(19.4)(5)	= 0.6(19.4)(5)	= 0.6(19.4)(5)
	= 43.65 kN	= 58.2 kN	= 58.2 kN	= 58.2 kN
Total load	89.65 kN	127.2 kN	121.45 kN	115.7 kN

Where side sway consideration becomes critical due to un-symmetry in geometry or loading, rigorous analysis will be required. For lateral loads, simplified methods may be used to obtain the moments and shear forces for structures that are symmetrical. For unsymmetrical or very tall structures. More rigorous methods should be used.

2D Substitute Frame Analysis Using Software

Considering, Dead load = 30 kN/m and Live load = 10 kN/m

The longest span and the shortest span differ more than 15%. Therefore, Tables 12 and 13 of IS 456 are not allowed to use design bending moment and shear force. Different load cases are to be analyzed to get the worst value of bending moment and shear force, i.e., to get design bending moment and shear force at different locations, i.e., in span and at supports (Figures 7.5–7.19).

Design of continuous beam (A1 – F1) against design bending moment values obtained using Tables 12 and 13 of IS456 and detail of reinforcements (Tables 7.3–7.6).

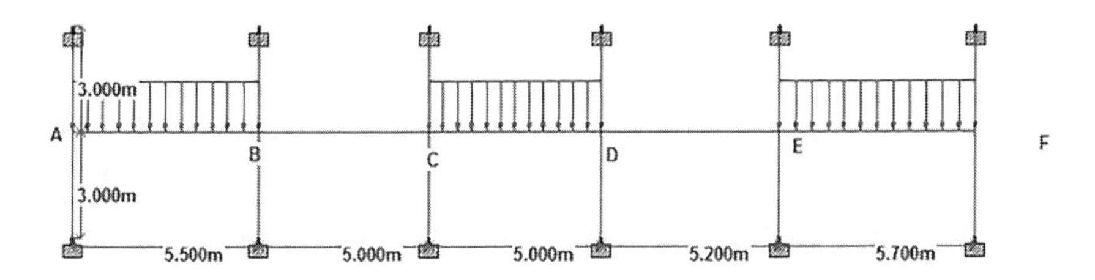

FIGURE 7.5 Case – I – live load on span AB, CD, and EF only.

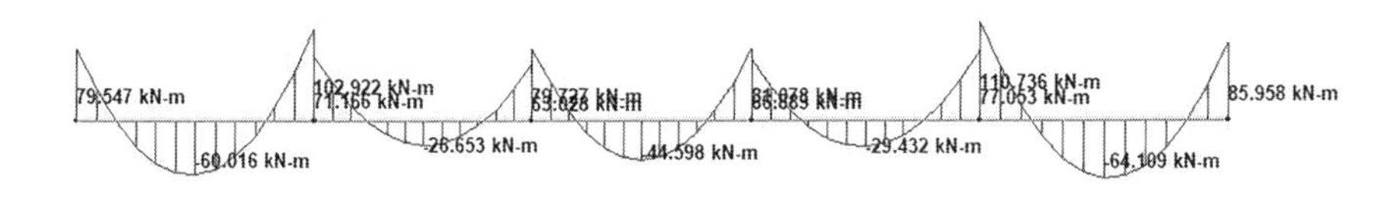

FIGURE 7.6 Bending moment diagram of continuous beam (case I).

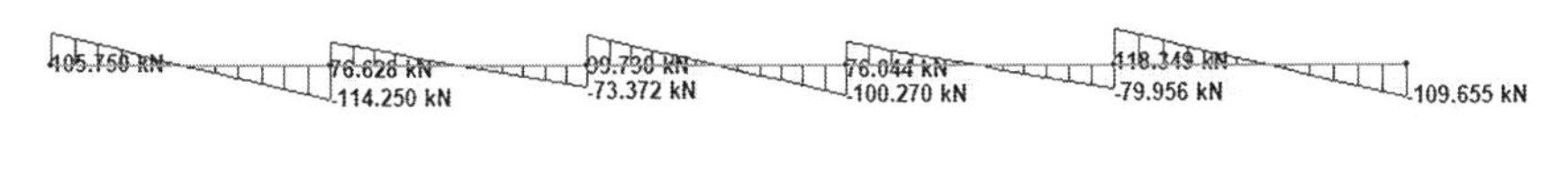

FIGURE 7.7 Shear force diagram of continuous beam (case I).

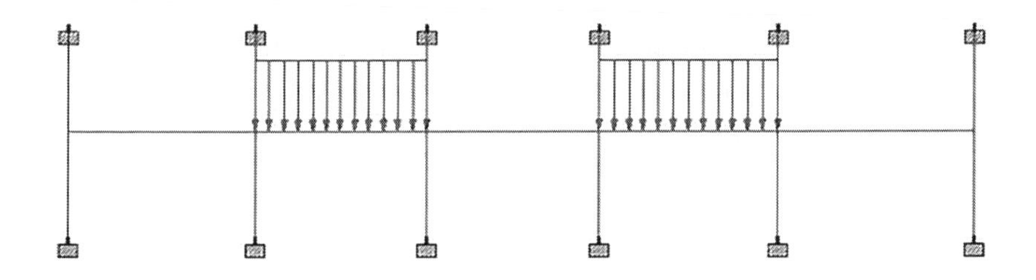

FIGURE 7.8 Case – II – live load on span BC and DE only.

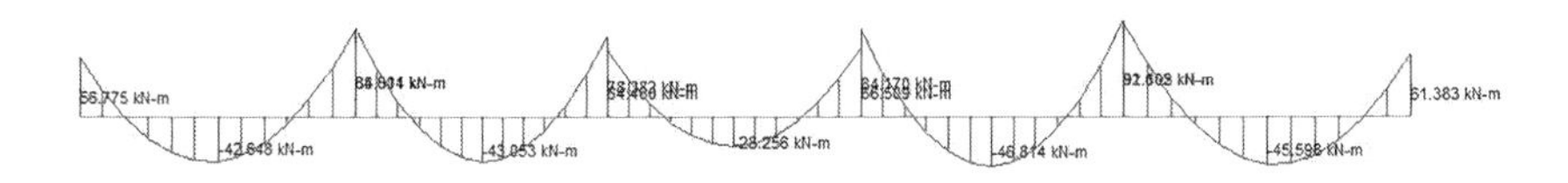

FIGURE 7.9 Bending moment diagram of continuous beam (case II).

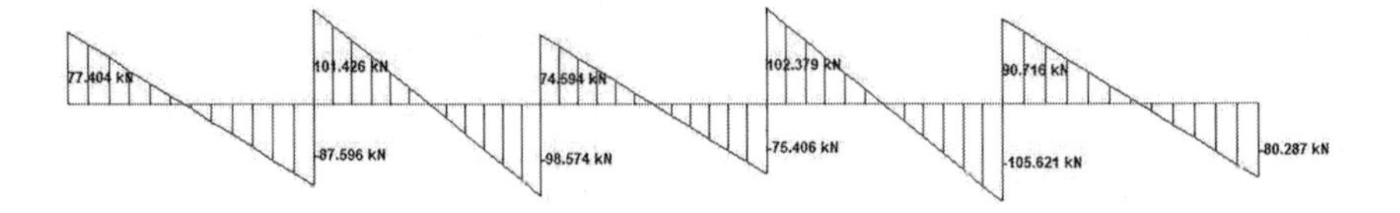

FIGURE 7.10 Shear force diagram of continuous beam (case II).

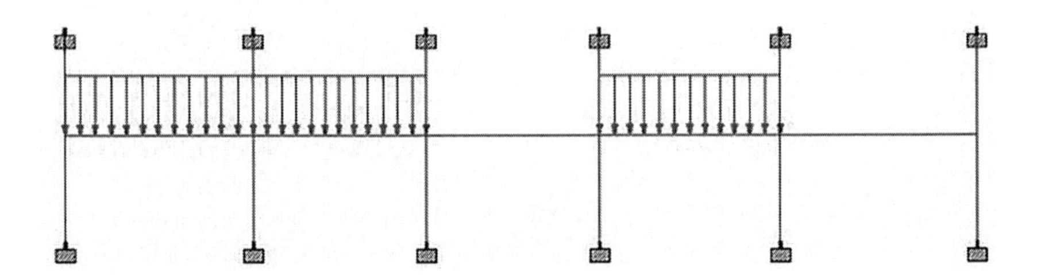

FIGURE 7.11 Case – III – live load on span AB, BC, and DE only.

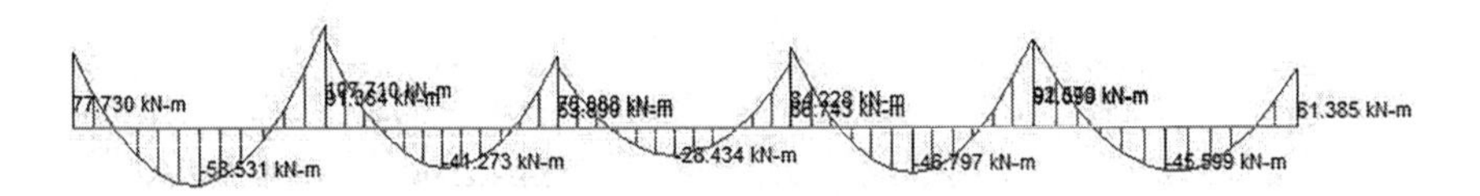

FIGURE 7.12 Bending moment diagram of continuous beam (case III).

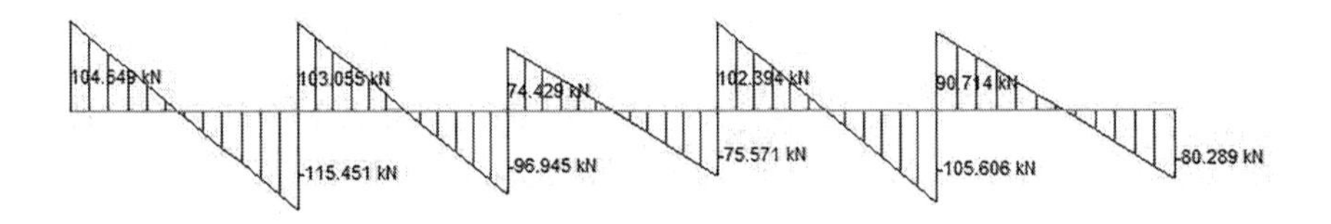

FIGURE 7.13 Shear force diagram of continuous beam (case III).

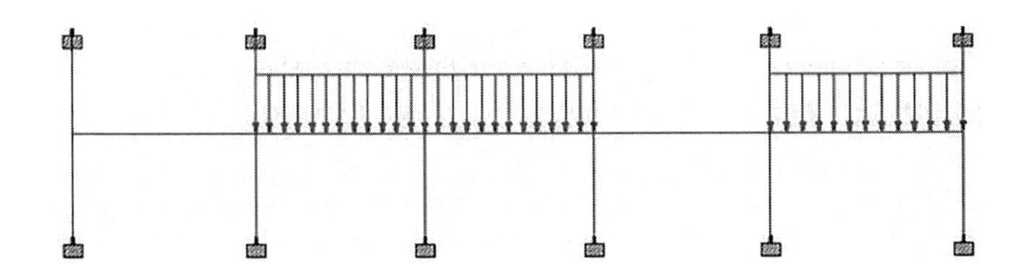

FIGURE 7.14 Case – IV – live load on span BC, CD, and EF only.

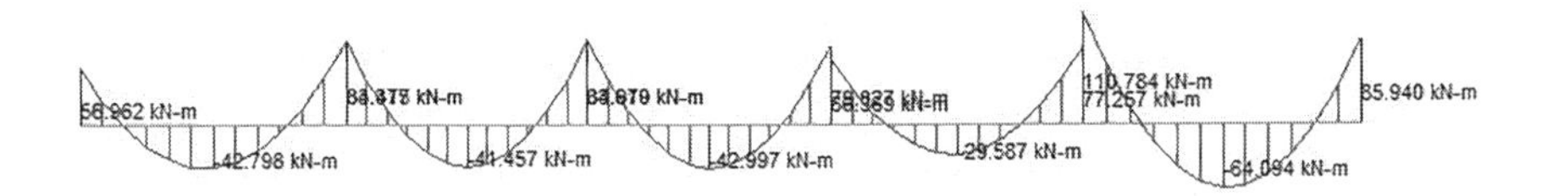

FIGURE 7.15 Bending moment diagram of continuous beam (case IV).

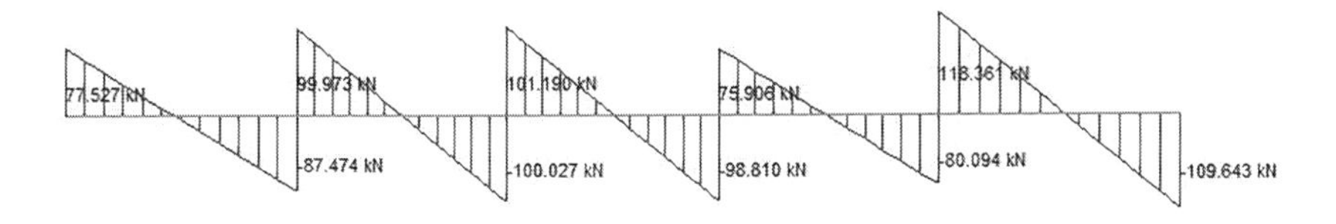

FIGURE 7.16 Shear force diagram of continuous beam (case IV).

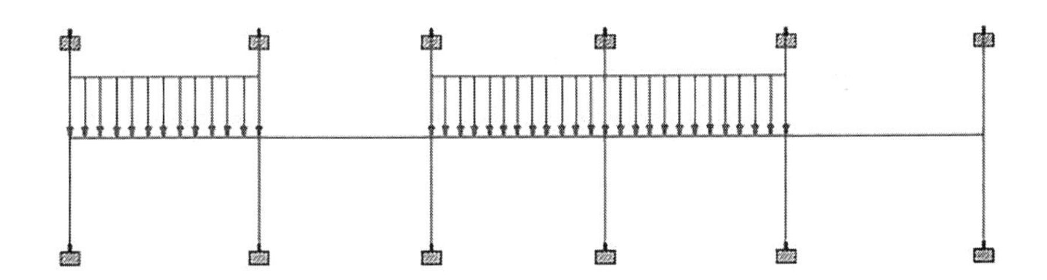

FIGURE 7.17 Case – V – live load on span AB, CD, and DE only.

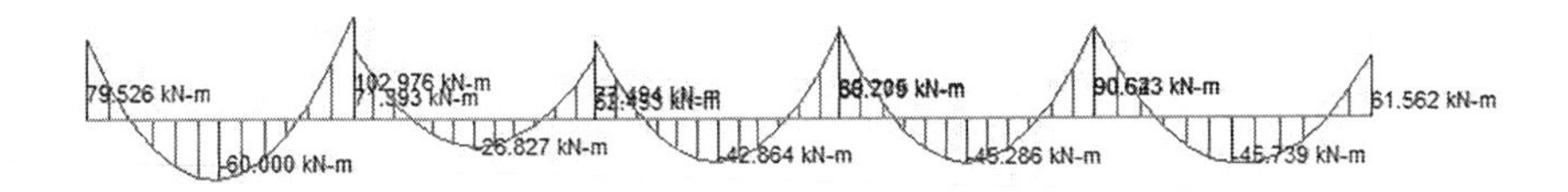

FIGURE 7.18 Bending moment diagram of continuous beam (case V).

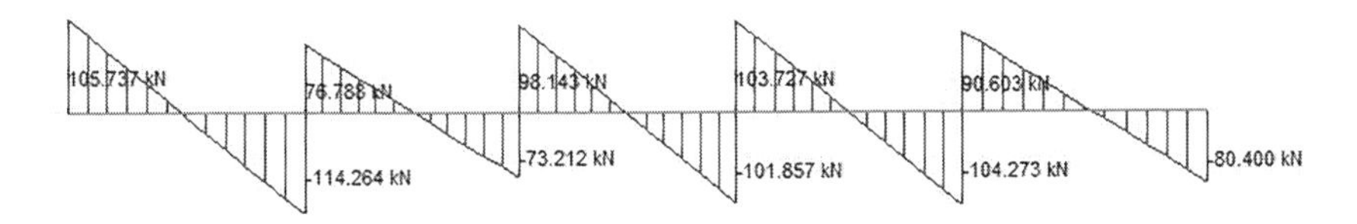

FIGURE 7.19 Shear force diagram of continuous beam (case V).

Let us assume beam section 250 mm × 450 mm. Clear cover = 25 mm, Using 16F – Fe 500 HYSD bars and. M25 grade concrete. $b = 250$ mm, $d = 400 - 25 - 8 = 417$ mm. $d'/d = 33/417 = 0.08$ i.e., 0.1, Minimum p_t (as per IS456) $= 0.85 \times 100/f_y = 0.17\%$,

Minimum $A_{st} = 0.17 \times 250 \times 417/100 = 177\ \text{mm}^2$

Applying limit state method of design as per IS456

γ_f as per IS456 = 1.5 for (DL+LL) combination.

TABLE 7.3
Comparison of Design Bending Moment Values Using Table 12 of IS 456 with 3D Frame Analysis Values

Sign Convention - Sagging Positive

	Span Moments		Support Moments	
	Near Middle of End Span	At Middle of Interior Span	At Support Next to End Support	At Other Interior Support
Design bending moment (as per IS456)	[a]+ 96.42 kN-m	[a]+ 76.36 kN-m	[a]111.4 kN-m	[a]–101.82 kN-m
Design bending moment (as per 3D analysis)	[b]+ 79.19 kN-m	[b]+ 60.63 kN-m	[b]–109.96 kN-m	[b]–90.56 kN-m

TABLE 7.4
Comparison of Design Shear Force Values Using 13 of IS 456 with 3D Frame Analysis Values

	At End Support	At Support Next to End Support		At all Other Interior Support
		Out Side	Inner Side	
Design Shear force (as per IS456)	[1]89.65 kN	[1]127.2kN	[1]121.45 kN	[1]115.7 kN
Design Shear force (as per 3D analysis)	[2]76.47 kN	[2]103.76 kN	[2]90.04 kN	[2]93.3 kN

TABLE 7.5
Design against Bending Moment (M)

	Span Moments		Support Moments	
	Near Middle of End Span	At Middle of Interior Span	At Support Next to End Support	At Other Interior Support
M	96.42 kN-m	76.36 kN-m	−111.4 kN-m	−101.82 kN-m
$M_u = \gamma_f M$	144.63 kN-m	114.54 kN-m	−167.1 kN-m	−152.73 kN-m
M_u/bd^2	3.32 kN/m^2	2.634 kN/m^2	3.84 kN/m^2	3.51 kN/m^2
p_t from SP16	0.945%	0.705%	1.069%	0.989%
p_c from SP16	0.001%	-	0.13%	0.047%
$A_{st} = p_t\,bd/100$	985 mm^2	735 mm^2	1115 mm^2	1031 mm^2
$A_{sc} = p_c bd/100$	p_c (minimum)	p_c (minimum)	p_c (minimum)	p_c (minimum)
No. of bars on tension face (16F)	4–20ϕ	3–20ϕ	4–20ϕ	4–20ϕ
No. of bars on compression face (16F)	2–20ϕ	2–20ϕ	2–20ϕ	2–20ϕ

TABLE 7.6
Design against Shear Force (V)

	At End Support	At Support Next to End Support: Out Side	At Support Next to End Support: Inner Side	At all Other Interior Support
V	89.65 kN	127.2 kN	121.45 kN	115.7 kN
$V_u = \gamma_f V$	134.475 kN	190.8 kN	182.175 kN	173.55 kN
Percentage of tension steel provided (p_{tp})	For 3–16 ϕ (mim) 0.58%	For 6–16ϕ 1.16%	For 6–16ϕ 1.16%	For 6–16ϕ 1.16%
7_c from SP16	0.52 N/mm²	0.68 N/mm²	0.68 N/mm²	0.68 N/mm²
V_c =7c bd	54.21 kN	70.89 kN	70.89 kN	70.89 kN
$V_{us} = V_u - V_c$	80.265 kN	119.91 kN	111.285 kN	102.66 kN
V_{us}/d	1.92 kN/cm	2.88 kN/cm	2.67 kN/cm	2.46 kN/cm
Diameter and spacing of stirrups from SP16-Table 62	[a]2L -8 ϕ @ 100c/c	2L -8 ϕ @ 100c/c	2L -8 ϕ @ 100c/c	2L -8 ϕ @ 100c/c

[a] Fe 500 stirrups, Requirement of spacing of stirrups obtained using Table 62 of IS456 and then adjusted from practical consideration.

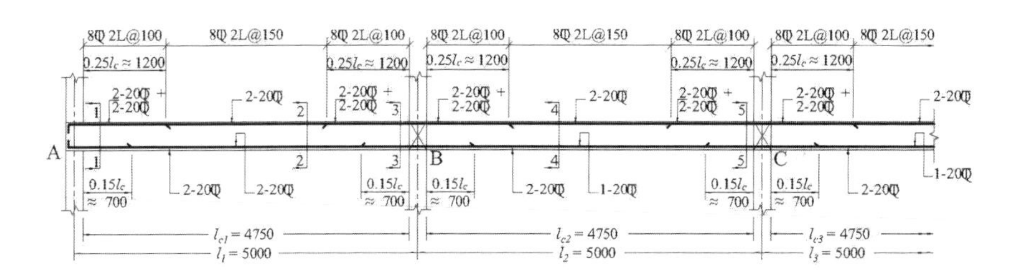

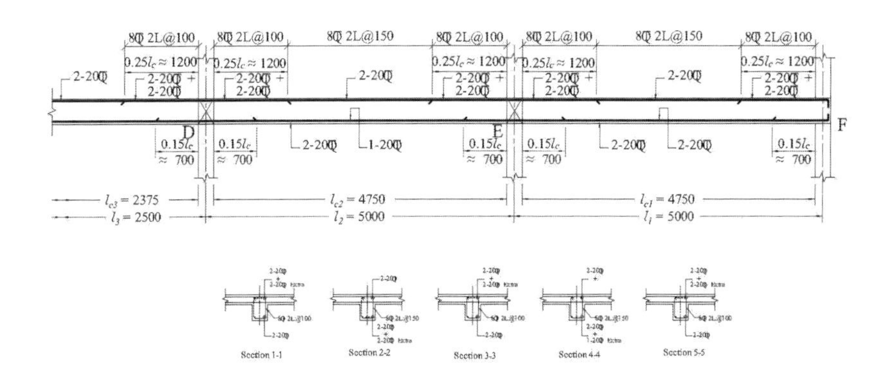

FIGURE 7.20 Reinforced detailing of continuous beams.

3D Frame Analysis Using Software

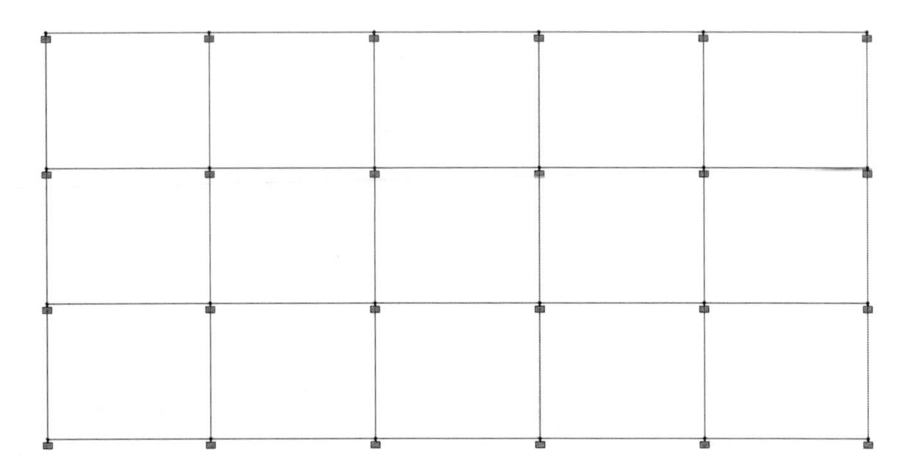

FIGURE 7.21 Structural plan showing columns and beams.

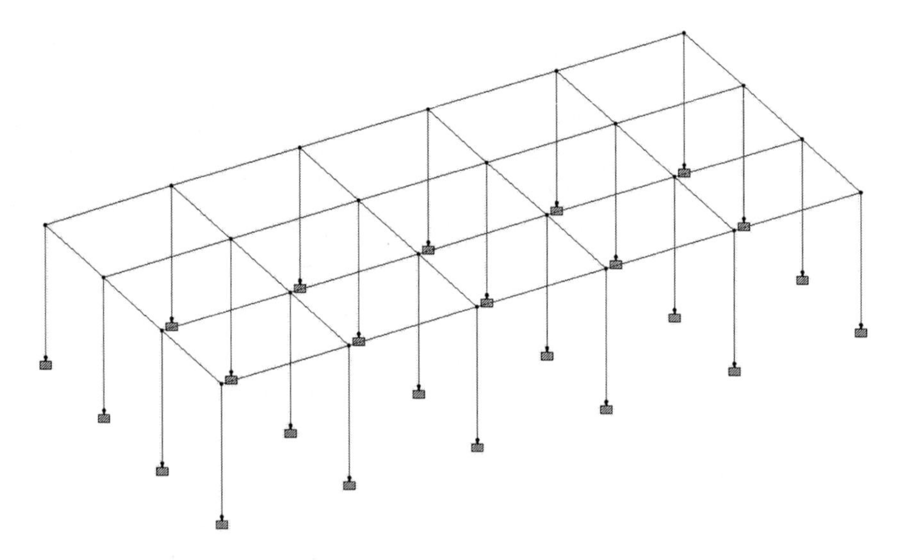

FIGURE 7.22 3D view of beam-column frame. Beam – 250 mm × 400 mm, column – 250 mm × 400 mm. The longer direction of the column is oriented along the shorter direction of the frame.

Case I – Dead Load and Live Load on All Slab Panels

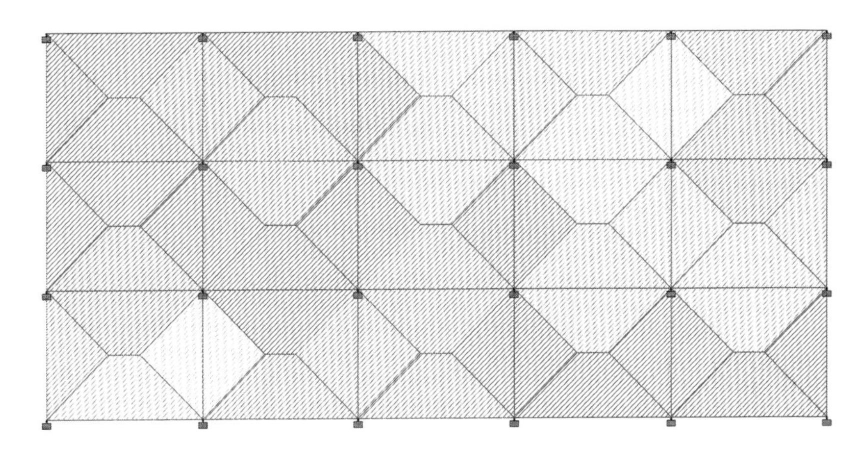

FIGURE 7.23 Load distribution ((DL+LL) distribution shown) (Case I).

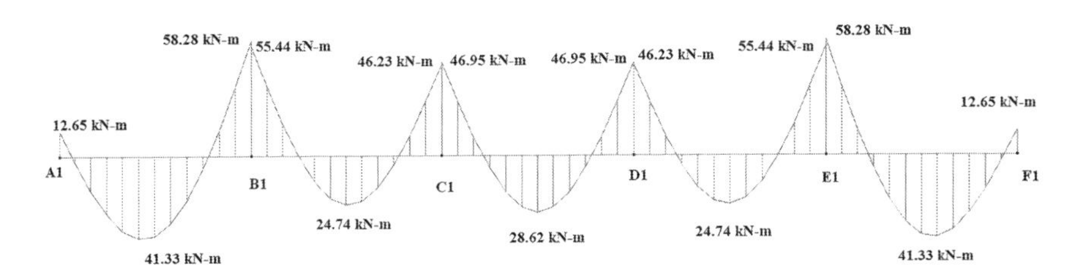

FIGURE 7.24 Bending moment diagram of continuous beam A1-F1 (Case I).

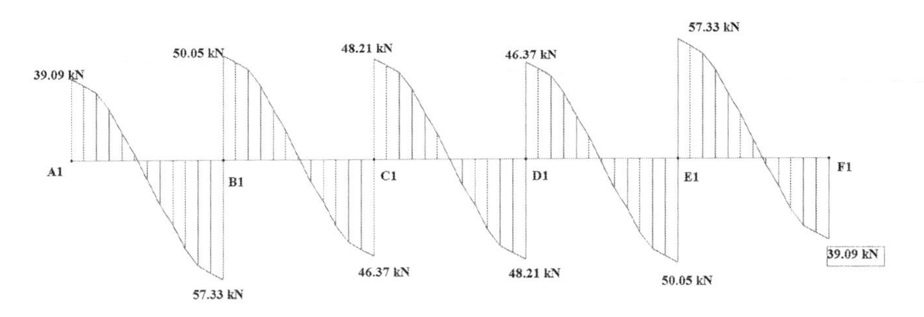

FIGURE 7.25 Shear force diagram of continuous beam A1-F1 (Case I).

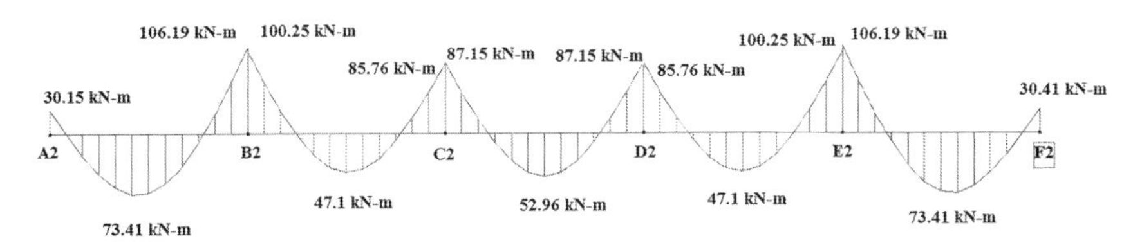

FIGURE 7.26 Bending moment diagram of continuous beam A2-F2 (Case I).

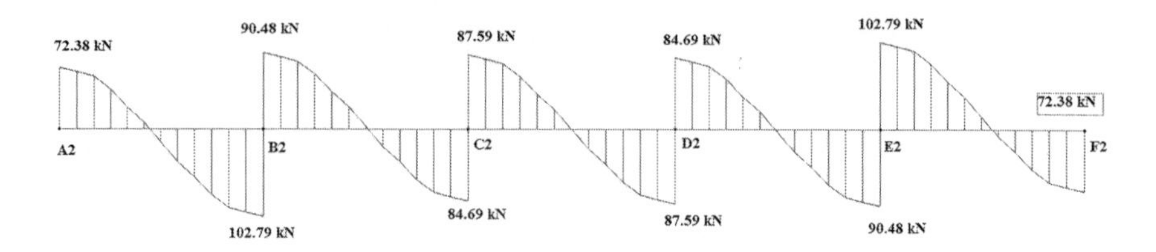

FIGURE 7.27 Shear force diagram of continuous beam A2-F2 (Case I).

Case II – Dead Load on All Slab Panels and Live Load on A1-B1, C1-D1, and E1-F1

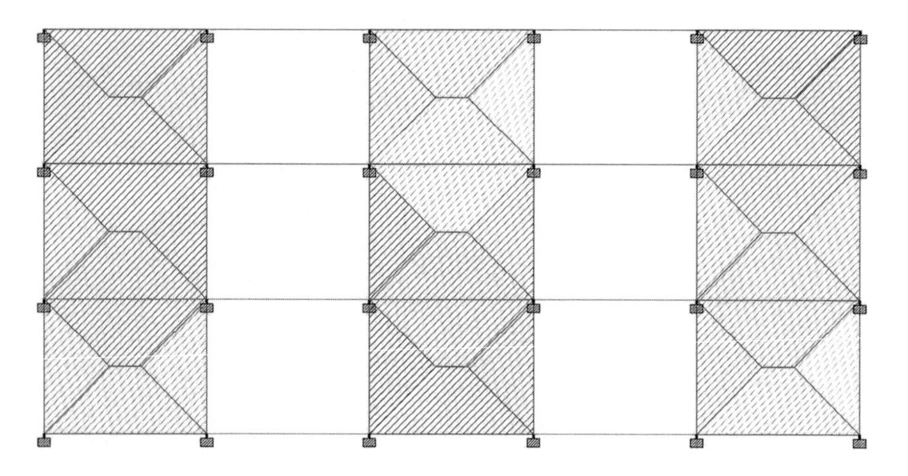

FIGURE 7.28 Load distribution (only LL distribution shown) (Case II).

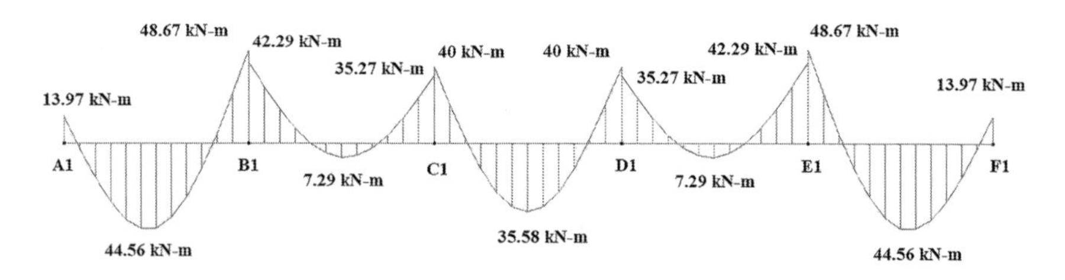

FIGURE 7.29 Bending moment diagram of continuous beam A1-F1 (Case II).

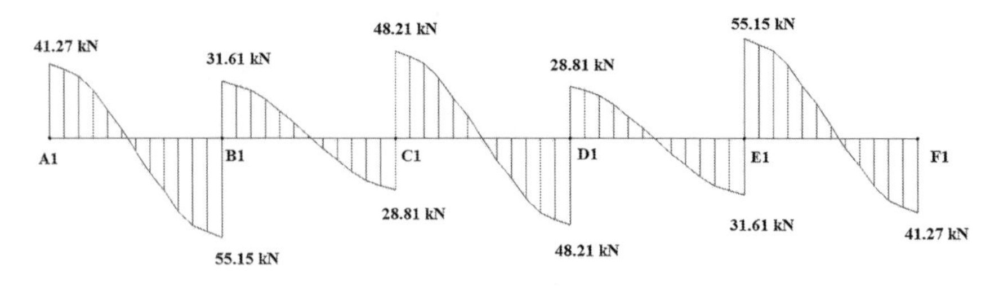

FIGURE 7.30 Shear force diagram of continuous beam A1-F1 (Case II).

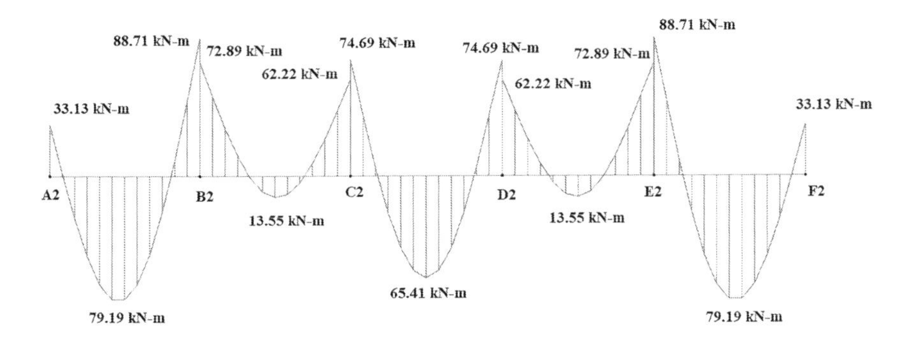

FIGURE 7.31 Bending moment diagram of continuous beam A2-F2 (Case II).

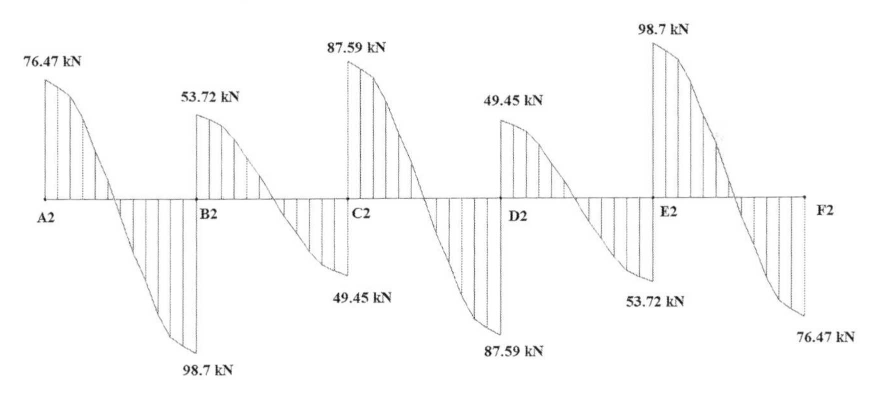

FIGURE 7.32 Shear force diagram of continuous beam A2-F2 (Case II).

Case III – Dead Load on All Slab Panels and Live Load on B1-C1 and D1-E1

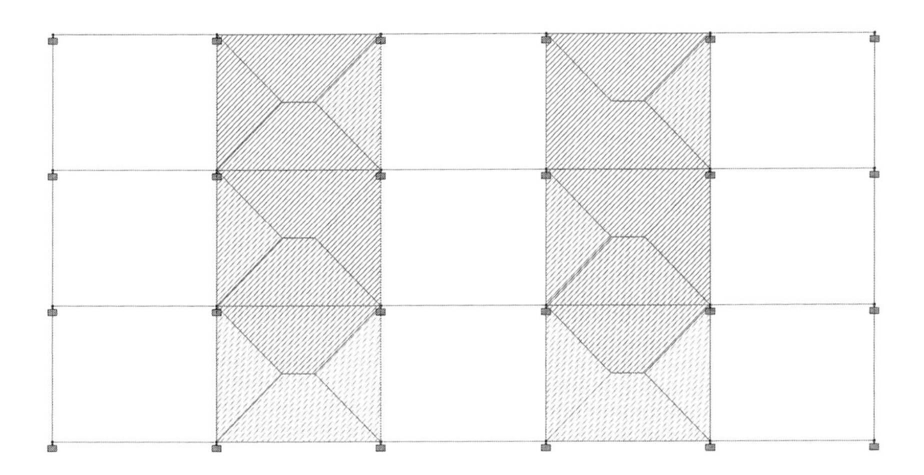

FIGURE 7.33 Load distribution (only LL distribution shown) (Case III).

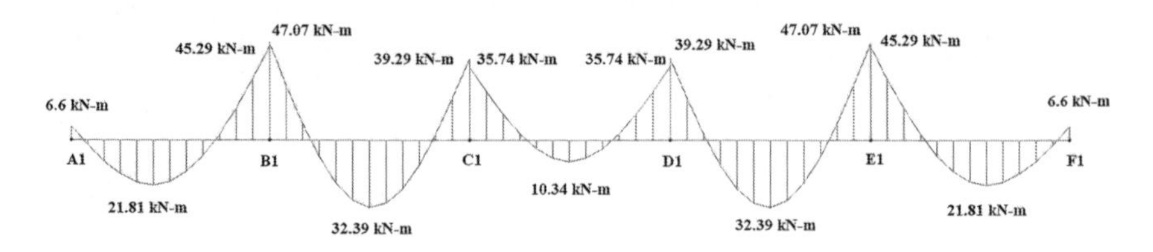

FIGURE 7.34 Bending moment diagram of continuous beam A1-F1 (Case III).

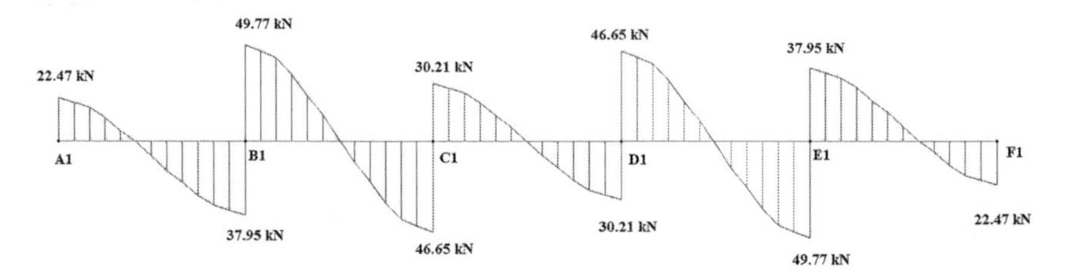

FIGURE 7.35 Shear force diagram of continuous beam A1-F1 (Case III).

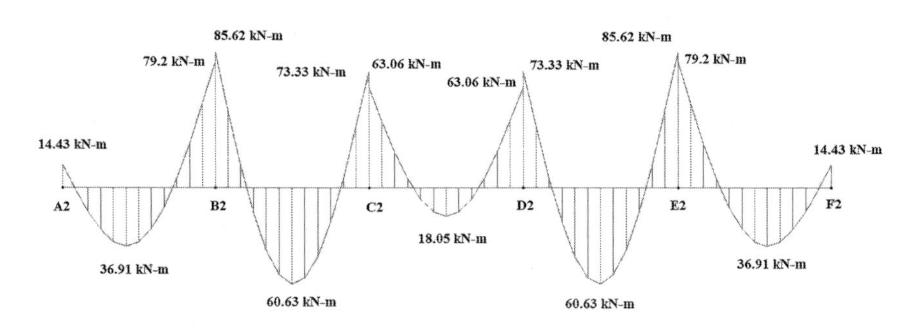

FIGURE 7.36 Bending moment diagram of continuous beam A2-F2 (Case III).

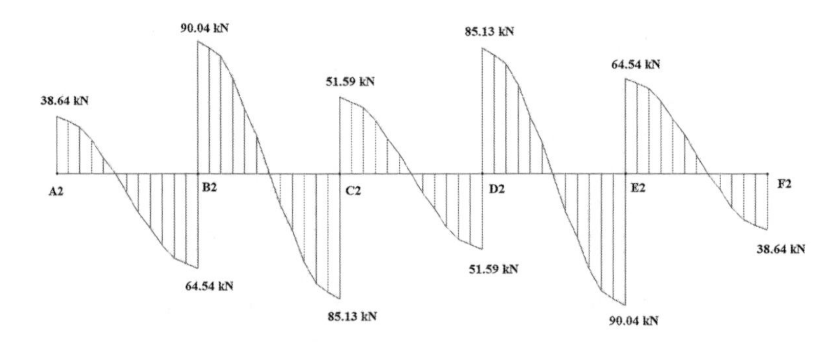

FIGURE 7.37 Shear force diagram of continuous beam A2-F2 (Case III).

Case IV – Dead Load on All Slab Panels and Live Load on A1-B1, B1-C1, and D1-E1

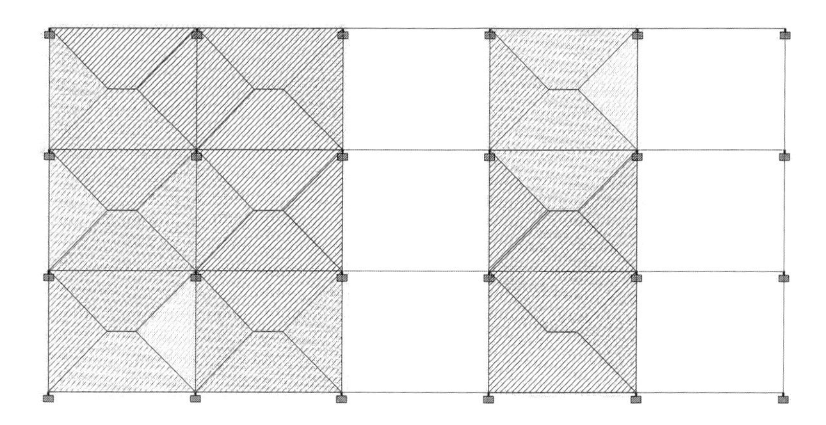

FIGURE 7.38 Load distribution (only LL distribution shown) (Case IV).

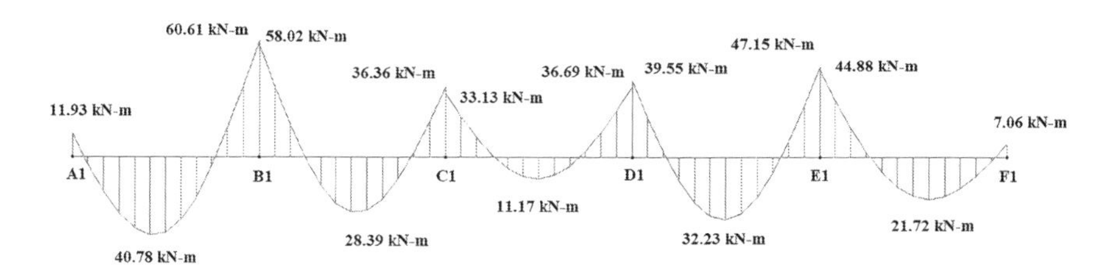

FIGURE 7.39 Bending moment diagram of continuous beam A1-F1 (Case IV).

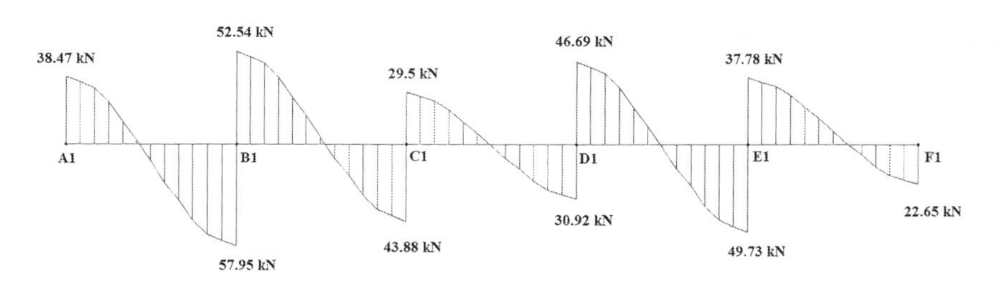

FIGURE 7.40 Shear force diagram of continuous beam A1-F1 (Case IV).

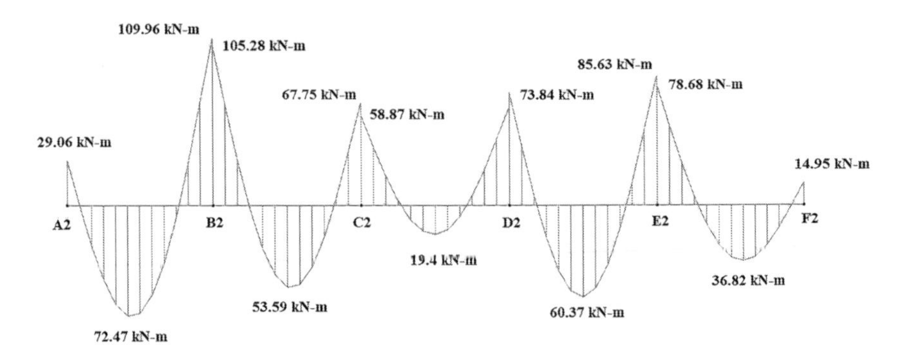

FIGURE 7.41 Bending moment diagram of continuous beam A2-F2 (Case IV).

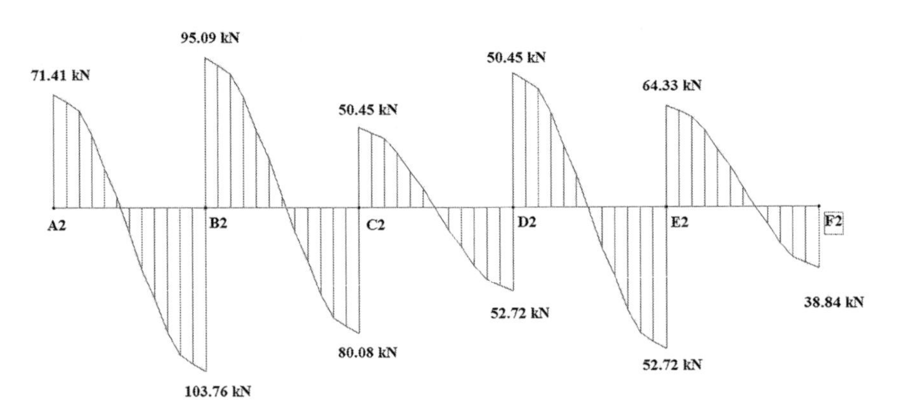

FIGURE 7.42 Shear force diagram of continuous beam A2-F2 (Case IV).

Case V – Dead Load on All Slab Panels and Live Load on B1-C1, C1-D1, and E1-F1

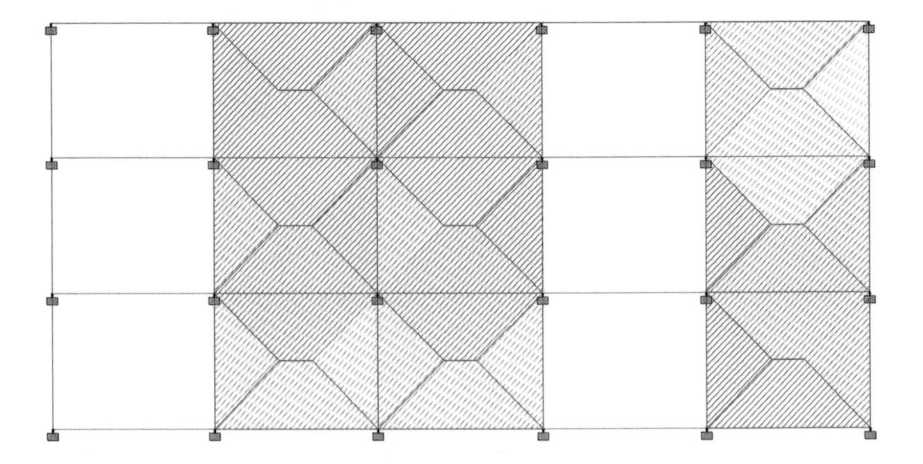

FIGURE 7.43 Load distribution (only LL distribution shown) (Case V).

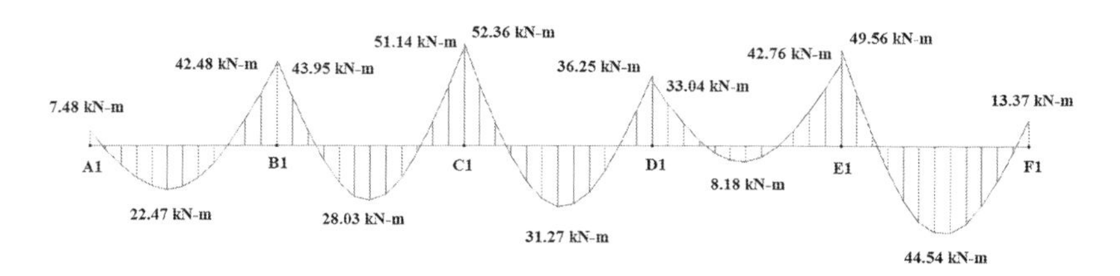

FIGURE 7.44 Bending moment diagram of continuous beam A1-F1 (Case V).

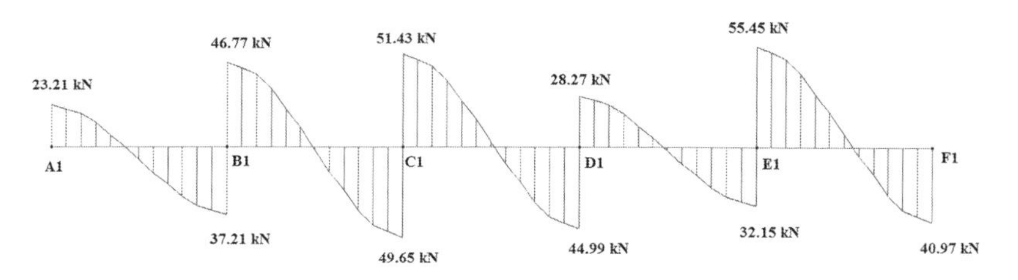

FIGURE 7.45 Shear force diagram of continuous beam A1-F1 (Case V).

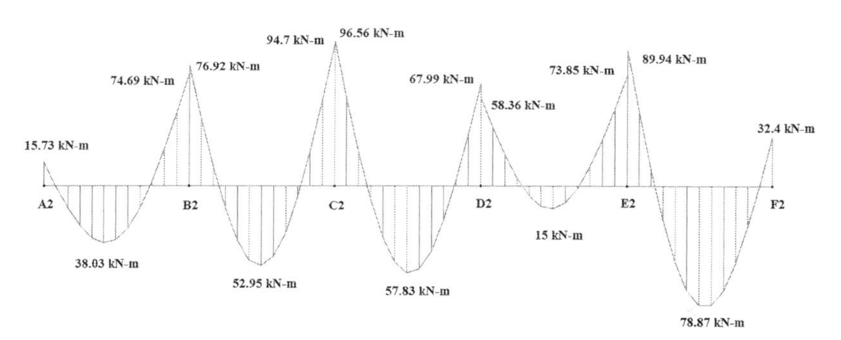

FIGURE 7.46 Bending moment diagram of continuous beam A2-F2 (Case V).

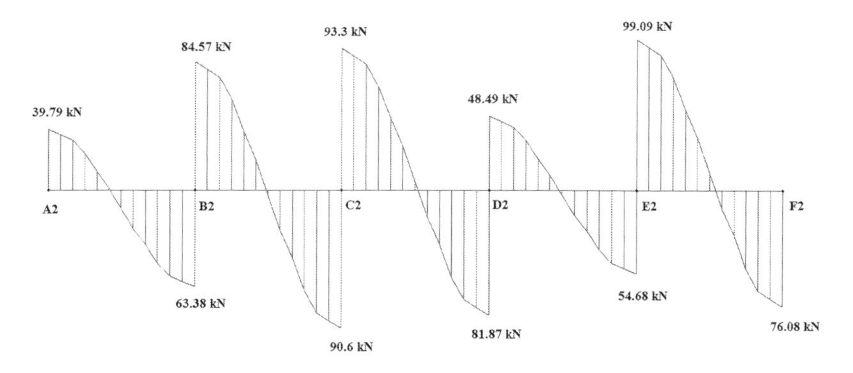

FIGURE 7.47 Shear force diagram of continuous beam A2-F2 (Case V).

Design bending moment and shear force have to be selected from the above bending moment diagram and shear force diagram as shown above, and accordingly, RCC design and detailing have to be made (Figures 7.20–7.47).

8 Design of Slab System

PREAMBLE

The slab system may be as follows:

1. Slab system with beams and columns or walls
2. Flat slab system with or without drops, supported generally without beams, only by columns with or without flared column heads

SLAB SYSTEM WITH BEAMS AND COLUMNS

The slab panel may be a One-way or Two-way slab and edge conditions may be continuous (edges may continue) or may be discontinuous (edge may be ended). Shorter span may be designated by L_x and longer span may be designated by L_y. If the L_y / L_x ratio is 2 or more, the slab is termed as 'One way' otherwise it is termed as 'Two-way'. Around 95% of the load is transferred in shorter direction in case of 'One way' slab panels. Naturally, main tension reinforcements are to be placed in shorter direction and minimum reinforcement as recommended in IS456, may be placed in longer direction. For two-way slab panels bending in both directions is equally important in both the direction and design bending moments in both directions may be calculated as recommended in Table 26 of IS456. In two-way slab panels, special attention is to be paid for torsional moments at corners and special care has to be taken at the corner of the slab panels where adjacent edges are discontinuous or one edge discontinuous and the other one is continuous, per clause D-1.8 to D-1.11. However, at corner, where both the edges are continuous, no special care has to be taken. Reinforcement against flexure/bending has to be provided as recommended in D-1.0 to D-1.7.

DEFLECTION CONTROL CRITERIA

CONTROL OF DEFLECTION

As per clause 23.2 of IS456, the deflection of structure or part thereof shall not adversely affect the appearance or efficiency of the structure or finishes or partitions. The deflection shall generally be limited to the following:

1. The final deflection due to all loads including the effects of temperature, creep, and shrinkage and measured from the as-cast level of the supports of floors, roofs, and all other horizontal members, should not normally exceed span/250.

DOI: 10.1201/9781003208204-8

2. The deflection including the effects of temperature, creep, and shrinkage occurring after the erection of partitions and the application of finishes should not normally exceed span/350 or 20 mm whichever is less.

The vertical deflection limits may generally be assumed to be satisfied provided that the span to depth is not greater than the values obtained as recommended below

Basic values of span to effective depth ratios for spans up to 10 m: for cantilever structural members = 7, simply structural members supported = 20, and continuous structural members = 26. For structural members having spans above 10m, the values above-mentioned values need to be multiplied by 10/span in meters, except for cantilever in which case deflection calculations should be made. Depending on the area and the stress of steel for tension reinforcement, the above-mentioned values shall be modified by multiplying with the modification factor obtained as per Figure 4 of IS456. Depending on the area of compression reinforcement, the value of span to depth ratio be further modified by multiplying with the modification factor obtained as per Figure 5 of IS456. For flanged beams, the above-mentioned values be modified as per Figure 6 of IS456 and the reinforcement percentage for using Figures 4 and 5 should be based on area of section equal to $b_f d$ (b_f = flange width). However, when deflections are required to be calculated, the method given in Annex C of IS 456 may be used.

Again, as per clause 24.1 of IS456, for slabs spanning in two directions, the shorter of the two spans should be used for calculating the span to effective depth ratios. For two-way slabs of shorter spans(up to 3.5 m) with mild steel reinforcement, the span to overall depth ratios given below may generally be assumed to satisfy vertical deflection limits for loading class up to 3 kN/m^2. For Simply supported slabs, the said ratio is 35 and for continuous slabs is 40. For high-strength deformed bars (HYSD bars) of grade Fe415, the above-mentioned values should be multiplied by 0.8.

As per clause 24.4 of IS456, the slabs spanning in two directions at right angles and carrying uniformly distributed load may be designed by any acceptable theory or by using coefficients given in Annex D of IS456. For determining bending moments in slabs spanning in two directions at right angles and carrying concentrated load, any accepted method approved by the engineer-in-charge may be adopted. The commonly used elastic methods are based on Pigeaud's, or Wester-guard's theory and the most commonly used limit state of collapse method is based on Johansen's yield line theory.

As per clause 24.4.1 of IS456, for a restrained slab with unequal conditions at adjacent panels, in some cases the support moments calculated from Table 26 for adjacent panels may differ significantly, the following procedure may be adopted to adjust them ie. calculate the sum of moments at mid-span and supports (neglecting signs) and treat the values from Table 26 as fixed end moments. Then, according to the relative stiffness of adjacent spans, distribute the fixed end moments across the supports, giving new support moments.

Then adjust mid-span moment such that, when added to the support moments from (neglecting signs), the total should be equal to mid-span moment.

If the resulting support moments are significantly greater than the value from Table 26, the tension steel over the supports will need to be extended further. Consider, the span moment as parabolic between supports: its maximum value and determine the points of contra-flexure of the new support moments with the span moment. Extend half the support tension steel at each end to at least an effective depth or 12 bar diameters beyond the nearest point of contra-flexure. Extend the full area of the support tension steel at each end to half the distance obtained.

DESIGN OF TWO-WAY SLAB SUPPORTED BY BEAM USING CODAL MOMENT COEFFICIENTS

Choice of Slab Thickness

Slab panels P1, P2, P3, P4, P5, P6 and P7 (Figure 8.1).

Maximum shorter span (l_x) of slab panels (l_x) = 3.145 m and corresponding longer span (l_y) of slab panels = 3.215 m

$L_y/l_x = 3.215/3.145 = 1.022 < 2$, Therefore, it is a two-way slab panel

Referring clause no. 24.1 of IS 456-2000

All slab panels are bounded with framed beams, therefore, it may be considered as continuous slab panels.

Then, Shorter span (l_x) / Effective depth (d) ratio = $(40 \times 0.8) = 32$, considering HYSD/TMT bars/ deformed bars

Hence, Effective depth (d) = 3145/32 = 99 mm

Referring to clause no. 26.4.2 and 26.4.3, Table 16 and 16A of IS 456-2000, recommended nominal cover of slab panels = 20 mm

Therefore, overall depth = 99 + 20 + 4 =123 mm, say **125 mm**

*considering 8mm diameter reinforcing bars

Similarly, for floor slab panel – P2 & P6, maximum shorter span (l_x) = 2556 mm and requirement of overall depth = **125 mm**

Calculation of Dead Load

(for slab panels marked – P1 to P3, P6)

For a Typical Floor Slab Panel

Self-weight of slab (125mm thick) = (0.125) (25) = 3.125 kN/m²

Floor finish:

Self weight of 25 mm thick concrete below tiles	= (0.025) (24)	= 0.6 kN/m²
Self weight of 9.8 mm thick Vitrified tiles		= 0.185 kN/m²
Self weight of 12 mm ceiling plaster	= (0.012) (20.4)	= 0.24 kN/m²
Light partition load directly on slab panels		= 1.0 kN/m²*
		≈ 5.15 kN/m²

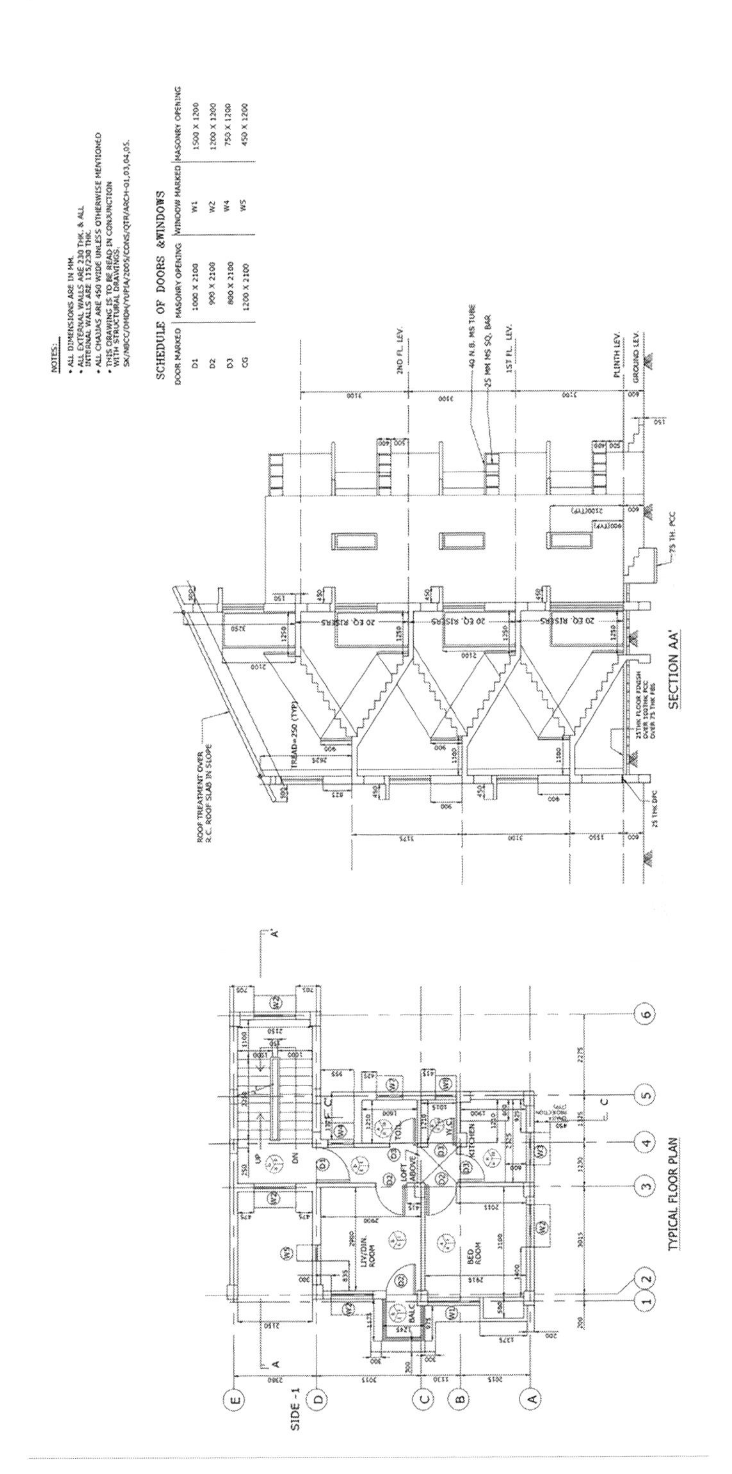

FIGURE 8.1 Architectural plan and sectional view.

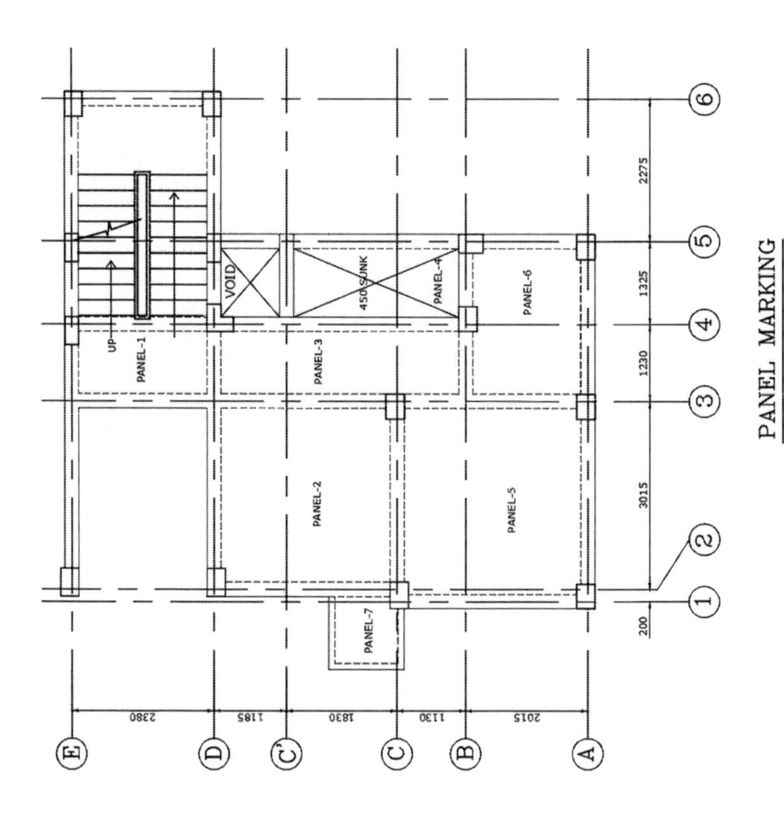

FIGURE 8.2 Marking of slab panels on structural plan.

* In case brick walls are placed directly on a slab panels, an approximate equivalent distributed load has to be added as follows (though brick walls need to be considered as line load and accordingly moment calculations need to be made). This is an approximate method. Rigorous plate theory may be applied in this regard. Calculations may done to make it simpler.

$$*w_{\text{bke}} = \frac{1.25\times \text{ Total weight of brick walls on a particular slab panel}}{\text{Area of the slab panel}}$$

Dead Load of RCC Slab

Self weight of slab (kN/m^2)	$= (0.1)(25)$	$= 2.5$ kN/m^2
Self weight of 25 mm thick concrete layer below tiles etc.	$= (0.025)(24)$	$= 0.6$ kN/m^2
Self weight of 9.8 mm thick Vitrified tiles		$= 0.185$ kN/m^2
Self weight of ceiling plaster 12 mm thick	$= (0.012)(20.4)$	$= 0.24$ kN/m^2
		$w_d \approx 3.5$ kN/m^2

Live Load

As per IS 875 Part 2, Table 1 (i), for residential building

Live load on all rooms, kitchen, toilet & bathroom = 2 kN/m^2

Corridors, passages, staircases including fire escapes and store rooms, balconies = 3 kN/m^2

As per IS 875 (Part 2) Table 1 (V) – for Business and office building

Live load on a typical floor = 4.0 kN/m^2 (Rooms without storage facility)

Live load on a corridor, staircase, etc. = 4.0 kN/m^2

Live load on a typical roof slab panel as per IS 875 (Part 2) Table 2 (1) (i) = 1.5 kN/m^2

(Considering accessible roof)

Design of Slab Panels

Let us use clause no. 37.1.2 and ANNEX- D of IS 456-2000

Shorter span moment coefficient:

Positive coefficient = α_x Negative coefficient = $\alpha_{x'}$

Longer span moment coefficient:

Positive coefficient = α_y Negative coefficient = $\alpha_{y'}$

Slab panels are marked according to edge condition as per Table 26 of IS 456, presented in Figure 8.2.

TABLE 8.1
Slab panels are marked according to edge condition (as per Table 26 of IS-456)

Sl. No.	Panel Marked	Description	Case Number
1	P3	Interior panel	Case 1
2	P1, P4	One long edge discontinuous	Case 3
3	P2, P5, P6	Two adjacent edges are discontinuous	Case 4
4	P7	Three edges discontinuous	Case 8

For Panel Marked P1 to P3 & P5 to P7

Dead load for a typical floor $= 5.15\ \text{kN/m}^2$

Live load for a typical floor $= 4.00\ \text{kN/m}^2$

$= 9.15\ \text{kN/m}^2$

For Panel Marked P4

Load due to drop slab in the toilet $= (0.125)\ (19) = 2.4\ \text{kN/m}^2$

Dead load for typical floor $= 5.15\ \text{kN/m}^2$

Live load for typical floor $= 4.00\ \text{kN/m}^2$

$= 11.55\ \text{kN/m}^2$

Bending moment coefficients according to Table 26 of IS 456 and corresponding design bending moments for slab panels are presented in Table 8.1.

Design calculations based on Tables 8.2 and 8.3.

REINFORCEMENT DETAIL

Few Important Aspects of Design and Detailing of RCC Slab

- Nominal cover should as per Table 16A for continuous slab with 2 hours fire resistance = 20 mm
- Minimum reinforcement of slab as per clause no. 26.5.2.1 = 0.12% (for HYSD/TMT deformed bar) but as per clause no. 6.2.1- IS 13920, 2016, for flexural member is 0.24%, considering slab as a flexural member.
- Maximum spacing of reinforcements in tension as per clause no. 26.3.3 = $3d$ or 300 mm, whichever is less.

TABLE 8.2

Design Bending Moment for Slab Panels (as per Table 26 of IS 456)

Slab Panel Marked	l_y (m)	l_x (m)	l_y/l_x	(DL + LL) (kN/m²)	Case No.	Bending Moment Coefficient α_x	α_y	$\alpha x'$	$\alpha y'$	Positive Bending Moment in Shorter Span M_{ux} (kNm)	Positive Bending Moment in Longer Span M_{uy} (kNm)	Negative Bending Moment in Shorter Span M_{ux}' (kNm)	Negative Bending Moment in Longer Span M_{uy}' (kNm)
P1	2.380	1.230	1.94	9.15	3	0.064	0.028	0.083	0.037	1.33	0.58	1.72	0.77
P2	3.015	3.015	1	9.15	4	0.035	0.035	0.047	0.047	4.37	4.37	5.86	5.86
P3	4.145	1.230	3.37	9.15	1	0.049	0.024	0.065	0.032	1.02	0.5	1.35	0.66
P4	2.96	1.325	2.234	11.55	3	0.065	0.028	0.085	0.037	1.96	0.844	2.56	1.12
P5	3.215	3.145	1.022	9.15	4	0.0361	0.035	0.0483	0.047	4.9	4.75	6.56	6.38
P6	2.556	2.015	1.27	9.15	4	0.0478	0.035	0.0635	0.047	2.66	1.95	3.54	2.62
P7	1.239	1.145	1.08	9.15	8	0.0494	0.043	-	0.057	0.89	0.77	-	1.03

TABLE 8.3
Design Calculations for Reinforcement for Slab Panels

Panel Marked	Revised Overall Depth (D)	Revised Effective Depth (d)	M_{ux}/bd^2	M_{uy}/bd^2	M_{ux}/bd^2	M_{uy}/bd^2	Percentage of Steel along Shorter Bottom	Percentage of Steel along Longer Bottom	Percentage of Steel along Shorter Top	Percentage of Steel along Longer Top
	mm	mm	N/mm^2	N/mm^2	N/mm^2	N/mm^2	p_{tx}	p_{ty}	ptx'	pty'
P1	125.00	101	0.13	0.06	0.17	0.075	0.085	0.085	0.085	0.085
P2	125.00	101	0.43	0.43	0.575	0.575	0.1224	0.1224	0.165	0.165
P3	125.00	101	0.10	0.049	0.132	0.065	0.085	0.085	0.085	0.085
P4	125.00	101	0.192	0.083	0.251	0.11	0.085	0.085	0.085	0.085
P5	125.00	101	0.48	0.466	0.643	0.63	0.137	0.133	0.185	0.181
P6	125.00	101	0.26	0.191	0.347	0.26	0.085	0.085	0.0982	0.085
P7	125.00	101	0.087	0.075	–	0.101	0.085	0.085	–	0.085

TABLE 8.3 (*Continued*)
Design Calculations for Reinforcement for Slab Panels

Panel	Area of Steel				Spacing of Reinforcement Required				Bar dia.	Spacing of Reinforcement Provided			
	Shorter Bottom	Longer Bottom	Shorter Top	Longer Top	Shorter Bottom	Longer Bottom	Shorter Top	Longer Top		Shorter Bottom	Longer Bottom	Shorter Top	Longer Top
	A_{stx}	A_{sty}	A_{stx}'	A_{sty}'	S_x	S_y	S_x'	S_y'		S_x	S_y	S_x'	S_y'
Marked	mm²	mm²	mm²	mm²	mm	mm	mm	mm		mm	mm	mm	mm
P1	86	86	86	86	300	300	300	300	8T	250	250	250	250
P2	124	124	167	167	300	300	299	299	8T	250	250	250	250
P3	86	86	86	86	300	300	300	300	8T	250	250	250	250
P4	86	86	86	86	300	300	300	300	8T	250	250	250	250
P5	139	135	187	183	300	300	267	274	8T	250	250	250	250
P6	86	86	100	86	300	300	300	300	8T	250	250	250	250
P7	86	86	-	86	300	300	-	300	8T	250	250	250	250

8T – Fe 415 TMT steel reinforcement & M20 concrete is used

- Maximum diameter of reinforcing bar as per clause no. 26.5.2.2 should be less than D/8.
- As per clause no. D-1.4 of Annex D, IS 456-2000, tension reinforcement provided at mid-span in the middle strip shall extend in the lower part of the slab to within $0.25L$ of a continuous edge, or $0.15\ L$ of a discontinuous edge. $L=$ clear span ie. face-to-face distance between columns and curtailment distances should be measured from the face of the column.
- As per clause no. D-1.5 of Annex D, IS 456-2000, over the continuous edges of a middle strip, the tension reinforcement shall extend in the upper part of the slab a distance of $0.15\ L$ from the support and at least 50% shall extend a distance of $0.3\ L$.
- As per clause no. D-1.6 of Annex D, IS 456-2000, at a discontinuous edge, negative moments may arise and depend on the degree of fixity at the edge of the slab panel but, in general, tension reinforcement equal to 50% of that provided at mid-span extending $0.1L$ into the span will be sufficient.

The above aspects are considered and accordingly, a rational detailing of slab reinforcements is shown in Figure 8.3

DESIGN OF FLAT SLABS SYSTEM

INTRODUCTION

Common practice of design and construction is to support the slabs with beams and support the beams with columns. This may be called as beam-slab construction. The beams reduce the available net clear ceiling height. Hence in warehouses, offices, and public halls sometimes beams are avoided and slabs are directly supported by columns. This type of construction is aesthetically appealing also. These slabs which are directly supported by columns are called Flat Slabs. Figure 8.4 shows a typical flat slab.

The column head is sometimes widened so as to reduce the punching shear in the slab. The widened portions are called column heads. The column heads may be provided with any angle from the consideration of architecture but for the design, concrete in the portion at 45° on either side of vertical only is considered as effective for the design [Ref. Figure 8.5]

Moments in the slabs are more near the column. Hence the slab is thickened near the columns by providing the drops as shown in Figure 8.6. Sometimes the drops are called as capital of the column. Thus we have the following types of flat slabs:

i. Slabs without drop and column head (Figure 8.4).
ii. Slabs without drop and column with column head (Figure 8.5).
iii. Slabs with drop and column without column head (Figure 8.6).
iv. Slabs with drop and column head as shown in Figure 8.7.

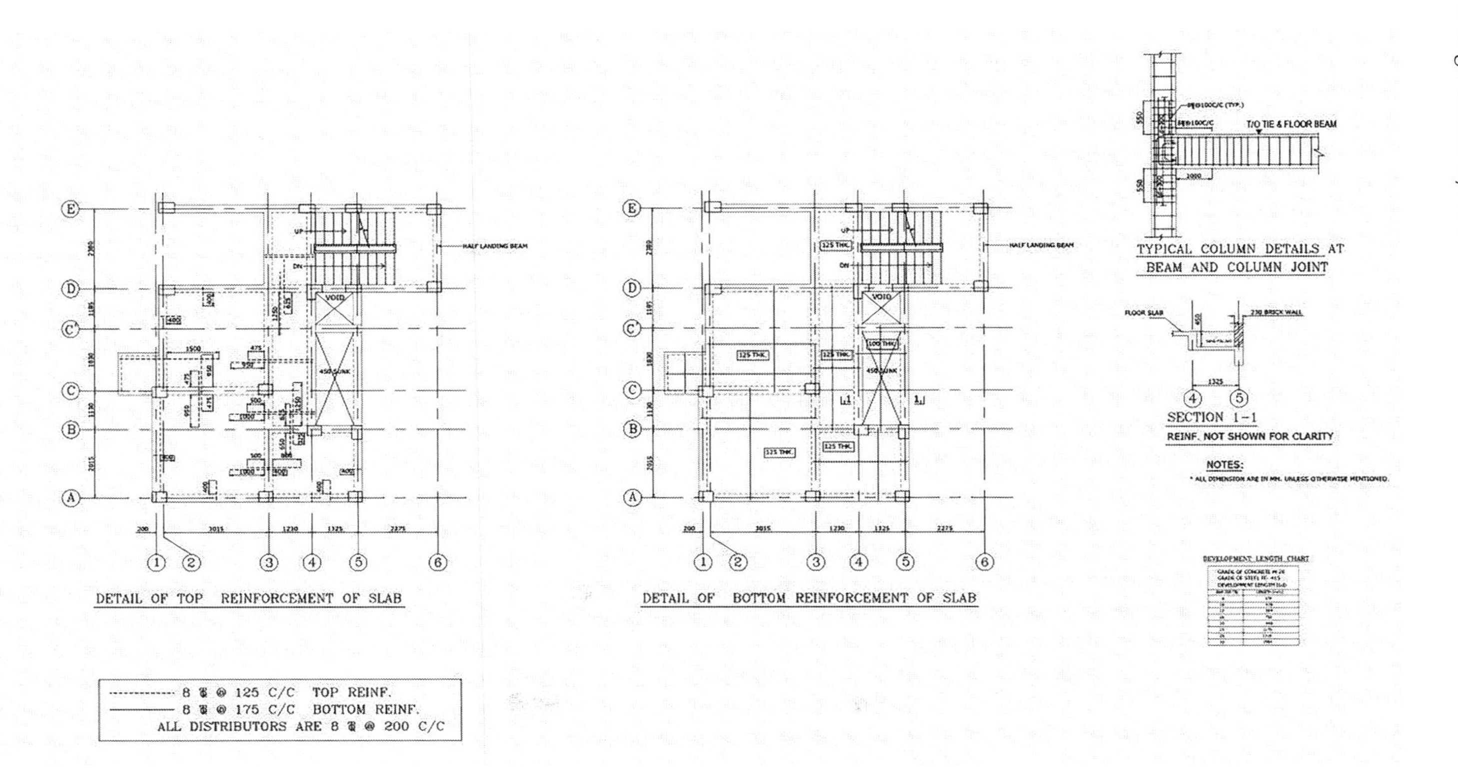

FIGURE 8.3 Reinforcement detailing.

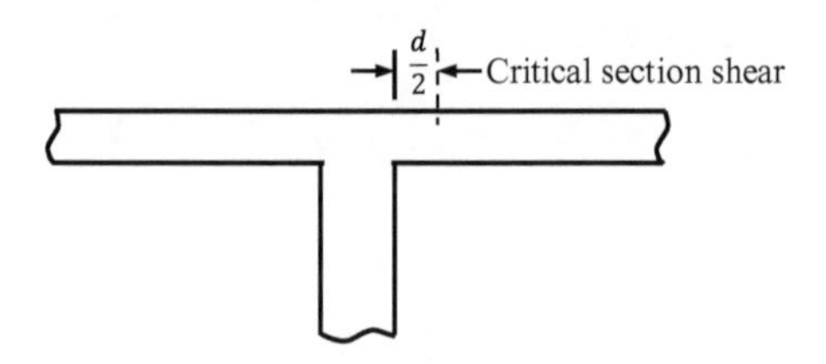

FIGURE 8.4 A typical flat slab (without drop and column head).

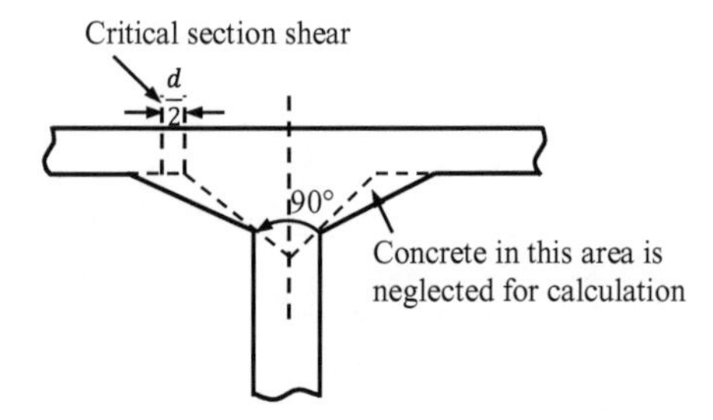

FIGURE 8.5 Slab without drop and column with column head.

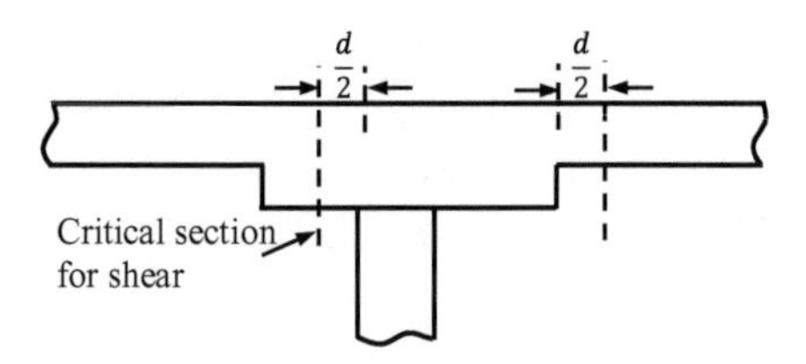

FIGURE 8.6 Slab with drop and column without column head.

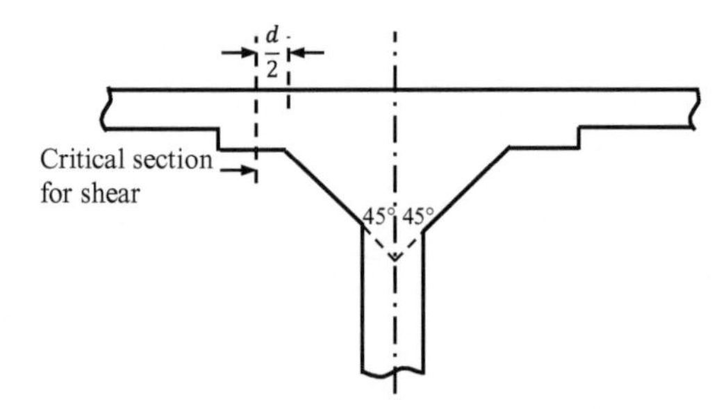

FIGURE 8.7 Slab with drop and column with column head.

The portion of flat slab that is bound on each of its four sides by center lines of adjacent columns is called a panel. The panel shown in Figure 8.8 has size $L_1 \times L_2$. A panel may be divided into column strips and middle strips. Column strip means a design strip having a width of 0.25 L_1 or 0.25 L_2, whichever is less.

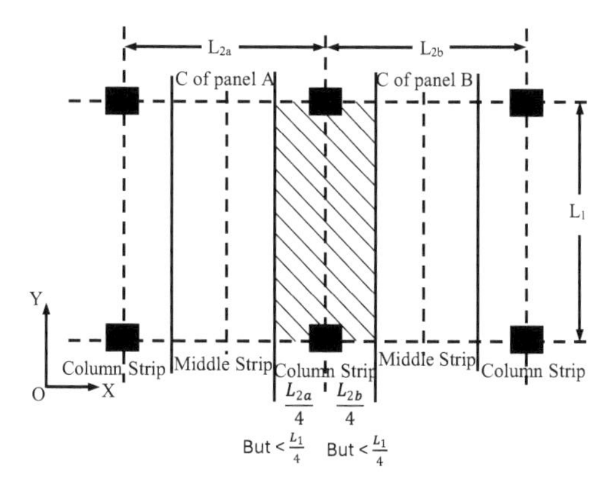

FIGURE 8.8 Panels, column strips, and middle strips are *y*-direction.

The remaining middle portion which is bound by the column strips is called middle strip. Figure 8.8 shows the division of flat slab panel into column and middle strips in the direction *y*.

Proportioning of Flat Slabs

IS 456-2000 [Clause 31.2] gives the following guidelines for proportioning:

- Drops: The drops when provided shall be rectangular in plan, and have a length in each direction not less than one-third of the panel in that direction. For exterior panels, the width of drops at right angles to the non-continuous edge and measured from the center-line of the columns shall be equal to one half of the width of drop for interior panels.
- Column Heads: Where column heads are provided, that portion of the column head which lies within the largest right circular cone or pyramid entirely within the outlines of the column and the column head, shall be considered for design purposes as shown in Figures 8.5 and 8.7.
- Thickness of Flat Slabs: From the consideration of deflection control IS 456-2000 specifies minimum thickness in terms of span to effective depth ratio. For this purpose larger span is to be considered. If drop as specified in 1.2.1 is provided, then the maximum value of ratio of larger span to thickness shall be
  = 40, if mild steel is used
  = 32, if Fe 415 or Fe 500 steel is used

  If drops are not provided or size of drops does not satisfy specification 1.2.1, then the ratio shall not exceed 0.9 times the value specified above i.e.,
  $= 40 \times 0.9 = 36$, if mild steel is used.
  $= 32 \times 0.9 = 28.8$, if HYSD bars are used

  It is also specified that in no case, the thickness of flat slab shall be less than 125 mm.

For Determination of Bending Moment and Shear Force

For this IS 456-2000 permits use of any one of the following two methods:

a. The Direct Design Method
b. The Equivalent Frame Method

The Direct Design Method

This method has the limitation that it can be used only if the following conditions are fulfilled:

i. There shall be minimum of three continuous spans in each directions.
ii. The panels shall be rectangular and the ratio of the longer span to the shorter span within a panel shall not be greater than 2.
iii. The successive span length in each direction shall not differ by more than one-third of longer span.
iv. The design live load shall not exceed three times the design dead load.
v. The end span must be shorter but not greater than the interior span.

It shall be permissible to offset columns a maximum

Total Design Moment

The absolute sum of the positive and negative moment in each direction is given by –

$$M_0 = \frac{WL_n}{8}$$

where

M_0 = Total moment
W = Design load on the area $L_2 \times L_n$
L_n = Clear span extending from face to face of columns, capitals brackets or walls but not less than $0.65L_1$
L_1 = Length of span in the direction of M_0; and
L_2 = Length of span transverse to L_1

In taking the values of L_n, L_1 and L_2, the following clauses are to be carefully noted:

a. Circular support shall be treated as square supports having the same area i.e., squares of size $0.886D$.
b. When the transverse span of the panel on either side of the center line of support varies, L_2 shall be taken as the average of the transverse spans. In Figure 8.8 it is given by $\frac{L_{2a} + L_{2b}}{2}$.

When the span adjustment and panel to an edge are being considered, the distance from the edge to the center-line of the panel shall be substituted for L_2

Distribution of Total Design Moment into Negative and Positive Moments

The total design moment M_0 in a panel is to be distributed into negative moment and positive moment as specified below:

In an Interior Span

Negative Design Moment $0.65\,M_0$
Positive Design Moment $0.35\,M_0$

In an End Span

$$\text{Interior negative design moment} = \left[0.75 - \frac{0.10}{1+\frac{1}{\alpha_c}}\right] M_0$$

$$\text{Interior positive design moment} = \left[0.63 - \frac{0.28}{1+\frac{1}{\alpha_c}}\right] M_0$$

$$\text{Exterior negative design moment} = \left[\frac{0.65}{1+\frac{1}{\alpha_c}}\right] M_0$$

Where α_c is the ratio of flexural stiffness at the exterior column to the flexural stiffness of the slab at a joint taken in the direction moments are being determined and is given by

$$\alpha_c = \frac{\Sigma K_c}{\Sigma K_s}$$

Where,

K_c = Sum of the flexural stiffness of the exterior columns meeting at the joint; and

K_s = Flexural stiffness of the slab, expressed as moment per unit rotation.

Distribution of the Bending Moment across the Panel Width

The negative moment and positive moments found are to be distributed across the column strip in a panel as shown in Table 8.4. The moment in the middle strip shall be the difference between panel and the column strip moments.

Moments in Columns

In this type of constructions column moments are to be modified as suggested in IS 456-2000 [Clause No. 31.4.5].

TABLE 8.4
Distribution Moments across the Panel Width in a Column Strip

Sl. No.	Description	Case number
a	Negative BM at the exterior support	100
b	Negative BM at the interior support	75
c	Positive BM	60

ONE-WAY SHEAR AND TWO-WAY SHEAR

Shear Force

The critical section for shear shall be at a distance *d*/2 from the periphery of the column/capital drop panel. Hence if drops are provided there are two critical sections near columns. These critical sections are shown in Figures 8.4–8.7. The shape of the critical section in plan is similar to the support immediately below the slab as shown in Figure 8.9.

For columns sections with re-entrant angles, the critical section shall be taken as indicated in Figure 8.10.

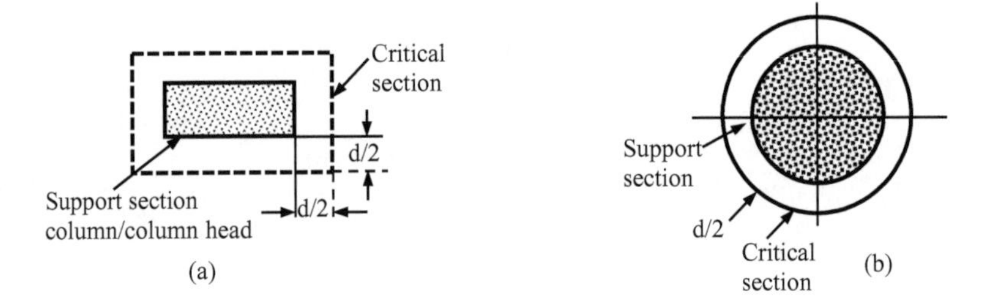

FIGURE 8.9 Critical sections for two way shear of rectangular and circular column.

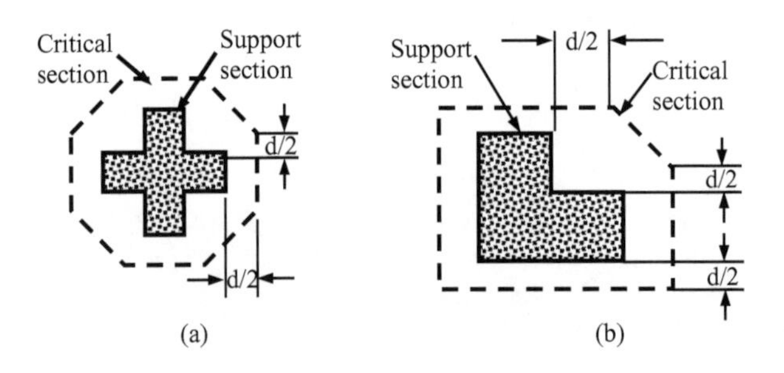

FIGURE 8.10 Critical sections for two way shear of plus-type and L-type column.

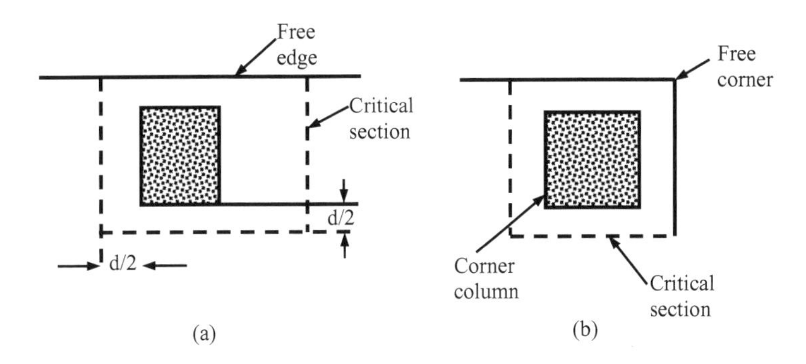

FIGURE 8.11 Critical sections for two way shear of columns near free edges of the slab.

In case of column near the free edge of a slab, the critical section shall be taken as shown in Figure 8.11

The nominal shear stress may be calculated as

$$\tau_v = \frac{V}{b_o d}$$

where

V – is shear force due to design

b_o – is the periphery of the critical section

d – is the effect depth

The permissible shear stress in concrete may be calculated as $k_s\ \tau_c$, where $k_s = 0.5 + \beta_c$ but not greater than 1, where β_c is the ratio of short side to long side of the column/capital; and

$$\tau_c = 0.25\sqrt{f_{ck}}$$

If shear stress $\tau_v < \tau_c$ – no shear reinforcement are required. If $\tau_c < \tau_v < 1.5\ \tau_c$, shear reinforcement shall be provided. If shear stress exceeds $1.5\ \tau_c$ flat slab shall be redesigned.

Equivalent Frame Method

IS 456-2000 recommends the analysis of flat slab and column structure as a rigid frame to get design moment and shear forces with the following assumptions:

a. Beam portion of frame is taken as equivalent to the moment of inertia of flat slab bounded laterally by center line of the panel on each side of the center line of the column. In frames adjacent and parallel to an edge beam portion shall be equal to flat slab bounded by the edge and the center line of the adjacent panel.

b. Moment of inertia of the members of the frame may be taken as that of the gross section of the concrete alone.
c. Variation of moment of inertia along the axis of the slab on account of provision of drops shall be taken into account. In the case of recessed or coffered slab which is made solid in the region of the columns, the stiffening effect may be ignored provided the solid part of the slab does not extend more than 0.15 l_{ef} into the span measured from the center line of the columns. The stiffening effect of flared columns heads may be ignored.
d. Analysis of frame may be carried out with substitute frame method or any other accepted method like moment distribution or matrix method.

Limitations of Equivalent Frame Method

The bending moments and shear forces may be determined by an analysis of the structure as a continuous frame and the following assumptions may be made:

- The structure shall be considered to be made up of equivalent frames on column lines taken longitudinally and transversely through the building. Each frame consists of a row of equivalent columns or supports, bounded laterally by the center-line of the panel on each side of the center-line of the columns or supports. Frames adjacent and parallel to an edge shall be bounded by the edge and the center line of the adjacent panel.
- Each such frame may be analyzed in its entirety, or, for vertical loading, each floor thereof and the roof may be analyzed separately with its columns being assumed fixed at their remote ends. Where slabs are thus analyzed separately, it may be assumed in determining the bending moment at a given support that the slab is fixed at any support two panels distant therefrom, provided the slab continues beyond the point.
- For the purpose of determining relative stiffness of members, the moment of inertia of any slab or column may be assumed to be that of the gross cross-section of the concrete alone.

Variations of moment of inertia along the axis of the slab on account of provision of drops shall be taken into account. In the case of recessed or coffered slab which is made solid in the region of the columns. The stiffening effect may be ignored provided the solid part of the slab does not extend more than 0.15 l_{ef} in to the span measured from the center-line of the columns.

The stiffening effect of flared column heads may be ignored.

One-way shear and Two-way shear have to be checked at slab-column junction as per clause 31.6 of IS456. Transfer of moments from slab to column has to be assessed by clause 31.4.5.

A design example is provided below considering and indicating above discussions as per recommendation of IS456.

Loading Pattern

When the live load does not exceed ¾th of dead load, the maximum moments may be assumed to occur at all sections when full design live load is on the entire slab. If live load exceeds ¾th dead load analysis is to be carried out for the following pattern of loading also:

i. To get maximum moment near mid-span – ¾th of live load on the panel and full live load on alternate panel
ii. To get maximum moment in the slab near the support – ¾th of live load is on the adjacent panel only

It is to be carefully noted that in no case design moment shall be taken to be less than those occurring with full design live load on all panels.

The moments determined in the beam of frame (flat slab) may be reduced in such proportion that the numerical sum of positive and average negative moments is not less than the value of total design moment

$$M_0 = WL_n / 8.$$

The distribution of slab moments into column strips and middle strips is to be made in the same manner as specified in direct design method.

Slab Reinforcement

- Spacing: The spacing of bars in a flat slab, shall not exceed 2 times the slab thickness.
- Area of Reinforcement: When the drop panels are used, the thickness of drop panel for determining area of reinforcement shall be the lesser of the following:
  - Thickness of drop, and
  - Thickness of slab plus one-quarter the distance between edge of drop and edge of capital.
  - The minimum percentage of the reinforcement is the same as that in solid slab i.e., 0.12% if HYSD bars used and 0.15% if mild steel is used.

Minimum length of Reinforcement: At least 50% of bottom bars should be from support to support. The rest may be bent up. The minimum length of different reinforcement in flat slabs should be as shown in Figure 8.9 (Figure 16 in IS 456-2000). If adjacent spans are not equal, the extension of the –ve reinforcement beyond each face shall be based on the longer span. All slab reinforcement should be anchored properties at discontinuous edges.

NUMERICAL EXAMPLES

Example 1

Design a flat slab system of 20 × 30 m, Column interval = 5 m c/c, Live load on slab = 5 kN/m^2

Calculate the positive and negative bending moments in column and middle strips of interior. Materials used are Fe415 HYSD bars and M20 concrete

PROPORTIONING OF FLAT SLAB

Assume l/d as 32 d = 5000/32, d = 156.25 mm

d = 175 mm (assumed), D = 175 + 20 + 10/2 = 200 mm

As per ACI code, the thickness of drop > 100 mm and > (Thickness of slab)/4

Therefore, 100 mm or 200/4 = 50 mm

PROVIDE A COLUMN DROP OF 100 MM

Overall depth of slab at drop = 200 + 100 = 300 mm, Length of the drop > L/3 = 5/3 = 1.67 m

Provide length of drop as 2.5 m. For the panel, 1.25 m is the contribution of drop. Column head = L/4 = 5/4 = 1.25 m.

$$L_1 = L_2 = 5 \text{ m}$$

$$L_n = L_2 - D = 5 - 1.25 = 3.75 \text{ m}$$

As per IS456, $M_o = WL_o/8$

Loading on slab: (Average thickness = (300 + 200)/2 = 250 mm)

$$\text{Self-weight of slab } = 25 \times 0.25 = 6.25 \text{ kN/m}^2$$

$$\text{Live load } = 5 \text{ kN/m}^2$$

$$\text{Floor finish } = 0.75 \text{ kN/m}^2$$

$$\text{Total} = 12 \text{ kN/m}^2$$

$$\text{Factored load } = 1.5 \times 12 = 18 \text{ kN/m}^2$$

$$W = w_u \times L_2 \times L_n = 18 \times 5 \times 3.75 = 337.5 \text{ kN}$$

$$\text{Total moment on slab panel } = (337.5 \times 3.75)/8 = 158.203 \text{ kNm}$$

DISTRIBUTION OF MOMENT

See Table 8.5

TABLE 8.5
Bending moment in interior panel

	Column Strip	Middle Strip
Negative Moment (65%)	$65 \times 0.75 = 49\%$	$65 \times 0.2 = 49\%$
	$0.49 \times 158.2 = 77.52$ kNm	$0.15 \times 158.2 = 23.73$ kNm
Positive Moment (35%)	$35 \times 0.6 = 21\%$	$35 \times 0.4 = 15\%$
	$0.21 \times 158.2 = 33.22$ kNm	$0.15 \times 158.2 = 23.73$ kNm

CHECK FOR DEPTH

Column Strip

Applying Limit state method of design approach as per IS456

$$M_u = 0.138 \cdot f_{ck} \cdot b \cdot d^2$$

$$b = 2.5 \text{ m}$$

$$77.5 \times 10^6 = 0.138 \times 20 \times 2.5 \times 1000 \times d^2$$

$d = 105.98$ mm ~ 106 mm < 275 mm Middle strip:

$$M_u = 0.138 \cdot f_{ck} \cdot b \cdot d^2, b = 2.5 \text{ m}$$

$$23.73 \times 10^6 = 0.138 \times 20 \times 2.5 \times 1000 \times d^2$$

$$d = 58.68 \text{ mm} \sim 59 \text{ mm} < 175 \text{ mm}$$

However, check against one-way and two-way shear needs to be checked.

Example 2

Design an exterior panel of a flat slab system of size 24 m × 24 m, divided into panels of 6 m × 6 m. Live load = 5 kN/m^2 and the columns diameter = 400 mm. Height of each story is 3 m. Use M20 concrete and Fe415 steel.

Let, $l/d = 32$, $d = 6000/32 = 187.5$ mm, Length of drop = Column strip = 3 m

Assume effective depth, $d = 175$ mm, $D = 200$ mm As per ACI, Assume a drop of 100 mm

Depth of slab at the drop is 300 mm Diameter of column head = $l/4$ = 6/4 = 1.5 m

LOADING ON SLAB

$$\text{Self weight of slab} = (0.2+0.3)/2\times 25 = 0.25\times 25 = 6.25\ \text{kN/m}^2$$

$$\text{Live load } = 5\,\text{kN}/\text{m}^2$$

$$\text{Floor finish } = 0.75\,\text{kN}/\text{m}^2$$

$$\text{Total } = 12\ \text{kN/m}^2$$

$$\text{Factored load} = 1.5\times 12 = 18\ \text{kN/m}^2$$

Let us apply Cl. 31.4.3.3, IS456-2000

$$\text{Stiffness of slab and column } = 4EI/L$$

α_c need to be obtained and checked with $\alpha_{c\ min}$ given in Table 17 of IS456.

Applying Cl.31.4.3.3, the interior and exterior negative moments and the positive moments are obtained.

$$\text{Interior negative design moment} = 0.75 - 0.10/(1+1/\alpha_c)\ \text{where, } \alpha_c = \Sigma K_c / K_s$$

$$\text{Interior positive design moment } = 0.63 - 0.28/(1+1/\alpha_c)$$

$$\text{Exterior negative design moment } = 0.65/(1+1/\alpha_c)$$

The distribution of interior negative moment for column strip and middle strip is in the ratio 3:1 (0.75 : 0.25)

The exterior negative moment is fully taken by the column strip. The distribution of positive moment in column strip and middle strip is in the ratio (0.6 : 0.4).

$$\alpha_c = \Sigma K_c / K_s\ \text{as per Cl.3.4.6.,}$$

$$\alpha_c = \text{Ratio of flexural stiffness of column and slab}$$

$$\Sigma K_c = \text{Summation of flexural stiffness of columns above and below}$$

$$\Sigma K_s = \text{summation of flexural stiffness of slab on both sides}$$

$$E = 5000\sqrt{f_{ck}} = 22.3606\times 10^3\ \text{N/mm}^2, I = \pi d^4/64 = \pi\times 400^4/64 = 1.25\times 10^9\ \text{mm}^4$$

$$\Sigma K_c = 2((4EI)/L) = 2\left((4EI_c)/L_c\right) = 2\times\left(4\times 22.3606\times 10^3\times 1.25\times 10^9\right)/3000$$

$$= 7.453\times 10^{10}\ \text{N/mm}$$

$$\Sigma K_s = 2\times\left(4\times 22.3606\times 10^3\times 6000\times (250)^3/(12)/6000\right) = 2.329\times 10^{11}\ \text{N/mm}$$

$$\alpha_c = 0.32$$

From Table 17 of IS456-2000

$$\alpha_{c\,\min} = L_2 / L_1 = 6/6 = 1$$

$$L_L / D_L = 5/(6.25 + 0.75) = 0.71 \text{ (approx)}$$

$$\alpha_{c\,\min} = 0.28 < \alpha_c = 0.32, \text{ So ok.}$$

Total moment on slab = $WL_n/8$ = 273.375 kNm

$$\mathrm{W = W_u \times L_2 \times L_n = 18 \times 60 \times 4.5 = 486\ kN}$$

$$L_n = 6 - 1.5 = 4.5 \text{ m}$$

As per Cl.31.4.3.3 of IS456-2000.
Exterior negative design moment is,

$$0.65/(1+1/\alpha_c) \times M_o = 43 \text{ kNm where, } \alpha_c = 0.32$$

Interior negative design moment is.

$$0.75 - 0.1/(1+1/\alpha_c) \times M_o \text{ where, } \alpha_c = \Sigma K_c / K_s$$
$$= 198.4 \text{ kNm}$$

For column strip (75%),

$$= 0.75 \times 198.4 = 148.8 \text{ kNm}$$

$$\text{For middle strip (25\%).} = 0.25 \times 198.4 = 49.6 \text{ kNm}$$

Interior positive design moment is,

$$0.63 - (0.28/1 + 1/\alpha_c) M_o$$
$$= 153.67 \text{ kNm}$$

For column strip (60%),

$$= 0.6 \times 153.67 = 92.2 \text{ kNm}$$

For middle strip (40%).

$$= 0.4 \times 153.67 = 61.47 \text{ kNm}$$

However, check against bending moment, one-way and two-way shear need to be checked.

Example 3

Design an interior panel of a flat slab of size 6.5 m × 6.5 m without providing drop and column head. Size of columns is 600 × 600 mm and live load on the panel is 4 kN/m². Take floor finishing load as 1 kN/m². Use M25 concrete and Fe 415 steel.

THICKNESS

Since drop is not provided and HYSD bars are used span to thickness ratio shall not exceed $= \frac{1}{32}$

$$\therefore \text{ Minimum thickness required } = \frac{\text{Span}}{32} = \frac{6500}{32} = 203.125 \text{ mm}$$

Let, $d = 225$ mm and $D = 250$ mm.

LOADS

Self-weight of slab	$= 0.25 \times 25$	$= 6.25 \text{ kN/m}^2$
Finished load		$= 1.0 \text{ kN/m}^2$
Live load		$= 4.0 \text{ kN/m}^2$
∴ Total working load		$= 11.25 \text{ kN/m}^2$
Factored load	$= 1.5 \times 11.25$	$= 16.875 \text{ kN/m}^2$

Now, $L_n = 6.5 - 0.6 = 5.9$ m

∴ Total design load in a panel $W = 15\ L_2\ L_n = 15 \times 6.5 \times 5.9 = 575.25$ kN

MOMENTS

Panel moment $M_0 = \frac{WL_n}{8} = 575.25 \times \frac{5.9}{8} = 424.25 \text{ kNm}$

Panel negative moment = 0.65 × 424.25 = 275.76 kNm

Panel positive moment = 0.35 × 424.25 = 148.49 kNm

Checking the thickness selected:

Since Fe 415 steel is used,

$$M_{u\,\text{lim}} = 0.138 f_{ck} bd^2$$

Width of column strip = 0.5 × 6500 = 3250

$$M_{u\,\text{lim}} = 0.138 \times 6500 \times 25 \times 3250 \times 225^2 = 567.632 \times 10^6 \text{ Nmm}$$

$$= 567.632 \text{ kNm}$$

Hence singly reinforced section can be designed i.e., thickness provided is satisfactory from the consideration of bending moment (Table 8.6).

TABLE 8.6
Distribution of Moment into Column Strips and Middle Strip

	Column Strips in kNm	Middle Strip in kNm
Negative moment	$0.75 \times 275.76 = 206.82$	$0.25 \times 275.76 = 68.94$
+ve moment	$0.60 \times 148.49 = 89.09$	$0.40 \times 148.49 = 59.40$

CHECK FOR SHEAR

The critical section for shear is at a distance $\frac{d}{2}$ from the column face. Hence periphery of critical section around a column is square of a size = $600 + d = 600 + 225 = 825$ mm

Shear to be resisted by the critical section

$$V = 16.875 \times 6.5 \times 6.5 - 16.875 \times 0.825 \times 0.825 = 701.483 \text{ kN.}$$

$$\therefore \ \tau_v = \frac{701.483 \times 1000}{4 \times 825 \times 225} = 0.945 \text{ N/mm}^2$$

$$k_s = 1 + \beta_c \text{ subjected to maximum of 1.}$$

$$\beta_c = \frac{L_1}{L_2} = \frac{6.5}{6.5} = 1$$

$$\therefore \ k_s = 1$$

$$\tau_c = 0.25\sqrt{f_{ck}} = 0.25\sqrt{25} = 1.25 \text{ N/mm}^2$$

safe in shear since $\tau_v < \tau_c$ (Figure 8.12).

REINFORCEMENT

For negative moment in column strip:

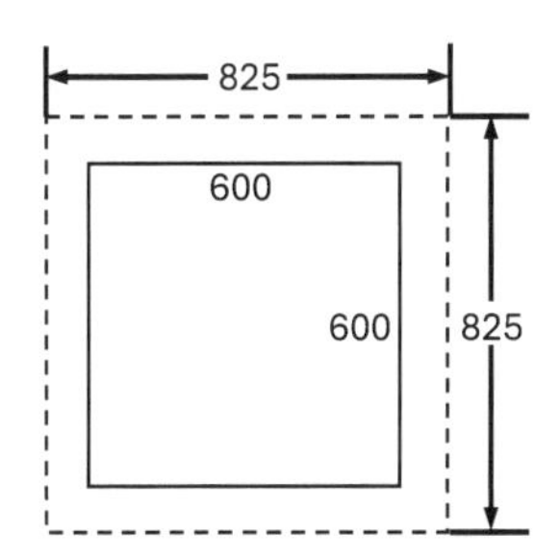

FIGURE 8.12 Critical sections for two way shear of square column.

$$M_u = 206.82 \text{ kNm}$$

$$206.82\times10^6 = 0.87\, f_y\, A_{st} d\left[1-\frac{A_{st}}{bd}\times\frac{f_y}{f_{ck}}\right]$$

$$= 0.87\times415\times A_{st}\ \times225\times\left[1-\frac{A_{st}}{3250\times225}\times\frac{415}{25}\right]$$

$$\frac{206.82\times10^6}{0.87\times415\times225} = A_{st}\times\left[1-\frac{A_{st}}{44{,}051.2}\right]$$

$$\text{i.e., } 2545.91\times44{,}051.2 = 44{,}051.2\, A_{st} - A_{st}^{\ 2}$$

$$\text{i.e., } A_{st}^{\ 2} - 44{,}051.2\, A_{st} + 112{,}150{,}390.6 = 0$$

$$\therefore\ \text{A}_{st} = 2713 \text{ mm}^2$$

This is to be provided in a column strip of width 3250 mm. Hence using 12 mm bars, spacing required is given by

$$s = \frac{\pi/4\times12^2}{2713}\times3250 = 135.48 \text{ mm} \approx 136 \text{ mm}$$

Provided 12 mm bars at 125 mm c/c.

For positive moment in column strip:

$$M_u = 89.09 \text{ kNm}$$

$$89.09\times10^6 = 0.87\times415\times A_{st}\times225\times\left[1-\frac{A_{st}}{3250\times225}\times\frac{415}{25}\right]$$

$$\frac{89.09\times10^6}{0.87\times415\times225} = A_{st}\times\left[1-\frac{A_{st}}{44{,}051.2}\right]$$

$$\text{i.e., } 1096.68\times44{,}051.2 = 44{,}051.2\,\text{A}_{st} - \text{A}_{st}^{\ 2}$$

$$\text{i.e., } \text{A}_{st}^{\ 2} - 44{,}051.2\text{A}_{st} + 48{,}310{,}070.016 = 0$$

$$\therefore\ \text{A}_{st} = 1125 \text{ mm}^2$$

This is to be provided in a column strip of width 3250 mm.

Using 10 mm bars, spacing required is

$$\text{s} = \frac{\pi/4\times10^2}{1125}\times3250 = 226.89 \text{ mm} < 2\times \text{ thickness of slab}$$

Hence provided 10 mm bars at 225 mm c/c.

Provide 10 mm diameter bars at 225 mm c/c in the middle strip to take up negative and positive moments. Since span is same in both directions, provided similar reinforcement in other direction also.

Example 4

Design the interior panel of a flat slab in Example 3, providing a suitable column head, if columns are of 600 mm diameter.

Let the diameter of column head be = $0.25L = 0.25 \times 6.5 = 1.625$ m

Its equivalent square has side 'a' where

$$\frac{\pi}{4} \times 1.625^2 = a^2$$

$$a = 1.44 \text{ m}$$

$$\therefore L_n = 6.5 - 1.44 = 5.06 \text{ m}$$

$$W_0 = 20.625 \times 6.5 \times 5.06 = 678.36 \text{ kN}$$

$$M_0 = \frac{W_0 L_n}{8} = \frac{678.36 \times 5.06}{8} = 429.06 \text{ kNm}$$

$$\therefore \text{ Total negative moment} = 0.65 \times 429.06 = 278.889 \text{ kNm}$$

$$\therefore \text{ Total positive moment} = 0.35 \times 429.06 = 150.171 \text{ kNm}$$

Width of the column strip = width of middle strip = 3250 mm

$M_{u\,\text{lim}} = 0.138 f_{ck} b d^2 = 0.138 \times 25 \times 3250 \times 325^2 = 1184.32 \times 10^6$ Nmm $= 1184.32$ kNm

Thus $M_{u\,\text{lim}} > M_u$, Hence thickness selected is sufficient (Table 8.7).

CHECK FOR SHEAR

The critical section is at a distance $\frac{d}{2} = \frac{325}{2} = 162.5$ mm from the face of column

∴ Diameter of critical section = 1625 + 325 = 1950 mm = 1.95 m

∴ Perimeter of critical section = $\pi D = 1.95\,\pi$ m

Shear on this section

$$V = 20.625 \times 6.5 \times 6.5 - 20.625 \times \frac{\pi}{4} \times 1.95^2 = 809.80 \text{ kN.}$$

TABLE 8.7
The above Moments Are to Be Distributed into Column Strip and Middle Strip

	Column Strips in kNm	Middle Strip in kNm
Negative moment	$0.75 \times 278.889 = 209.167$ kNm	$0.25 \times 278.889 = 69.72$ kNm
+ve moment	$0.60 \times 150.171 = 90.7$ kNm	$0.40 \times 150.171 = 60.47$ kNm

$$\therefore \text{ Nominal shear } = \tau_v = \frac{809.80 \times 1000}{4 \times 1950 \times 325} = 0.40 \text{ N/mm}^2$$

$$\text{Shear strength } = k_s \tau_c$$

where $k_s = 1 + \beta_c$ subjected to maximum of 1 (Figure 8.13).

where $\beta_c = \frac{L_1}{L_2} = \frac{6.5}{6.5} = 1$

$$\therefore \; k_s = 1$$

$$\tau_c = 0.25\sqrt{f_{ck}} = 0.25\sqrt{25} = 1.25 \text{ N/mm}^2$$

Design shear stress permitted = 1.25 N/mm² > τ_v, there is no need to provide shear reinforcement.

DESIGN OF REINFORCEMENT

a. For negative moment in column strip:

$$M_u = 209.167 \text{ kNm}$$

$$\text{Thickness d} = 325 \text{ mm}$$

$$\therefore \; M_u = 0.87 f_y A_{st} d \left[1 - \frac{A_{st}}{bd} \times \frac{f_y}{f_{ck}} \right]$$

$$209.167 \times 10^6 = 0.87 \times 415 \times A_{st} \times 325 \times \left[1 - \frac{A_{st}}{3250 \times 325} \times \frac{415}{25} \right]$$

$$\frac{209.167 \times 10^6}{0.87 \times 415 \times 325} = A_{st} \times \left[1 - \frac{A_{st}}{63629.52} \right]$$

$$\text{i.e., } 1782.55 \times 63{,}629.52 = 63{,}629.52 A_{st} - A_{st}^2$$

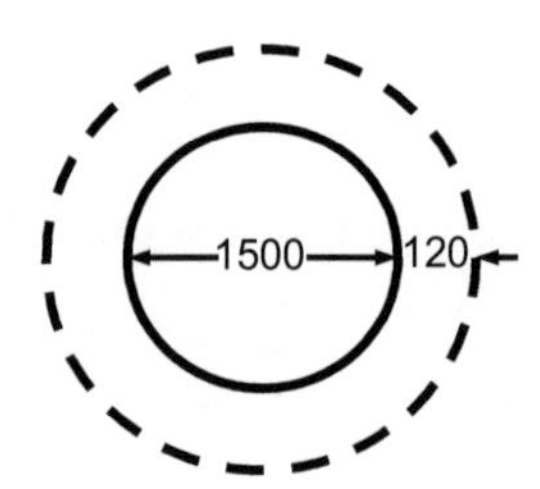

FIGURE 8.13 Critical sections for two way shear of circular column.

$$\text{i.e.,}\ A_{st}^2 - 63{,}629.52A_{st} + 1782.55 \times 63{,}629.52 = 0$$

$$\therefore\ A_{st} = 1836\ \text{mm}^2\ \text{in}\ 3250\ \text{mm width}$$

Using 12 mm bars, spacing required is

$$s = \frac{\pi/4 \times 12^2}{1836} \times 3250 = 200.199\ \text{mm} \approx 200\ \text{mm}$$

Provided 12 mm bars at 175 mm c/c.

b. For positive moment in column strip:

$$M_u = 90.7\ \text{kNm} = 90.7 \times 10^6\ \text{Nmm. Thickness}\ d = 325\ \text{mm.}$$

$$\therefore\ 90.7 \times 10^6 = 0.87 \times 415 \times A_{st} \times 325 \times \left[1 - \frac{A_{st}}{3250 \times 325} \times \frac{415}{25}\right]$$

$$\frac{90.7 \times 10^6}{0.87 \times 415 \times 325} = A_{st} \times \left[1 - \frac{A_{st}}{63{,}629.52}\right]$$

$$\text{i.e.,}\ 772.96 \times 63{,}629.52 = 63{,}629.52A_{st} - A_{st}^2$$

$$\text{i.e.,}\ A_{st}^2 - 63{,}629.52\,A_{st} + 49{,}183{,}073.78 = 0$$

$$\therefore\ A_{st} = 782\ \text{mm}^2$$

Using 10 mm bars, spacing required is

$$s = \frac{\pi/4 \times 10^2}{782} \times 3250 = 326\ \text{mm c/c}$$

Hence provided 10 mm bars at 300 mm c/c.

c. For negative moment in middle strip:

$$M_u = 69.72\ \text{kNm}$$

$$\text{Thickness}\ d = 225\ \text{mm}$$

$$M_u = 0.87 f_y A_{st} d\left[1 - \frac{A_{st}}{bd} \times \frac{f_y}{f_{ck}}\right]$$

$$69.72 \times 10^6 = 0.87 \times 415 \times A_{st} \times 225 \times \left[1 - \frac{A_{st}}{3250 \times 225} \times \frac{415}{25}\right]$$

$$\frac{69.72 \times 10^6}{0.87 \times 415 \times 225} = A_{st} \times \left[1 - \frac{A_{st}}{44,051.2}\right]$$

$$\text{i.e., } 858.24 \times 44,051.2 = 44,051.2\, A_{st} - A_{st}^2$$

$$\text{i.e., } A_{st}^2 - 44,051.2\, A_{st} + 37,806,501.89 = 0$$

$$\therefore\ A_{st} = 599.8 \text{ mm}^2 \text{ in } 3250 \text{ mm width}$$

Using 10 mm bars, spacing required is

$$s = \frac{\pi/4 \times 10^2}{599.80} \times 3250 = 425.57 \text{ mm} \approx 426 \text{ mm}$$

Provided 10 mm bars at 300 mm c/c.

d. Provided 10 mm bars at 300 mm c/c for positive moment in middle strip also.

As the span is same in both directions, provide similar reinforcement in both directions. Reinforcement detail may be shown as was done in the previous problem.

Example 5

A flat slab system consists of 5.5 m × 6.5 m panels and is without drop and column head. It has a live load of 4 kN/m² and a finishing load of 1 kN/m². It is to be designed using M25 grade concrete and Fe 415 steel. The size of the columns supporting the system is 600 × 600 mm and floor-to-floor height is 4 m. Calculate design moments in interior and exterior panels at column and middle strips in both directions (Tables 8.8–8.10).

THICKNESS

Since Fe 415 steel is used and no drops are provided, longer span to depth ratio is not more than 32.

$$\therefore\ d = \frac{\text{Span}}{32} = \frac{6500}{32} = 203.125 \text{ mm}$$

Let us select $d = 225$ mm and $D = 250$ mm.

LOADS

Self-weight of slab	$= 0.25 \times 1 \times 1 \times 25$	$= 6.25\ \text{kN/m}^2$
Finished load		$= 1.0\ \text{kN/m}^2$
Live load		$= 4.0\ \text{kN/m}^2$
Total working load		$= 11.25\ \text{kN/m}^2$
Factored load	$= 1.5 \times 11.25$	$= 16.875\ \text{kN/m}^2$

Along Length

$L_1 = 6.5\text{m}$ and $L_2 = 5.5\text{ m}$

Width of column strip = $0.25L_1$ or L_2 whichever is less.

$= 0.25 \times 5.5 = 1.375\text{ m}$ on either side of column center line

∴ Total width of column strip = $1.375 \times 2 = 2.75\text{ m}$

Width of middle strip = $5.5 - 2.75 = 2.75\text{ m}$

Along Width

$L_1 = 5.5\text{ m}$ and $L_2 = 6.5\text{ m}$

Width of column strip = $0.25 \times 5.5 = 1.375\text{ m}$ on either side

∴ Total width of column strip = $2 \times 1.375 = 2.75\text{ m}$

Hence, width of middle strip = $6.5 - 2.75 = 3.75\text{ m}$

INTERIOR PANELS

Moments along Longer Size

$L_1 = 6.5\text{m}$ and $L_2 = 5.5\text{ m}$

$L_n = 6.5 - 0.6 = 5.9\text{ m}$ subjected to minimum of $0.65 \times L_1 = 4.225\text{ m}$

$$\therefore\ L_n = 5.9\text{ m}$$

Load on panel $W_0 = 16.875 \times L_2 \times L_n = 16.875 \times 5.5 \times 5.9 = 547.59\text{ kN}$

Panel moment $M_0 = \dfrac{W_0 L_n}{8} = 547.59 \times \dfrac{5.9}{8} = 403.85\text{ kNm}$

APPROPRIATION OF MOMENT

Total negative moment = $0.65 \times 403.85 = 262.5\text{ kNm}$

∴ Total positive moment = $403.85 - 262.5 = 141.35\text{ kNm}$

TABLE 8.8

The Moment in Column Strip and Middle Strip along Longer Direction in Interior Panels

	Column Strips (in kNm)	Middle Strip (in kNm)
Negative moment	$0.75 \times 262.5 = 195.875$	$262.5 - 196.875 = 65.625$
+ve moment	$0.60 \times 141.35 = 84.81$	$141.35 - 84.81 = 56.54$

TABLE 8.9
Moments in Column Strip and Middle Strip

	Column Strips (in kNm)	Middle Strip (in kNm)
Negative moment	$0.75 \times 213.98 = 160.485$	$0.25 \times 213.98 = 53.495$
+ve moment	$0.60 \times 115.22 = 69.132$	$0.40 \times 115.22 = 46.088$

MOMENTS ALONG WIDTH

$L_1 = 5.5\,\text{m}$ and $L_2 = 6.5\,\text{m}$

$$L_n = 5.5 - 0.6 = 4.9\text{m}$$

Load on panel $W_0 = 16.875 \times L_2 \times L_n = 16.875 \times 6.5 \times 4.9 = 537.47$ kN

$$\text{Panel moment } M_0 = \frac{W_0 L_n}{8} = 537.47 \times \frac{4.9}{8} = 329.20 \text{ kNm}$$

APPROPRIATION OF MOMENT

Total negative moment = 0.65 × 329.20 = 213.98 kNm
∴ Total positive moment = 329.20 – 213.98 = 115.22 kNm

INTERIOR PANELS

Length of column = 4.0 – 0.25 = 3.75 m

The building is not restrained from lateral sway. Hence as per Table 28 in IS 456-2000, effective length of column = 1.2 × length of column = 1.2 × 3.75 = 4.5 m

Size of column = 600 mm × 600 mm

$$\text{Moment of inertia of column } = \frac{1}{12} \times 600^4$$

$$\therefore \quad k_c = \frac{I}{L} = \frac{1}{12} \times \frac{600^4}{4500} = 2.4 \times 10^6 \,\text{mm}^4$$

ALONG LONGER SPAN

Moment of Inertia of Beam

$$I_s = \text{Moment of inertia of slab}$$

$$= \frac{1}{12} \times 6500 \times 250^3 \,\text{mm}^4$$

TABLE 8.10
Appropriation of Moment in kNm

	Total	Column Strip	Middle Strip
Interior – ve	$0.674 \times 403.85 = 145.00$	$0.75 \times 145.00 = 108.75$	$145.00 - 108.75 = 36.25$
Exterior – ve	$0.492 \times 403.85 = 198.69$	$1.00 \times 198.69 = 198.69$	$198.69 - 198.69 = 0$
+ Moment	$0.418 \times 403.85 = 168.80$	$0.60 \times 168.80 = 101.28$	$168.80 - 101.28 = 67.52$

TABLE 8.11
Appropriation of Moment in kNm

	Total	Column Strip	Middle Strip
Interior – ve	$0.669 \times 329.20 = 220.235$	$0.75 \times 220.235 = 165.176$	$220.235 - 165.176 = 55.059$
Exterior – ve	$0.528 \times 329.20 = 173.818$	$1.00 \times 173.818 = 173.818$	$173.818 - 173.818 = 0$
+ve Moment	$0.400 \times 329.20 = 131.680$	$0.60 \times 131.680 = 79.008$	$131.680 - 79.008 = 52.672$

Its length $= L_2 = 5500\,\text{mm}$

$$\therefore \quad k_c = \frac{I_s}{L_2} = \frac{1}{12} \times \frac{6500 \times 250^3}{5500} = 1{,}538{,}825.758\ \text{mm}^4$$

$$\frac{\text{Live load}}{\text{Dead load}} = \frac{4}{7} < 0.75$$

$$\therefore \text{ Relative stiffness ratio is } \alpha_c = \frac{k_{C1} + k_{C2}}{k_S} = \frac{2 \times 2.4 \times 10^6}{1{,}538{,}825.758} = 3.119$$

$$\alpha = 1 + \frac{1}{\alpha_c} = 1 + \frac{1}{3.119} = 1.32$$

Hence various moment coefficients are:

$$\text{Interior} - \text{ve moment coefficient } = 0.75 - \frac{0.1}{\alpha} = 0.674$$

$$\text{Exterior } - \text{ve moment coefficient } = \frac{0.65}{\alpha} = 0.492$$

$$\text{Positive moment coefficient } = 0.63 - \frac{0.28}{\alpha} = 0.418$$

$$\text{Total moment } M_0 = 403.85\ \text{kNm}$$

ALONG SHORTER SPAN

$$\therefore \ k_s = \frac{1}{12} \times \frac{5500 \times 250^3}{6500} = 1{,}101{,}762.821\ \text{mm}^4$$

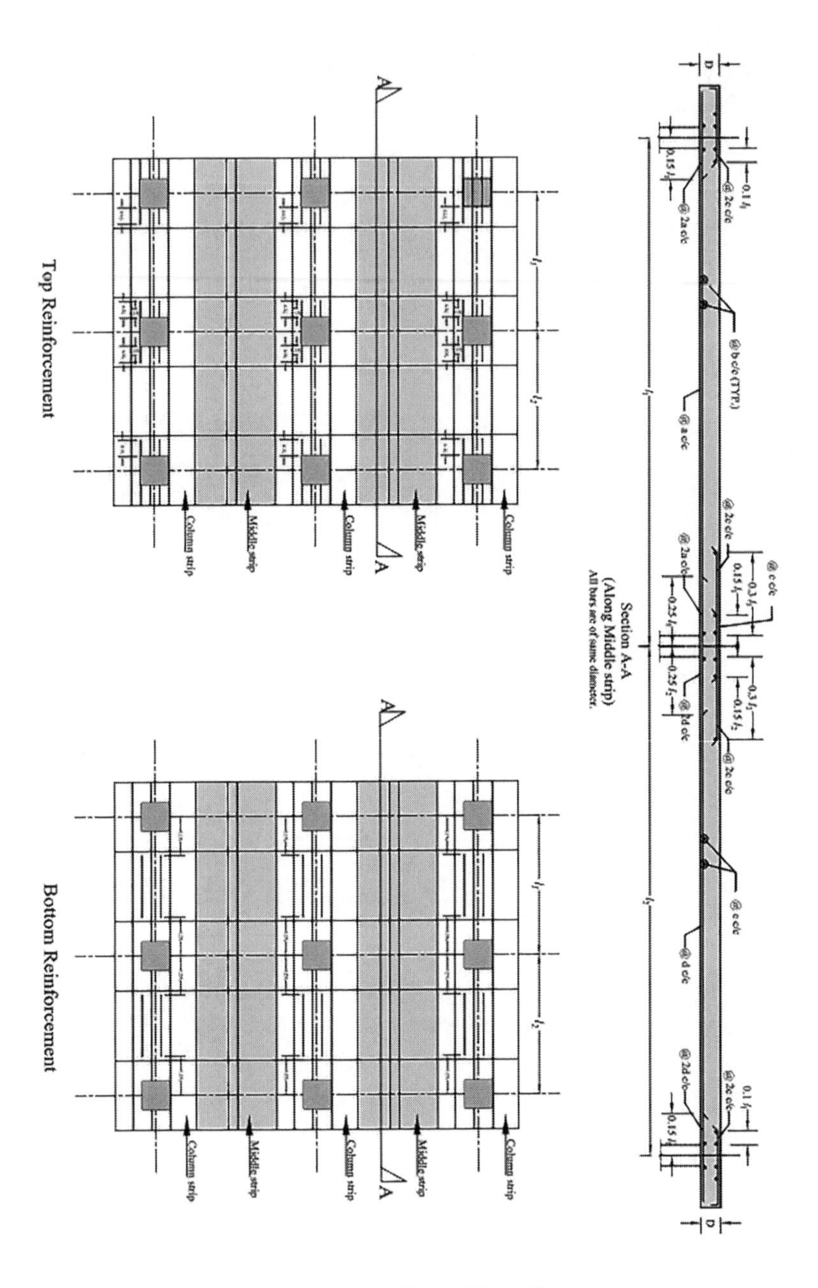

FIGURE 8.14 Typical reinforcement detailing of flat slab system.

$$\alpha_c = \frac{k_{C1} + k_{C2}}{k_S} = \frac{2 \times 2.4 \times 10^6}{1{,}101{,}762.821} = 4.357$$

$$\alpha = 1 + \frac{1}{\alpha_c} = 1 + \frac{1}{4.357} = 1.230$$

Hence various moment coefficients are (Figure 8.14 and Table 8.11):

$$\text{Interior} - \text{ve moment coefficient} = 0.75 - \frac{0.1}{\alpha} = 0.75 - \frac{0.1}{1.230} = 0.669$$

$$\text{Exterior} - \text{ve moment coefficient} = \frac{0.65}{\alpha} = \frac{0.65}{1.230} = 0.528$$

$$\text{Positive moment coefficient} = 0.63 - \frac{0.28}{\alpha} = 0.63 - \frac{0.28}{1.230} = 0.40$$

$$\text{Total moment } M_0 = 329.20 \text{ kNm}$$

9 Design of Staircases

PREAMBLE

Staircase is an important component of a building, providing access to different floors and roof of the building. Stairs are a structure designed to bridge a large vertical distance between lower and higher levels by dividing it into smaller vertical distances. It consists of a flight of steps and one or more intermediate landing slabs between the floor levels. Different types of staircases can be made by arranging stairs and landing slabs. The design of the main components of a staircase-stair, landing slabs, and supporting beams or walls are covered in this chapter. In fact, staircase is a folded plate structure. It consists of a flight of steps and one or more intermediate landing slabs between the floor levels as shown in Figures 9.1 and 9.2. Different types of staircases can be made by arranging flight and landing slabs. The angle of a staircase, often measured in degrees from the horizontal, determines the steepness of the stairs. Common angles range from 30° to 50°. However, a widely accepted standard for the most comfortable and safe stair angle is around 35°. In terms of width in a normal residential building, it should be 1200 mm and riser 150 mm. The minimum width of tread without nosing should be 250 mm, and if nosing is required, it may be adjusted. Riser may be 175 mm, if required. The standard size for a stair room in a residential home is typically between 2.5 m × 5 m, although, this depends on floor to floor height. The definitions of some technical terms, which are used in connection with design of stairs, are given. (i) Tread or Going: horizontal upper portion of a step. (ii) Riser: vertical portion of a step. (iii) Rise: vertical distance between two consecutive treads. (iv) Flight: a series of steps provided between two landings. (v) Landing: a horizontal slab provided between two flights. (vi) Waist: the least thickness of a stair slab. (vii) Winder: radiating or angular tapering steps. (viii) Soffit: the bottom surface of a stair slab. (ix) Nosing: the intersection of the tread and the riser. (x) Headroom: the vertical distance from a line connecting the nosing of all treads and the soffit above. The following are some of the general guidelines to be considered while planning a staircase: Rise = 150–175 mm, Tread = 250–275 mm, for residential buildings. Rise may be 125–150 mm, Tread = 250–300 mm for public buildings. Number of riser = (Floor/Floor Height)/Rise. Number of risers in one Flight = 0.5*(Number of risers). Number of tread = (Number of risers − 1). Going distance = (No. of tread) × (tread width). Width of landing ≥ width of stair. For structural design, Effective span = c/c distance between supports. If not given, the width of support can be taken in between 250 and 300 mm. Trial depth of waist slab may be according to is 456:2000 article 23.2.1 by calculating ratio of span to effective depth and after that ratio is multiplying by the modification factor. Modification factor can be calculated by assuming percentage of tension reinforcement.

DOI: 10.1201/9781003208204-9

Effective span of stairs should be as per stipulations of clause 33 of IS 456. Three different cases are given to determine the effective span of stairs without stringer beams. (i) The horizontal center-to-center distance of beams should be considered as the effective span when the slab is supported at top and bottom risers by beams spanning parallel with the risers. (ii) The horizontal distance is equal to the going of the stairs, plus at each end either half the width of the landing or one meter, whichever is smaller when the stair slab is spanning on to the edge of a landing slab that spans parallel with the risers. It is worth mentioning that certain idealizations are made in the actual structures for the applicability of the simplified analysis. The designs based on the simplified analysis have been found to satisfy the practical needs.

TYPES OF STAIRCASES

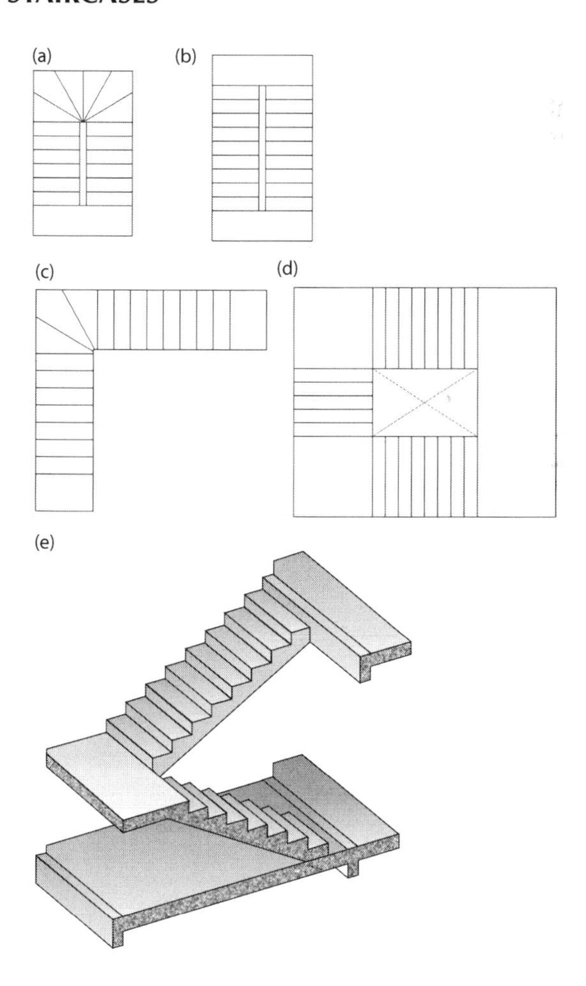

FIGURE 9.1 Plan and 3D view of staircase.

STRUCTURAL CLASSIFICATION

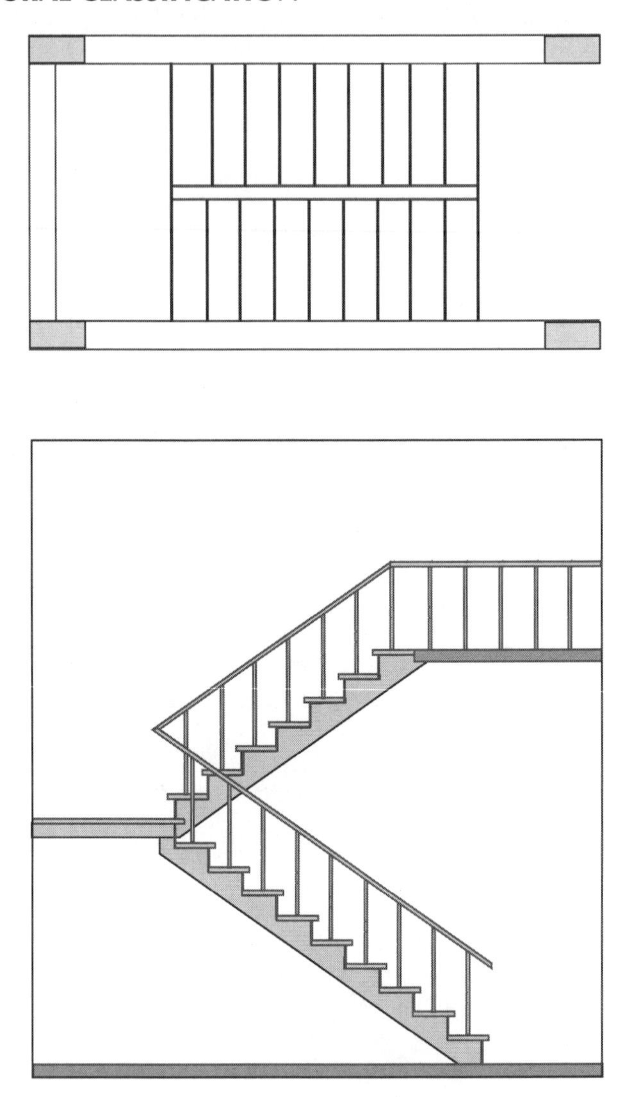

FIGURE 9.2 Detail of a typical dog-legged staircase.

DESIGN EXAMPLES ON DIFFERENT TYPES OF STAIRCASES

Example 1

Design a dog-legged stair for a building in which the vertical distance between floors is 3.0 m. The stair hall measures 2.5 m × 4.75 m. The live load may be taken as 3000 N/m². Use M 20 concrete, and grade of steel Fe415 HYSD bars. Assume the unit weight of concrete = 25 kN/m³.

Solution:

GENERAL ARRANGEMENT

Figure 9.3 shows the plan of stair hall. Let the rise be 150 mm and tread be 250 mm. Let us keep the width of each flight = 1.2 m.

$$\text{Height of each flight} = \frac{3.0}{2} = 1.5 \text{ m}$$

$$\text{No. of risers required} = \frac{1.5}{0.15} = 10 \text{ in each flight}$$

$$\text{No. of tread in each flight} = 10 - 1 = 9.$$

$$\text{Space occupied by treads} = 9 \times 250 = 2250 \text{ mm}$$

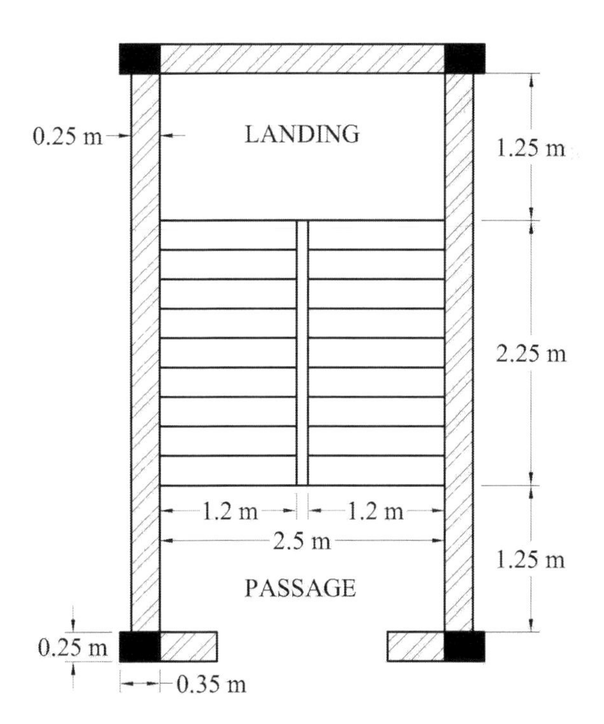

FIGURE 9.3 Plan of dog-legged stair.

Keeping width of landing equal to 1.25 m.

$$\text{Hence space left for passage} = (4.75 - 1.25 - 2.25) = 1 \text{ m}.$$

Applying "Limit state method" as per IS 456

For M 20 concrete, and Fe 415 HYSD bars having $\sigma_{st} = 230$ N/mm^2.

Loading on Each Flight

$$\text{Then effective span } = 2.25 + (1.25 \times 2) + 0.25 = 5 \text{ m}.$$

Let the thickness of the waist slab be equal to 175 mm

$$\text{Weight of slab } w' \text{ on slope } = \frac{175}{1000} \times 1 \times 1 \times 25{,}000 = 4375 \text{ N/m}^2.$$

$$\text{Dead weight on plan } w_1 = w' \times \frac{\sqrt{R^2 + T^2}}{T}$$

$$= 4375 \times \frac{\sqrt{150^2 + 250^2}}{250} = 5102 \text{ N/m}^2.$$

Dead weight of steps

$$w_2 = \frac{R}{2 \times 1000} \times 1 \times 1 \times 25{,}000 = \frac{150}{2000} \times 25{,}000 = 1875 \text{ N / m}.$$

Total dead weight per meter run = 5102 + 1875 = 6977 N/m

Weight of finishing, etc. = 100 N/m

Live load = 3000 N/m

Total w = 10,077 N/m.

The 'w' on the landing portion will be 10,077 – 1875 = 8202 N/m, since weight of steps will not come on it. However, a uniform value of 'w' has been adopted here.

Design of Waist Slab

$$M = \frac{wL^2}{10} = \frac{10{,}077 \times 5^2}{10} = 25{,}192.5 \text{ N-m.} = 25.1925 \times 10^6 \text{ N-mm}$$

$$M_u = 1.5 \times 25.1925 \times 10^6 \text{ N-mm} = 37.7888 \times 10^6 \text{ N-mm}.$$

From SP16, Table – D, for M20 concrete and Fe415 steel $\frac{M_{u\,\lim}}{bd^2} = 2.76 \text{ n/mm}^2$

$$\therefore\ d = \sqrt{\frac{M_u}{b \times 2.76}} = \sqrt{\frac{37.7888 \times 10^6}{2.76 \times 1000}} = 117 \text{ mm}$$

Let us provide 150 mm overall depth. Clear cover = 20 mm and using 12 mm ϕ Fe 415 HYSD bars,

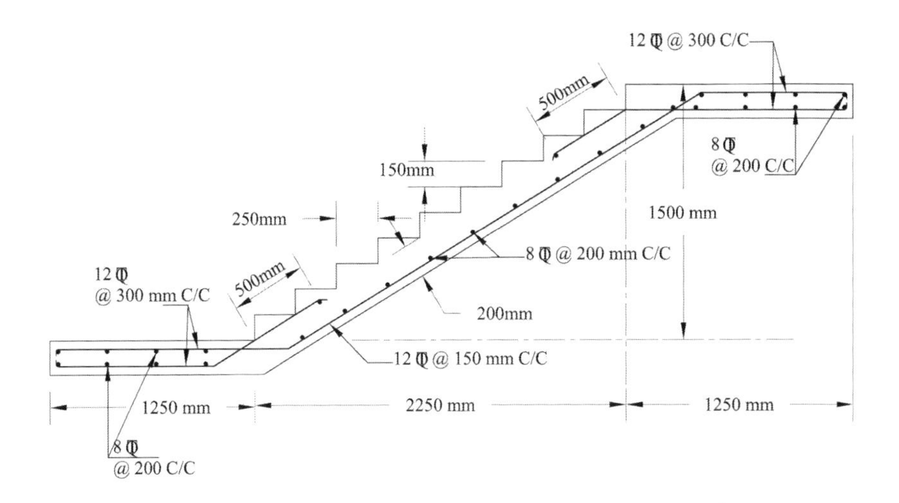

FIGURE 9.4 Reinforcement details of dog-legged stair.

Effective depth = 150 – 20 – 6 = 124 mm. > 117 mm

Reinforcement

$$\frac{M_u}{bd^2} = \frac{37.7888 \times 10^6}{1000 \times 124^2} = 2.46$$

From SP16, Table – 2, Page – 49, for M20 concrete and Fe415 steel, $P_t = 0.823$

$$A_{st} = \frac{P_t \times b \times d}{100} = \frac{0.823 \times 1000 \times 124}{100} = 1020.52 \text{ mm}^2$$

Using 12 mm ϕ bars $A_\phi = \frac{\pi}{4} \times (12)^2 = 113 \text{ mm}^2$

$$\text{Spacing of bars} = \frac{1000 \times 113}{1020.52} = 110.73 \text{ mm.}$$

Providing 12 mm ϕ bars @ 100 mm c/c.

Distribution reinforcement @ 0.12% of area.

$$A_{sd} = \frac{0.12 \times 150 \times 1000}{100} = 180 \text{ mm}^2$$

Using 8 mm ϕ bars, $A_\phi = \frac{\pi}{4} \times (8)^2 = 50.3 \text{ mm}^2$

$$s = \frac{1000 \times 50.3}{180} \approx 280 \text{ mm.}$$

Providing 8 mm ϕ bars @ 200 mm c/c (Figure 9.4).

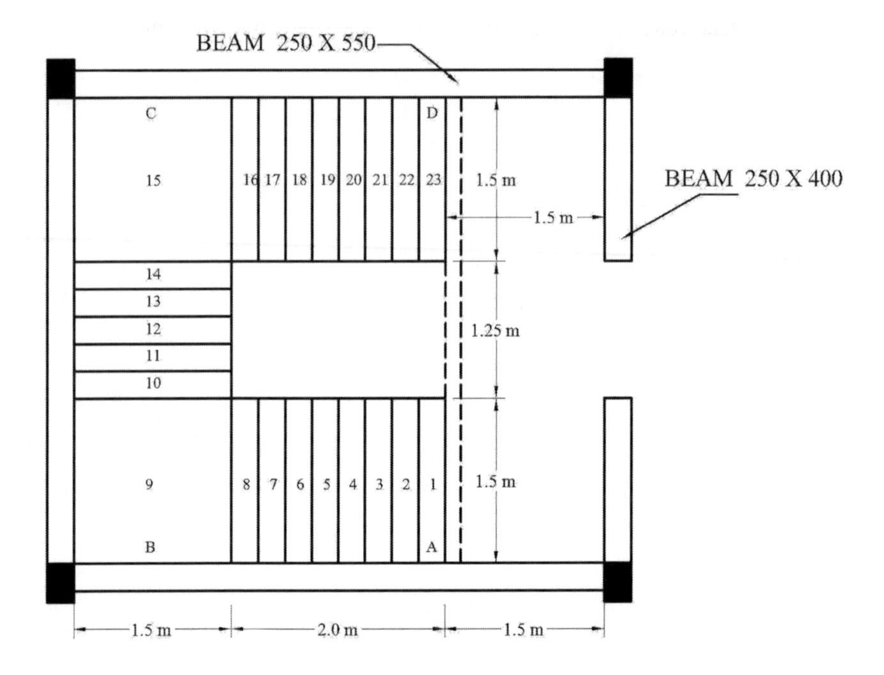

FIGURE 9.5 Plan of open newel stair.

Example 2

General arrangement of an open newel type staircase of a building, as shown in Figure 9.5. The risers are 150 mm and the treads are 250 mm. Design the staircase for a live load of 5000 N/m². The width of stair is 1.5 m and the width of wall is 400 mm. Adopt unit weight of concrete as 2500 N/m³. Assume size of the beam = 250 mm × 400 mm and 250 mm × 550 mm as shown in Figure 9.5

Using M20 concrete, Fe 415 steel.

Solution:

Applying "Limit state method of Design" as per IS456

M20, concrete, $f_{ck} = 20$ N/mm².

Fe 415, steel, $f_y = 415$ N/mm².

EFFECTIVE SPAN

For flight AB and CD, $L = 0.125 + 2 + 1.5 + 0.125 = 3.75$ m

For flight BC, $L = 0.125 + 1.5 + 1.25 + 1.5 + 0.125 = 4.5$ m.

FLIGHT PORTION

Let us consider waist slab and landing slab be 200 mm thick.

Considering a strip of slab of 1 m wide.

$$\text{Weight of slab } w' \text{ on slope } = \frac{200 \times 1 \times 1}{1000} \times 25{,}000 = 5000\ \text{N/m}^2$$

$$\text{Weight on horizontal area } = 5000 \times \frac{\sqrt{150^2 + 250^2}}{250} = 5830 \text{ N / m}^2$$

Dead weight of steps is given by

$$w_2 = \frac{R}{2 \times 1000} \times 1 \times 1 \times 25{,}000 = \frac{150}{2000} \times 25{,}000 = 1875 \text{ N}$$

Let the weight of finishing = 100 N/m²
Live load = 5000 N/m².
Hence total 1oad = 5830 + 1875 + 100 + 5000 = 12,805 N/m².

LANDING PORTION

$$\text{Dead load } = \frac{200}{1000} \times 1 \times 1 \times 25{,}000 = 5000 \text{ N/m}^2.$$

$$\text{Weight of finishing } = 100 \text{ N/m}^2 \text{ (assumed)}$$

$$\text{Live load } = 5000 \text{ N / m}^2.$$

$$\text{Total } = (5000 + 100 + 5000) = 10{,}100 \text{ N/m}^2.$$

However, since each quarter space landing is common to both the flights, only half of the above loading, i.e., 5050 N/m will be taken.

DESIGN OF FLIGHT AB

$$R_A = \frac{1}{3.75}[(5050 \times 1.625 \times 0.8125) + (12{,}805 \times 2.125 \times 2.6875)] = 21{,}279 \text{ N}$$

$$R_B = \frac{1}{3.75}\left[\frac{12{,}805 \times 2.125^2}{2} + (5050 \times 1.625 \times 2.9375)\right] = 14{,}138 \text{ N}$$

S.F. is zero at $\frac{21{,}279}{12{,}805} = 1.66$ m from A (Figure 9.6).

$$M_{\max} = \left[(21{,}279 \times 1.66) - \frac{12{,}805 \times 1.66^2}{2}\right] = 17.68 \times 10^6 \text{ N-mm.}$$

$$M_u = 1.5 \times 17.68 \times 10^6 \text{ N-mm.}$$

From SP16, Table – D, for M20 concrete and Fe415 steel $\frac{M_{u\lim}}{bd^2} = 2.76$

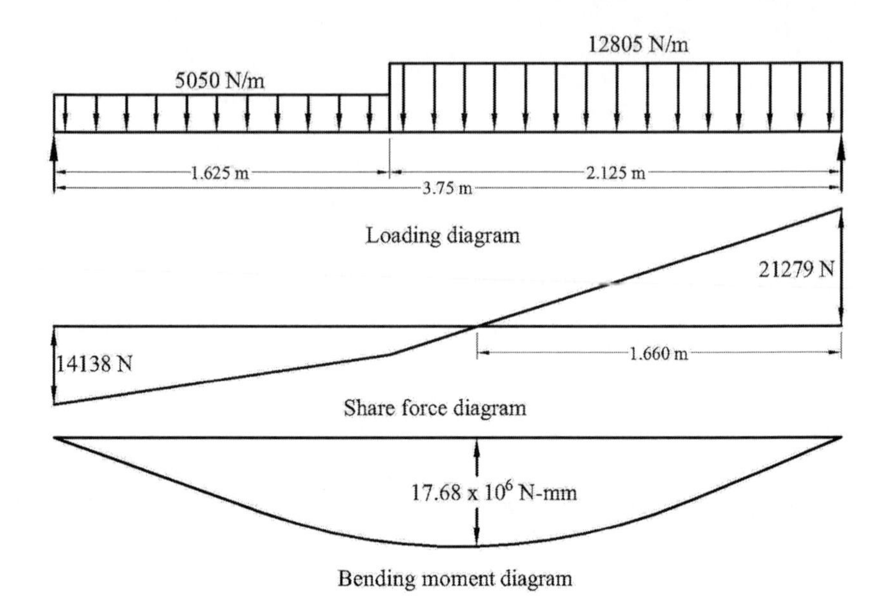

FIGURE 9.6 S.F. and B.M. diagram.

$$\therefore\ d = \sqrt{\frac{M_u}{b \times 2.76}} = \sqrt{\frac{1.5 \times 17.68 \times 10^6}{2.76 \times 1000}} = 98.02\ \text{mm}$$

Provide overall depth of 150 mm so that available $d = 124$ mm. > 98.02 mm.

$$\frac{M_u}{bd^2} = \frac{1.5 \times 17.68 \times 10^6}{1000 \times 124^2} = 1.72$$

From SP16, Table – 2, Page – 49, for M20 concrete and Fe415 steel, $P_t = 0.54$

$$A_{st} = \frac{P_t \times b \times d}{100} = \frac{0.54 \times 1000 \times 124}{100} = 669.6\ \text{mm}^2$$

Using 12 mm ϕ bars $A_\phi = \frac{\pi}{4} \times (12)^2 = 113\ \text{mm}^2$

$$\text{Spacing of bars} = \frac{1000 \times 113.1}{669.6} = 168.75\ \text{mm.}$$

Hence use 12 mm ϕ bars @ 150 mm C/C.

$$A_{st\ provided} = \frac{1000 \times 113}{150} = 753.33\ \text{mm}^2.$$

$$P_{t\ provided} = \frac{100 \times 753.33}{1000 \times 124} = 0.6\ \text{mm}^2.$$

Distribution reinforcement $A_{sd} = 1.5D = 1.5 \times 150 = 225\,mm^2$
Provide 8 mm @ 175 mm C/C.

$$\text{Nominal shear stress } \tau_v = \frac{v_u}{bd} = \frac{1.5 \times 21{,}279}{1000 \times 124} = 0.26\ \text{N/mm}^2 < \tau_c \text{ as per IS 456}$$

Provided $P_t = 0.6$, so $\tau_c = 0.49$ as per table – 19 of IS 456
So, there is no requirement of shear reinforcement.

DESIGN OF FLIGHT BC

$$R_B = R_C = \frac{1}{2} \times [5050 \times 1.625 + 12{,}805 \times 1.25 + 5050 \times 1.625] = 16{,}209.375\ \text{N}$$

$$M_{max} = \left[(16{,}209.375 \times 2.25) - (5050 \times 1.625 \times 1.4375) - \frac{12{,}805 \times 0.625^2}{2} \right]$$

$$\times 1000 = 22.174 \times 10^6\ \text{N-mm}$$

$$M_u = 1.5 \times 22.174 \times 10^6\ \text{N-mm}.$$

From SP16, Table – D, for M20 concrete and Fe415 steel $\frac{M_{u\,lim}}{bd^2} = 2.76\ \text{n/mm}^2$

$$\therefore\ d = \sqrt{\frac{M_u}{b \times 2.76}} = \sqrt{\frac{1.5 \times 22.174 \times 10^6}{2.76 \times 1000}} = 109\ \text{mm}$$

Let overall depth = 150 mm so $d = 124$ mm. > 109 mm (Figure 9.7).

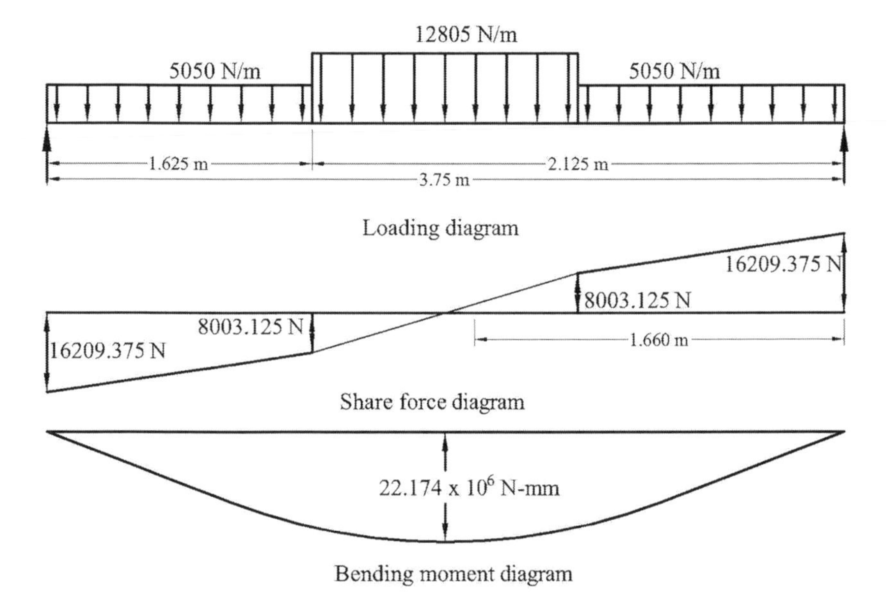

FIGURE 9.7 S.F. and B.M. diagram.

$$\frac{M_u}{bd^2} = \frac{1.5 \times 22.174 \times 10^6}{1000 \times 124^2} = 2.16$$

From SP16, Table – 2, Page – 49, for M20 concrete and Fe415 steel,

$$P_t = 0.701\%,\ A_{st} = \frac{P_t \times b \times d}{100} = \frac{0.701 \times 1000 \times 124}{100} = 869.24\ \text{mm}^2$$

Using 12 mm ϕ bars having $A_\phi = \frac{\pi}{4} \times (12)^2 = 113\ \text{mm}^2$

$$\text{Spacing of bars } = \frac{1000 \times 113}{869.24} = 130\ \text{mm.}$$

Hence provide 12 mm ϕ bars @ 125 mm c/c (Figure 9.8)

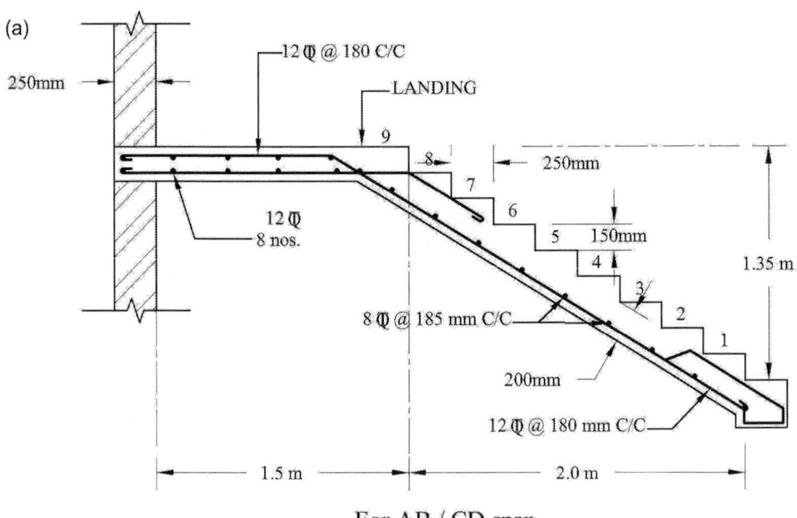

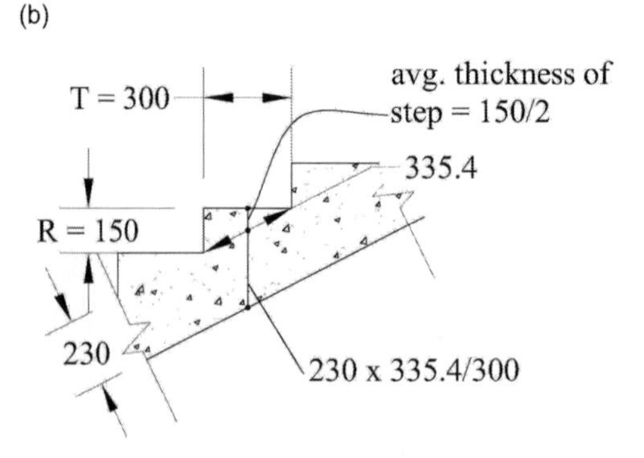

FIGURE 9.8 Reinforcement details of open newel stair.

Example 3

General layout of a staircase with central stringer beam supported on columns at B, C, and D. The rise and tread of the stairs are 150 and 250 mm respectively. The width of the steps is 1.40 m. Design the staircase for a live load of 5000 N/m². Adopt the unit weight of concrete as 25,000 N/m³.

Applying "Limit state method" and M20 concrete and Fe415 steel.

(Design of staircase with central stringer beam)

Solution:

M20, concrete, $f_{ck} = 20$ N/mm².

Fe 415, steel, $f_y = 415$ N/mm².

LOADING ON WAIST SLAB

Let the waist slab be 125 mm thick. The weight of waist slab on the slope should be multiplied by the factor $\dfrac{\sqrt{R^2+T^2}}{T}$

$$= \frac{\sqrt{150^2+250^2}}{250} = 1.166 \text{ to get the equivalent weight horizontal plane.}$$

Consider 1 m width of slab

Load per meter horizontal run will be as follows:

a. Self weight $= \left(\dfrac{125}{1000} \times 1 \times 1 \times 25{,}000\right) \times 1.166 \qquad = 3643.75 \text{ N}$

b. Wt. of steps $= \left(\dfrac{150}{2 \times 1000} \times 1 \times 1 \times 25{,}000\right) \qquad = 1875 \text{ N}$

c. Finishing etc. (given) $\qquad = 100$ N (say)

d. Live load (given) $\qquad = 5000$ N

Total = 10,675.75 N

The loading on the landing will be lesser; however, for simplicity, we will take the same loading throughout (Figure 9.9).

DESIGN OF WAIST SLAB

The waist slab is supported on the central stringer beam. Hence the worst condition may be when we consider the concentrated live load of 1300 N to act to one side only.

$$\text{Dead weight} = 3643.75 + 1875 + 100 = 5618.75 \text{ N}$$

Let the width of stringer beam be 250 mm.

$$\text{Projection of slab beyond the rib of beam} = \frac{1.4 - 0.25}{2} = 0.575 \text{ m}$$

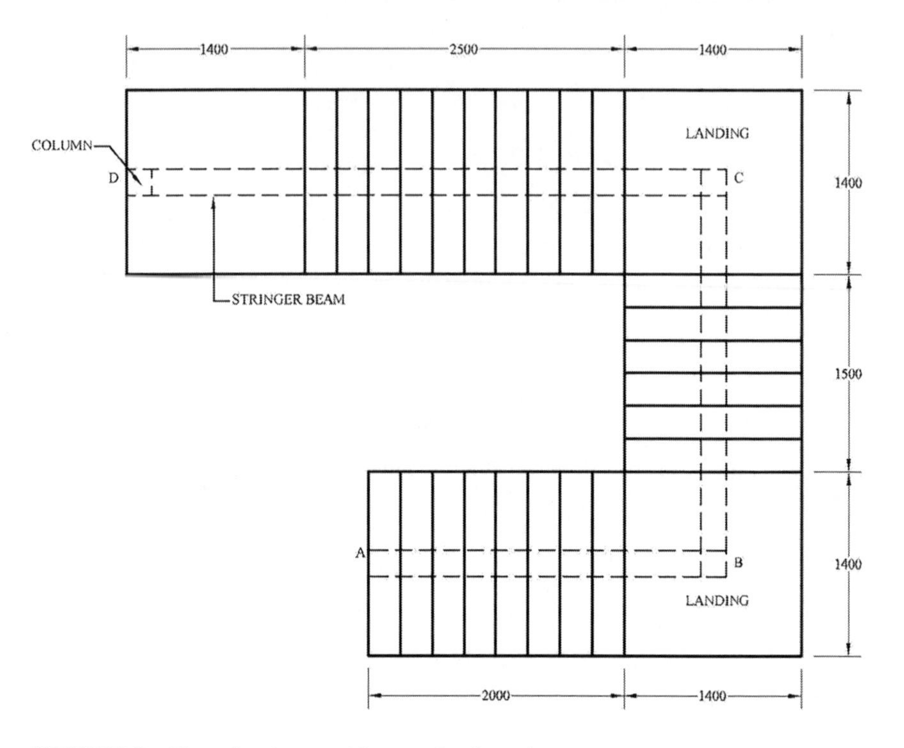

FIGURE 9.9 Plan of staircase with central stringer beam.

$$\text{Bending Moment due to dead load } = \frac{5618.75 \times 0.575^2}{2} = 929 \text{ N-m}$$

$$\text{Bending Moment due to U.D. live load } = \frac{5000 \times 0.575^2}{2} = 827 \text{ N-m}$$

Bending moment due to concentrated live load = $1300 \times 0.575 = 748$ N-m

$$\text{Max. B.M. } M = 929 + 748 = 1.677 \times 10^6 \text{ N-mm.}$$

$$M_u = 1.5 \times 1.677 \times 10^6 \text{ N-mm.}$$

From SP16, Table – D, for M20 concrete and Fe415 steel $\dfrac{M_{u\lim}}{bd^2} = 2.76$

$$d = \sqrt{\frac{M_u}{b \times 2.76}} = \sqrt{\frac{1.5 \times 1.677 \times 10^6}{2.76 \times 1000}} = 30.18 \text{ mm}$$

Keep overall depth of 125 mm so that available $d = 105$ mm. > 30.18 mm.

$$\frac{M_u}{bd^2} = \frac{1.5 \times 1.677 \times 10^6}{1000 \times 105^2} = 0.23$$

From SP16, Table – 2, Page – 49, for M20 concrete and Fe415 steel,

$$P_t = 0.085$$

As per IS 456 minimum percentage of steel reqd. = 0.12% of gross cross section

$$A_{st} = 0.12 \times 1000 \times 125 = 150 \text{ mm}^2$$

Using 10 mm ϕ bars $A_\phi = \frac{\pi}{4} \times (10)^2 = 78.5 \text{ mm}^2$

$$s = \frac{1000 \times 78.5}{150} \approx 523 \text{ mm.}$$

Let us keep keep the spacing = 250 mm

Distribution reinforcement = 1.5 × 125 = 187.5 mm².

∴ Provide 8 mm ϕ bars @ 200 mm c/c.

DESIGN OF STRINGER BEAM

The stringer beam will act as a T-beam. Flight CD is the longest, and hence we will design the stringer beam CD.

$$\text{Effective span } = \left(1.4 - \frac{0.2}{2}\right) + 2.5 + \frac{1.4}{2} = 4.5 \text{ m.}$$

The loading on the stringer beam will as follows assuming the web to be 250 mm width and 200 mm deep.

a. Weight of rib / m run $= \left(\frac{250 \times 200}{10^6} \times 25{,}000\right) \times 1.166$ = 1457.5 N.

b. Load from waist slab = 10,618.75 × 1 × 1.4 = 14,866.25 N.

Total = 16,323.75 N/m

Assuming partial fixity at the ends.

$$M = \frac{w \times L^2}{10} = \frac{16{,}323.75 \times 4.5^2}{10} \times 1000 = 33.05 \times 10^6 \text{ N-mm.}$$

$$M_u = 1.5 \times 33.05 \times 10^6 \text{ N-mm.}$$

Taking lever arm = 0.9*d*,

As per IS 456 Cl. G-2.2

$$M_u = 0.36 \cdot \frac{x_{u\,\max}}{d}\left(1 - 0.42\frac{x_{u\,\max}}{d}\right) f_{ck} \cdot b_w \cdot d^2 + 0.45 \cdot f_{ck} \cdot (b_f - b_w) D_f \cdot \left(\frac{d - D_f}{2}\right)$$

where b_f = flange width of isolated T-Beam given by

$$b_f = \frac{l_0}{\left(\frac{l_0}{b}\right) + 4} + b_w$$

where $l_0 \approx L = 4.5$ m; b = actual width of flange = 1.4 m; $b_w = 0.25$ m.

$$b_f = \frac{4.5}{\left(\frac{4.5}{1.5}\right)+4} + 0.25 = 0.893 \text{ m} = 893 \text{ mm.}$$

$$\frac{x_{u\max}}{d} = 0.48 \text{ for Fe 415 steel [As per IS 456]}$$

$$1.5 \times 33.05 \times 106 = 0.36 \times 0.48 \times (1 - 0.42 \times 0.48) \times 20 \times 250 \times d^2$$
$$+ 0.45 \times 20(893 - 250) \times 125 \times (d - 125/2)$$

$$49{,}620{,}210.94 = 689.8d^2 + 723.375d$$

Hence from

$$\text{or, } d = \frac{-723.37 \pm \sqrt{723.37^6 + 4 \times 689.8 \times 49{,}620{,}210.94}}{2 \times 689.8}$$

From which $d = 268$ mm

From shear consideration as per Cl. B-5.1 of IS 456

$$d = \frac{V_e}{b_w \tau_{c\max}}$$

where $\tau_{c\max} = 1.8$ N/mm^2 for M 20 concrete (Table 24 of IS 456)

As per Cl.B-6 3.1 of IS 456

$$V_e = V + 1.6\frac{T}{b_w}$$

$$V = \frac{wL}{2} = \frac{16{,}324 \times 4.5}{2} = 36{,}728 \text{ N}$$

T = Torsional moment, which will be induced due to live load acting only on one side of step.

$$T = \left[\frac{5000 \times 0.575^2}{2}\right] \times \frac{4.5}{2} \times 1000 \text{ N-mm} = 1.86 \times 10^6 \text{ N-mm}$$

or

$$T = (1300 \times 0.575) \times \frac{4.5}{2} \times 1000 \text{ N-mm} = 1.68 \times 10^6 \text{ N-mm}$$

whichever is more.

$$T = 1.68 \times 10^6 \text{ N-mm.}$$

$$V_e = 36,728 + 1.6 \times \frac{1.86 \times 10^6}{250} = 48,632 \text{ N}$$

$$\text{Hence } d = \frac{48,632}{250 \times 1.8} = 108 \text{ mm}$$

Let us provide total depth $(D) = 300$ mm.

Using 20 mm ϕ bars and 25 mm clear cover, available $d = 300 - 10 - 25 = 265$ mm.

Assuming, Lever arm = 0.9 d

$$A_{st} = \frac{1.5 \times 33.05 \times 10^6}{0.87 \times 230 \times 0.9 \times 265} = 1039 \text{ mm}^2$$

For 20 mm ϕ bars $A_\phi = \frac{\pi}{4} \times (20)^2 = 314 \text{ mm}^2$

$$\text{No. of bars } = \frac{1039}{314} = 3.3$$

Let us provide 4 no. of 20 mm ϕ. ie. A_{st} provided = 314 × 4 = 1256 mm².

The above reinforcement is for bending requirements only. Additional longitudinal reinforcement need to be calculated for torsion moment.

DESIGN FOR TORSION

$T = 1.86 \times 106$ N-mm, $V = 36,728$ N and $V_e = 48,632$ N.

$$\therefore \tau_{ve} = \frac{V_e}{b_w d} = \frac{48,632}{250 \times 265} = 0.734 \text{ N/mm}^2.$$

$$\frac{100 A_{st}}{bd} = \frac{100 \times 1256}{250 \times 265} = 1.9\%$$

Hence from Table 23 of IS 456, $\tau_c = 0.482$ N/mm².

Since $\tau_{ve} > \tau_c$, shear reinforcement is necessary.

LONGITUDINAL REINFORCEMENT

$$M_{e1} = M + M_T$$

where $M = 33.05 \times 10^6$ N-mm

$$M_T = T \frac{(1 + D / b_w)}{1.7}$$

$$= 1.86 \times 10^6 \times \frac{(1 + 300 / 250)}{1.7} = 2.4 \times 10^6 \text{ N-mm}$$

$$M_{e1} = 33.05 \times 10^6 + 2.4 \times 10^6 = 35.45 \times 10^6 \text{ N-mm}$$

$$V_e = 48{,}632 \text{ N.}$$

From SP16, $p_t = 1.047\%$ and $p_c = 0.095\%$

$$\text{against } \frac{1.5 \times M_{e1}}{bd^2} = 1.5 \times 33.45 \times 10^6 / 250 \times (265)^2 = 3.03 \text{ N/mm}^2$$

$$A_{st} = 1.047 \times 250 \times 265 / 100 = 692 \text{ mm}^2$$

$$A_{sc} = 0.095 \times 250 \times 265 / 100 = 63 \text{ mm}^2$$

Therefore, 4 no. 20 mm ϕ, giving $A_{st} = 1256 \text{ mm}^2$ is OK.

Provide 2 no. 12 mm ϕ at compression face giving $A_{sc} = 113 \text{ mm}^2$ is also OK.

TRANSVERSE REINFORCEMENT

Transverse reinforcement will be provided in the form of vertical stirrups. Let us provide 25 mm clear cover all round.

$$b_1 = \text{center to center distance between corner bars in the direction of width}$$

$$= 250 - 2 \times 25 - 10 = 190 \text{ mm}$$

$$d_1 = \text{center to center distance between corner bars in the direction of depth}$$

$$= 300 - 2 \times 25 - 10 = 240 \text{ mm}$$

Let us provide 10 mm ϕ 2-legged stirrups, $A_{sv} = 2\frac{\pi}{4} \times (10)^2 = 157 \text{ mm}^2$

$$\text{Now, } A_{sv} = \frac{T \cdot S_v}{b_1 d_1 \sigma_{sv}} + \frac{V \cdot S_v}{2.5 d_1 \sigma_{sv}}$$

$$\text{or, } 157 = \left[\frac{1.86 \times 10^6}{190 \times 240 \times 230} + \frac{36728}{2.5 \times 240 \times 230}\right] S_v$$

$$\text{or. } S_v = 354 \text{ mm.}$$

However, the spacing should not exceed the least of x_1, $\frac{x_1 + x_1}{4}$ and 300 mm, where

x_1 = short dimension of stirrup = 190 + 20 + 10 = 220 mm (Figure 9.10).

(a)

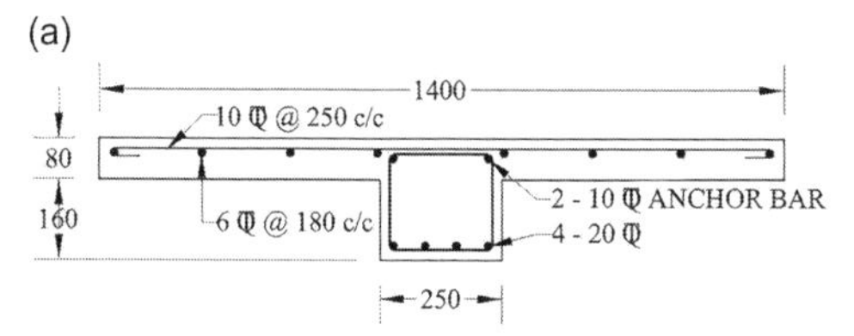

(b)

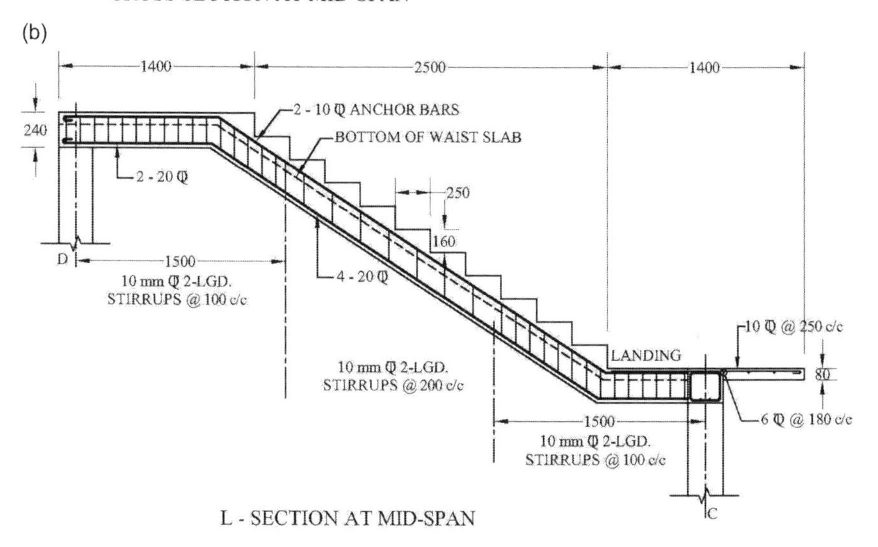

FIGURE 9.10 Reinforcement details of staircase with central stringer beam.

10 Design of Compression Members

CLASSIFICATION OF COLUMNS

Compression member is a structural element, which is subjected predominantly to axial compressive forces. These members are commonly found in reinforced concrete buildings as columns, sometimes walls, etc.

IS 456 defines the column as a compression member where the effective length exceeds three times the least lateral dimension. It may be noted here that the pedestal is described as a vertical compression member whose effective length is less than three times its least lateral dimension. This chapter primarily deals with tied columns, which are commonly used in reinforced concrete construction. All cross-sectional shapes, square, rectangle, *L*, etc., are dealt in this chapter. Shear effects in columns are generally neglected. However, shear resistance is high in the presence of axial compression and lateral ties.

Columns may be classified based on the nature of loading:

1. Columns with axial loading
2. Columns with uniaxial eccentric loading
3. Columns with biaxial eccentric loading

The combination of axial compression (P) with bending moment (M) at any column section is statically equivalent to a system consisting of the load P applied with an eccentricity $e = M/P$ with respect to the longitudinal centroidal axis of the column section. In a more general loading situation, bending moments (M_x and M_y) are applied simultaneously on the axially loaded column in two perpendicular directions – about the major axis (XX) and minor axis (YY) of the column section.

Again, depending on slenderness ratio, compression members may be classified into the following two categories: short columns and slender long columns.

l = unsupported length → length between the floor beams or foundation and another floor beam.

$l_e = Kl$. For uniform column – Table 28 of IS456, to be considered. If $l_e/b \leq 12$ → termed as **Short column**

If $l_e/b > 12$ → termed as **Long column or Slender column.**

Slenderness is defined as in IS 456, effective length to lateral dimension. It is important from elastic instability (buckling) point of view. Columns with very high slenderness ratios are accompanied with large lateral deflection under

DOI: 10.1201/9781003208204-10

relatively low compressive loads. When a column is subjected to flexure combined with axial compression, the action of the axial compression in the displaced geometry of the column introduces, secondary moments, usually ignored in first-order analysis. These secondary moments become significant with increasing column slenderness. According to the IS Code, sometimes, slenderness ratio is better expressed in terms of the radius of gyration† (r). IS 456, recommended in its working design methodology. The effective length depends on the unsupported length (l) and the ends conditions). Charts are given in Figs. 26 and 27 of IS 456 for determining the effective length ratios of braced columns and unbraced columns, respectively, in terms of coefficients β_1 and β_2 that represent the degrees of rotational freedom at the top and bottom ends of the column, which are need to be used. Slenderness effects in columns effectively result in reduced strength, on account of the additional, secondary moments, introduced. In the case of very slender columns, failure may occur suddenly under small loads due to instability (elastic buckling). The IS 456 recommended that the ratio of the unsupported length (l) to the least lateral dimension (d) of a column should not exceed 60.

Eccentricities arising out of structural analysis due to various reasons, like, lateral loads not considered in design, live load placements not considered in design, accidental lateral/eccentric loads, errors in construction, and slenderness effects underestimated in design. For this reason, IS456 recommended that every column needs to be designed for a minimum eccentricity e_{min} (in any plane) equal to the unsupported length/500 plus lateral dimension/30, subject to a minimum of 20 mm.

Minimum longitudinal bars must, in general, have a cross-sectional area not less than 0.8% of the gross area of the column section. In very large-sized columns (for architectural consideration) under axial compression, the limit of 0.8% of gross area may result in high area of reinforcement. In such cases, the IS456, permitting the minimum area of steel, may be calculated as 0.8% of the area of concrete required to resist the direct stress, and not the actual gross area. For pedestals, the minimum requirement of longitudinal bars may be taken as 0.15% of the gross area of cross-section. Maximum cross-sectional area of longitudinal bars should not exceed 6% of the gross area of the column section. However, a reduced maximum limit of 4% is recommended, in general, to avoid congestion of reinforcements. Longitudinal bars in columns (and pedestals) should not be less than 12 mm in diameter and should not be spaced more than 300 mm apart. At least four bars should be provided in a rectangular column section and at least six bars, equally spaced near the periphery, in a circular column. Minimum clear cover of 40 mm or bar diameter (whichever is greater) to the column ties is recommended by the IS456. Minimum diameter of lateral ties is generally considered 6 mm. IS456 recommends maximum lateral tie spacing recommended as 48φ or 300 mm, whichever is less. In the interest of durability, increased cover generally not greater than 75 mm.

The maximum compressive strain in concrete under axial loading at the limit state of collapse in compression is specified as $\varepsilon_c = 0.002$ as per IS456 (Limit State Design Methodology).

$$P_u = 0.4 f_{ck} A_c + 0.67 f_y A_{sc}$$

where P_u is the axial compression capacity (including the effect of minimum eccentricities as recommended by IS 456). Axial compression capacity may be enhanced by 5% when spiral reinforcement is provided as per IS 456.

Design of RCC column as per IS 456:2000, Clause 39, P-70, has to be followed.

- Axial strain $\varepsilon_{cc} \leq 0.0020$ under direct compression
- Maximum compressive strain $\varepsilon_{bc} \leq (0.0035 - 0.75 \times$ strain at least compressed fiber)
  (Under no tension)
- Minimum eccentricity to be considered as per clause – 25.4, p – 42 of Code.
  $e_{min} = \left(\frac{l}{500} + \frac{b}{30}\right)$ subjected to a minimum 20 mm, to be considered about one axis at a time.
- General reinforcement detailing to be followed as per clause 26, p – 42–49.
- For column clause 26.5.3, to be followed.
- Load factor as per Table 18
- For Effective length – Annex E, p – 92–94

DESIGN OF SHORT COLUMN UNDER AXIAL COMPRESSION

Approximate Load Assessment on a Column

Through "Substitute frame analysis", normally one can get P_x and P_y, i.e., load transfer from frame in x-direction and y-direction, respectively, on a particular column (Figure 10.1).

$$P = P_x + P_y$$

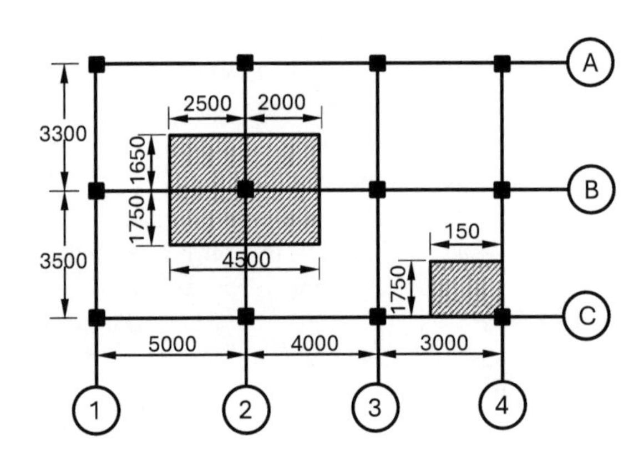

FIGURE 10.1 Influence area of a column.

Any column will be under biaxial moment. Similarly, M_x and M_y can be obtained

Then column has to be designed against P, M_x, and M_y

Approximate method to calculate axial load (P) from floor and floor beams on a particular column may be made, using **"Influence area concept"**

For column B-2, referring to Figure 10.1

$$A_f = \text{Influence area} = 3.4 \text{ m} \times 4.5 \text{ m}$$

Similarly for A-3, A_f = Influence area = 3.5 m × 1.65 m

For column C-4, A_f = Influence area = 1.5 m × 1.75 m

It is assumed that load within influence area, all the loads, will be transferred to that particular column (Figure 10.1).

Load Calculation on a Particular Column

w_d = Dead load per square meter, w_l = Live load per square meter

W_{BK} = Weight of brick work within influence zone

W_b = Weight of beam rib within influence zone only

Therefore, P_f = Load from a particular floor = $A_f(w_d + w_l) + W_{BK} + W_b$

If the floor planning is more or less same and the number of typical floor is 'n'.

Then, P = Axial force at foundation level = $nP_f + P_r + P_{TB}$

$$P = \text{Axial force at any floor level} = nP_f + P_r$$

where n = number of typical floors carrying by that particular column

P_r = Load from roof (Including water tank, stair room etc.)

P_{TB} = Load from Tie beam (brick work etc.)

Ultimate load carrying capacity (P_u) as per LSD (IS 456), If Slenderness ratio $l_e/b \leq 12$

$P_n = 0.4 f_{ck} A_c + 0.67 f_y A_{sc}$ as per clause 39.3, p – 71- IS456, if e_{min} as per clause 25.4 does exceed.

Design example of Axial loaded compression member/column as per IS 456:2000

P = 2000 kN, l = unsupported length = 3.0 m, M20 concrete, Fe 415 steel.

Choice of column section

Let us assume $l_e/b < 12$, i.e. a short column.

Let A_{sc} = 1% of A_g = 0.01 A_g

$$P_u = 1.5 \times 2000 = 3000 \text{ kN} = 3000 \times 10^3 \text{ N}$$

For a short column, $P_u = 0.4 f_{ck} A_c + 0.67 f_y A_{sc}$

$$A_g = bD, A_c = A_g - A_{sc} = A_g - 0.01A_g = 0.099A_g$$

$$3000 \times 10^3 = 0.4 \times 20 \times 0.99A_g + 0.67 \times 415 \times 0.01A_g$$

$$A_g = 280{,}360.73 \text{ mm}^2$$

Generally, minimum $\left(\dfrac{A_{sc}}{bD}\times 100\right) \geq 0.8$, maximum = 4%

We have to choose a section whose 'A_g' is more or less around 280,360.7 mm². It is may be slightly less or slightly more (Figure 10.2).

Let us provide a column section 450 mm × 600 mm, i.e., $A_g = 450 \times 600 =$ 270,000 mm² close to 280,360.7 mm², so okay.

$$e_{\min} \text{ as per clause } 25.4 = \frac{l}{500} + \frac{b}{30} = \frac{3000}{500} + \frac{450}{30} = 21 \text{ mm}$$

Again $0.05b = 0.05(450) = 22.5$ mm > $e_{\min}$ as per clause 25.4

Therefore, $P_u = 0.4 f_{ck} A_c + 0.67 f_y A_{sc}$, can be used

$$3000 \times 10^6 = 0.4 \times 20 \times (450 \times 600 - A_{sc}) + 0.67 \times 415 \times A_{sc}$$

$$A_{sc} = 3111 \text{ mm}^2$$

Let us provide 4–25ϕ + 4–20 ϕ

$$\text{i.e. } A_{sc} = \frac{\pi \times 25^2}{4} \times 4 + \frac{\pi \times 20^2}{4} = 3220 \text{ mm}^2 > 3111 \text{ mm}^2 \text{ So, Ok}$$

$$p = \frac{3220}{450 \times 600} \times 100 = 1.192\% > 0.8\% \text{ So Ok. } [\text{As per clause } 26.5.3.1(\text{a}) \text{ p} - 48]$$

$$\phi \geq 12 \text{ mm} \quad [\text{Ok as per clause } 26.5.3(\text{d})\text{p} - 48]$$

For circular columns minimum, six nos. longitudinal bars have to be provided.

Lateral Ties

Diameter of lateral ties $\phi_{Lt} > \phi/4 = 6.25$ mm or 6 mm whichever is higher

[As per clause 26.5.3.2(c), p – 49], Let us $16 \times \phi = 16 \times 20 = 320$ mm

Provide eight ϕ bars.

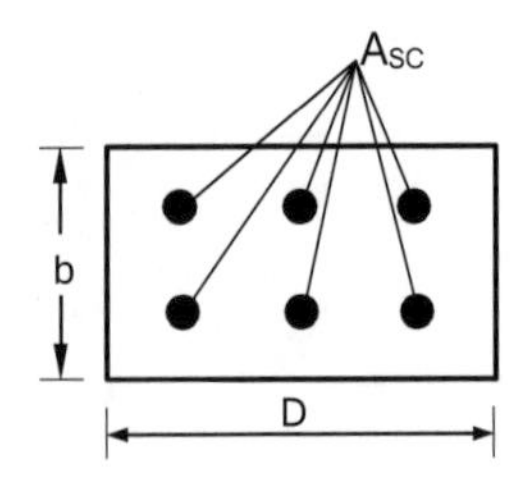

FIGURE 10.2 Column reinforcement.

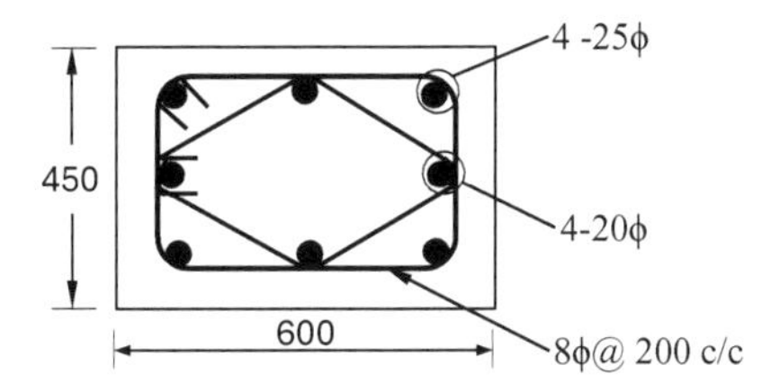

FIGURE 10.3 Lateral ties of column.

Spacing of lateral ties $S_{Lt} \leq b = 450\,\text{mm}$ or $16 \times \phi = 16 \times 20 = 320\,\text{mm}$ or 300 mm, whichever is lowest. i.e. 300 mm. Normally, minimum 200 mm c/c. Let us provide 8 ϕ @ 200 c/c

It always better to be provide closer spacing by adjusting diameter to be protect core concrete better and from ductility point of view (IS13920 may be followed from earthquake resistant design point of view) (Figure 10.3).

Design Examples

Design reinforcement in the column of size 400 × 600 mm subjected to an axial working load of 2000 kN. The column has an unsupported length of 3 m, and it is fixed against sideway in both directions. Adopt M20 and Fe 415.

We have, $f_{ck} = 20$ N/mm², $f_y = 415$ N/mm², P = Working load = 2000 kN, l = Unsupported Length = 3 m, $b = 400$, $D = 600$ mm

$l_e/b = 3000/400$ (D is the Least Lateral Dimension) = 7.5 < 12

Min. Eccentricity

$$e_{\min} = 3000 / 500 + 400 / 30 = 19.3 < 20 \text{ mm (Pg. No. 42 clause 25.4)}$$

$$e_{\max} = 3000 / 500 + 600 / 30 = 26 > 20 \text{ mm}$$

Check:

$$0.05 \times D = 30 > e_{y\,\min} \qquad 0.05 \times b = 20 > e_{x\min} \text{(Pg. No. 71)}$$

Ultimate load, P_u

$$P_u = \text{Working Load} \times \text{PSF} = 1.5 \times 2000$$

$$P_u = 3000 \text{ kN}$$

Longitudinal reinforcement

$$P_u = 0.4\, f_{ck} A_c + 0.67\, f_y A_{sc}, \text{ where, } A_c = A_g - A_{sc}$$

$$A_g = 400 \times 600(\text{Column area given}) = 240{,}000 \text{ mm}^2$$

$$3000 \times 10^3 = 0.4 \times 20 \times (240{,}000 - A_{sc}) + (0.67 \times 415 \times A_{sc})$$

$$A_{sc} = 3999.25\text{mm}^2 \text{ say } 4000 \text{ mm}^2$$

Use 25 mm dia. bar

$$n = A_{sc} / A_{\varnothing} = 4000 / \left(\pi \times 25^2 / 4\right) = 8.15 \text{ say } 10 \text{ rods}$$

In order to provide for column 10 nos., it can be split as 6 and 4 nos. with 25 and 20 mm, respectively

$$6 \times (\pi \times 25^2 / 4) + 4 \times (\pi \times 20^2 / 4) = 4202 \text{ mm}^2$$

Lateral ties

$$\text{Tie Diameter} < \text{Dia of bar } / 4 = 25 / 4 = 6.25 < 8 \text{ mm}$$

$$\text{Tie spacing } < 300 \text{ mm}, \; 16 \times \text{ dia } = 16 \times 25 = 400 \text{ mm} > 300 \text{ mm}$$

Provide 8 mm dia. bars with spacing of 250 mm c/c

Design Example

Design the reinforcement in the circular column of diameter 30 mm with helical reinforcement to support a factored load of 1500 kN; the columns have an unsupported length of 3 m and it is braced against side way. Adopt M20 and Fe 415 steel.

Given data
Diameter = 300 mm, Unsupported Length, $L = 3$ m = 3000 mm

$$P_u = 1500 \text{ kN}$$

$$f_{ck} = 20 \text{ N / mm}^2, f_x = 415 \text{ N / mm}^2.$$

Slenderness ratio

$$e_{\min} = [L / 500 + D / 300] = [3000 / 500 + 300 / 30] = 16 < 20 \text{ mm}$$

Also $0.05 \times D = 15 < 20$ mm

Longitudinal reinforcement

$$P_u = 0.4\, f_{ck} A_c + 0.67\, f_y A_{sc}, \text{ where, } A_c = A_g - A_{sc}$$

$$A_g = \pi \times 300^2 / 4 = 70{,}686 \text{ mm}^2,$$

$$\left(1500 \times 10^3\right) = \left[0.4 \times 20 \times \left(70{,}686 - A_{sc}\right)\right] + \left[0.67 \times 415 \times A_{sc}\right]$$

$$A_{sc} = 3460 \text{ mm}^2 > A_{sc\ min} = 0.8\% \text{ of Cross sectional area}$$

$$= 0.8 / 100 \times \pi \times 300^2 / 4 = 565 \text{ mm}^2$$

Use 25 mm diameter $n = A_{scc} / A_{\varnothing} = 3460/(\pi \times 25^2/4) = 7.046$ say 8 nos.

Provide nos. of bars of 25 mm diameter

Helical reinforcement

$$\text{Core diameter } = [300 - (2 \times 40)] = 220 \text{ mm}$$

Assuming clear cover of 40 mm

$$\text{Area of core, } A_c = (\pi \times 220^2 / 4) - (8 \times \pi \times 25^2 / 4) = 34{,}086 \text{ mm}^2$$

$$\text{Volume of core, } V_c = 34{,}086 \times 10^3 \text{ mm}^3$$

$$\text{Gross area, } A_g = (\pi \times 300^2 / 4) = 70{,}686 \text{ mm}^2$$

Use 8 mm diameter helical spirals at a pitch 'p' mm

The volume of helical/m length is given by

$$V_{ns} = \pi(300 - (2 \times 40) - 8) \times (\pi \times 8^2 / 4) \times 1000 / p = 33.48 \times 10^6 / p \text{ mm}^3$$

According to clause 39.4.1 (IS: 456) $\mathbf{V_{ns}/V_c < 0.36 \{(A_g/A_c) - 1\} f_{ck}/f_y}$ **(Pg. No 71)**

$$\left[\left(33.48 \times 10^6 / p\right) / \left(34086 \times 10^3\right)\right] = 0.36[(70686 / 34086) - 1](20 / 415)$$

$$p = 52.7 \text{ mm} < 75 \text{ mm}$$

Codal restriction on pitch according to **clause 26.5.3.2,** $p < 75$ mm (or) Core diameter/6 = 220/6 = 36.66 mm (or) $p > 25$ mm (or) $3 \times$ diameter of helical rod $3 \times 8 = 24$ mm

Therefore provide 8 mm diameter of spiral at pitch of 36.6 mm say 25 mm

AXIAL LOAD – MOMENT INTERACTION CURVE AS RECOMMENDED BY IS 456

The interaction curve is provided in SP16 is a graphical representation of the ultimate design axial load and moment capacities in mixed mode of a uniaxially eccentrically loaded column. Each point on the curve corresponds to the ultimate design capacity of P_u and M_u, associated with a specific eccentricity (e) of loading. If the load P is applied on a short column with an eccentricity e, the corresponding ultimate moment by $M_u = P_u.e$, then the coordinates (M_u, P_u) form a unique point on the interaction diagram. The interaction curve defines the different (M_u, P_u) combinations for all possible eccentricities of loading $0 \leq e < \infty$. For design purposes, the calculations of M_u and P_u are based on the design stress-strain curves (including the partial safety factors) and the resulting Axial Load–Moment interaction design curve. Design interaction curve serves as a failure envelope.

Design Charts (for Uniaxial Eccentric Compression) in SP: 16

The design charts (non-dimensional interaction curves) given in the Design Handbook, SP: 16 cover the following three cases of symmetrically arranged reinforcement:

a. Rectangular sections with reinforcement distributed equally on two sides, the 'two sides' refer to the sides parallel to the axis of bending; there are no inner rows of bars, and each outer row has an area of $0.5A_s$, this includes the simple four bars
b. Rectangular sections with reinforcement distributed equally on four sides; two outer rows and four inner rows have been considered in the calculations; however, the use of these charts can be extended, without significant error, to cases of not less than two inner rows.
c. Circular column sections, the charts are applicable for circular sections with at least six bars (of equal diameter) uniformly spaced circumferentially

However, proper interaction diagrams are required as indicated below to make the designs more economical.

- Unsymmetrically arranged reinforcement in rectangular sections;
- Non-rectangular and non-circular sections — such as *L*-shaped, *T*-shaped, *H*-shaped, cross-shaped sections, etc.

In such cases, it becomes necessary for proper interaction diagrams in order to obtain accurate and reliable solutions.

Design Example

Design the longitudinal and lateral reinforcement in rectangular reinforced concrete column of size 300×400 mm subjected to design ultimate load of 1200 kN and an ultimate moment of 200 kN-m with respect to major axis. Adopt M20 grade and Fe 415 steel.

Given Data: $P_u = 1200$ kN, $M_u = 200$ kNm, $f_{ck} = 20$ N/mm², $f_y = 415$ N/mm², Column dimention = 300×400 mm

Non-dimensional parameter

$$P_u/(f_{ck}\, bD) = (1200 \times 10^3) / (20 \times 300 \times 400) = 0.5$$

$$M_u/(f_{ck}\, b\, D^2) = (200 \times 10^6) / (20 \times 300 \times 400^2) = 0.208$$

Longitudinal reinforcement

Adopting an effective cover of 50 mm, $d' = 50$ mm, $d'/D = 50/400 = 0.125 = 0.15$

[Refer chart 33 sp 16 Page No. 118]

From chart 33 $P/f_{ck} = 0.2$ (Based on A and B)

$$P = 0.2 \times f_{ck} = 0.2 \times 20 = 4$$

$$A_{sc} = P \times b \times D / 100 = 4800 \text{ mm}^2$$

Use 25 mm dia bars and find out no. of bars

$$n = A_{st} / A_{\varnothing} = 4800 / (\pi \times 25^2 / 4) = 9.78 = 10 \text{ nos}$$

Provide 10 nos of 25 mm dia bars

Lateral ties

$$\text{Tie dia} = \text{diameter of rod}/4 = 25 / 4 = 6.25 \text{ mm} < 8 \text{ mm}$$

$$\text{Tie spacing} > 300 \text{ mm}, \quad 16 \times \text{d} = 16 \times 25 = 400 > 300 \text{ mm}$$

So, adopt 300 mm as spacing

Provide 8 mm dia. bars with spacing 250 mm c/c

Design Example

Design a short circular column of dia. 400 mm to support a factored axial load of 900 kN together with a factored moment of 100 kN-m. Adopt M20 Grade and Fe 415 Steel

$$P_u = 900 \text{ kN} = 900 \times 10^3 \text{ N}, \ M_u = 100 \text{ kN-m} = 100 \times 10^6 \text{ N-mm}$$

$$f_{ck} = 20 \text{ N/mm}^2, f_y = 415 \text{ N/mm}^2, \text{ Dia, } D = 400 \text{ mm}$$

Non-dimensional parameter

$$P_u/\left(f_{ck}\, D^2\right) = (900 \times 10^3) / (20 \times 400^2) = 0.28$$

$$M_u/\left(f_{ck}\, D^3\right) = (100 \times 10^6) / (20 \times 400^3) = 0.078$$

Longitudinal reinforcement
Adopting an effective cover of 50 mm

$$d' = 50 \text{ mm}, d'/D = 50/400 = 0.125 < 0.15$$

Use chart 57,
$P/f_{ck} = 0.12$ (Based on A and B)

$$P = 0.12 \times f_{ck} = 0.12 \times 20 = 2.4$$

$$A_{sc} = (P\pi D^2/4) \times (1/100)(\text{from same chart}) = 2.4 \times \pi \times 400^2/400 = 3015.9 \text{ mm}^2$$

Use 25 mm dia. bars

$$n = A_{st} \;/\; A_{\varnothing} = 3015.9 / (\pi \times 25^2 \;/\; 4) = 6.14 \text{ say } \ 6 \text{ nos.}$$

Provide six nos. of 25 mm dia. bars

Lateral ties

$$\text{Tie dia} = \text{diameter of rod } /4 = 25/4 = 6.25 \text{ mm} < 8 \text{ mm}$$

$$\text{Tie spacing } > 300 \text{ mm}, \quad 16 \times \text{d} = 16 \times 25 = 400 > 300 \text{ mm}$$

Adopt 250 mm as spacing.
Provide 8 mm dia. bars with spacing 250 mm c/c

DESIGN OF SHORT COLUMNS UNDER AXIAL COMPRESSION WITH BIAXIAL BENDING

Biaxial Eccentricities

As mentioned in Section 13.3.2, all columns are (in a strict sense) to be treated as being subject to axial compression combined with biaxial bending, as the design must account for possible eccentricities in loading (e_{min} at least) with respect to both major and minor principal axes of the column section. Uniaxial loading is an idealized approximation, which can be made when the e/D ratio with respect to one of the two principal axes can be considered to be negligible. Also, as mentioned in Section 13.4.1, if the e/D ratios are negligible with respect to both principal axes, conditions of axial loading may be assumed, as a further approximation.

In the Code, it is clarified (Cl. 25.4) that "where biaxial bending is considered, it is sufficient to ensure that eccentricity exceeds the minimum about one axis at a time". This implies that if either one or both the factored bending moments M_{ux} and M_{uy} (obtained from analysis) is less than the corresponding value, calculated from minimum eccentricity considerations, it suffices to ensure that at least one of the two minimum eccentricity conditions is satisfied. However, it also becomes necessary to check for the other biaxial bending condition wherein the minimum eccentricity in the other direction is also satisfied.

Code Procedure for Design of Biaxially Loaded Columns

The simplified method adopted by the Code (Cl. 39.6) is based on Bresler's formulation for the 'load contour' – whereby an approximate relationship between $M_{uR,x}$ and $M_{uR,y}$ (for a specified $P_u = P_{uR}$) is established. This relationship is conveniently expressed in a non-dimensional form as follows:

$$\left(\frac{\mathbf{M_{ux}}}{\mathbf{M_{ux1}}}\right)^{\alpha_n} + \left(\frac{\mathbf{M_{uy}}}{\mathbf{M_{uy1}}}\right)^{\alpha_n} \leq \mathbf{1.0}$$

where M_{ux} and M_{uy} denote the factored biaxial moments acting on the column, and (as explained earlier) M_{ux1} and M_{uy1} denote the uniaxial moment capacities with reference to the major and minor axes, respectively, all under an accompanying axial load $P_u = P_{uR}$. It may be noted that M_{ux}, M_{uy} (and P_u) are measures of the load effects due to external loading on the structure, whereas M_{ux1}, M_{uy1} (and P_{uR}) are measures of the inherent resistance of the column section.

α_n in Eq. 13.38 is a constant that depends on the factored axial compression P_u and which defines the shape of the 'load contour'. For low axial load levels, the load contour (in non-dimensional coordinates) is approximated as a straight line; accordingly, $\alpha_n = 1$. For high axial load levels, the load contour is approximated as the quadrant of a circle; accordingly, $\alpha_n = 2$. For moderate load levels, α_n takes a value between 1 and 2. In order to quantitatively relate α_n with P_u, it is convenient to normalize P_u with the maximum axial load capacity of the column (under 'pure compression').

$$P_{uz} = 0.45 f_{ck} A_c + 0.75 f_y A_{sc}$$

where A_g denotes the gross area of the section and A_{sc} the total area of steel in the section.

$\alpha_n = 1$ for $P_u/P_{uz} < 0.2$; $\alpha_n = 2$ for $P_u/P_{uz} > 0.8$; and α_n is assumed to vary linearly for values of P_u/P_{uz} between 0.2 and 0.8

This is usually achieved by increasing the percentage of reinforcement and/or improving the grade of concrete; the dimensions may also be increased, if required.

Generally, reinforced concrete columns are subjected to an axial force P_u, and biaxial moments M_{ux} and M_{uy} about x and y axes respectively. Such a section may be considered to be acted upon by an axial load P_u at eccentricities $e_x = M_{ux}/P_u$ and $e_y = M_{uy}/P_u$. If P_u, e_x and e_y are given, one can calculate $M_{ux} = e_x \times P_u$ and $M_{uy} = e_y \times P_u$. The method provided in IS 456: 2000 (clause 38.6) may be applied as follows:

$$\left(\frac{M_{ux}}{M_{ux1}}\right)^{\alpha_n} + \left(\frac{M_{uy}}{M_{uy1}}\right)^{\alpha_n} \leq 1.0 \tag{10.1}$$

where M_{ux} and M_{uy} are the moments about x and y axes, respectively, due to the design loads. M_{ux1} and M_{uy1} are the ultimate uniaxial moment capacities about x and y axes, respectively, under an axial load P_u.

α_n is an exponent which depends on the ratio on the P_u/P_{uz}, where $\mathbf{P_{uz} = 0.45 f_{ck} A_c + 0.75 f_y A_{sc}}$

P_u/P_{uz}	α_n
≤0.2	1.0
≤0.8	2.0

*Linear interpolation may be done. Alternatively, the value of α_n can be determined by the following equation as follows:

$$\alpha_n = \left(\frac{2}{3}\right)\left(1 + \left(\frac{5}{2}\right)\left(\frac{P_u}{P_{uz}}\right)\right)$$

Load P_{uz} may be evaluated from SP: 16 handbook also. Interaction curve between P_u and M_u are given in SP: 16, for different conditions in SP: 16.

Design Procedure

Step 1: Calculate area of concrete required using equation $P_u = 0.4 f_{ck} A_c + 0.67 f_y A_{sc}$, considering P_u only assuming A_{sc} = 1% of gross area. Gross area may be increased by (25% due to effect of M_{ux} and M_{uy}). Assume also type of reinforcement distribution (whether reinforcements are distributed uniformly along all four sides or provided on opposite ends along shorter or longer direction.

Step 2: Assume a value of (p/f_{ck}) or p for column section, where p = percentage of steel and f_{ck} is the characteristic direct compressive strength of concrete.

Step 3: Determine uniaxial ultimate moment capacities M_{ux1} and M_{uy1} considering axial load P_u using $P_u/f_{ck}\, b\, D$ vs. $M_u/f_{ck}\, bD^2$ interaction curve provided in SP16.

Step 4: Compute $P_{uz} = P_u = 0.45f_{ck}A_c + 0.75f_yA_{sc}$ and then find ratio P_u/P_{uz}.
Step 5: Compute α_n from IS456 or SP16.

Step 6: Calculate $\left(\frac{M_{ux}}{M_{ux1}}\right)^{\alpha_n} + \left(\frac{M_{uy}}{M_{uy1}}\right)^{\alpha_n}$ and check whether it is less than 1 or not.

If value is more than 1, then section is to be revised. This ratio near and less than 1 provides economic solution (Figure 10.4).

Design Example

$P = 2000$ kN, $M_x = 150$ kN-m, $M_y = 100$ kN-m, Design a square section
Grade of concrete M25 i.e., $f_{ck} = 25$ N/mm^2
Grade of steel Fe500 ie $f_y = 500$ N/mm^2
Partial safety factor for load, $\gamma_f = 1.5$
Let us initially consider only the effect of 2000 kN

$$A_{sc} = 1\% \text{ of } A_g = 0.01A_g, A_c = (A_g - A_{sc}) = 0.99A_g$$

$$P_u = 0.4f_{ck}A_c + 0.67f_yA_{sc}$$

$$\Rightarrow 1.5 \times 2000 \times 10^3 = (0.4 \times 25 \times 0.99A_g) + (0.67 \times 500 \times 0.01A_g) \text{ i.e., } A_g$$

$$= 226{,}415 \text{ mm}^2$$

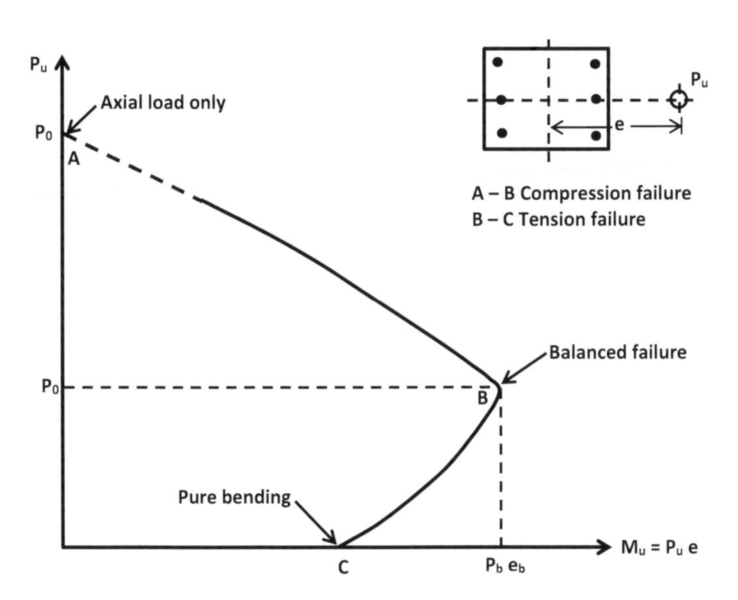

FIGURE 10.4 Axial compressive force vs. bending moment interaction curve.

$$B = 500 \text{ mm and } D = 500 \text{ mm}, \left(A_g\right)_{\text{prov.}} = B \times D = 500 \times 500 = 250{,}000 \text{ mm}^2$$

Let us now consider percentage of steel, $p = 2.125\%$

$$p / f_{\text{ck}} = (2.125/25) = 0.085, \ d' = 50 \text{ mm}, \ d' / D = 50 / 500 = 0.10$$

Let us consider that the reinforcements are assumed to be equally distributed on all four sides

$$P_u / f_{\text{ck}} BD = (1.5 \times 2000 \times 1000) / (25 \times 500 \times 500) = 0.48$$

From chart 48 of SP: 16, $M_{\text{ux1}}/f_{\text{ck}}BD^2 = 0.094 \Rightarrow M_{\text{ux1}} = 0.094 \times 25 \times 500 \times 500^2$ $= 293.75$ kN-m

$$M_{\text{ux}} = 1.5 \times M_x = 1.5 \times 150 = 225 \text{ kN-m}, \quad M_{\text{ux1}} > M_{\text{ux}}, \text{ So Ok.}$$

As it is a square column

$$M_{\text{uy1}} = 293.75 \text{ kN-m} > M_{\text{uy}}, \text{ So Ok.}$$

From chart 63 of SP: 16, $P_{\text{uz}}/A_g = 19.15$, $P_{\text{uz}} = 19.15 \times 250{,}000 \text{ N} = 4787.5$ kN

$$P_u / P_{\text{uz}} = (1.5 \times 2000) / 4787.5 = 0.627$$

$$\alpha_n = \left(\frac{2}{3}\right)\left(1 + \left(\frac{5}{2}\right)\left(\frac{P_u}{P_{\text{uz}}}\right)\right) = 1.711$$

$$\left(\frac{M_{\text{ux}}}{M_{\text{ux1}}}\right)^{\alpha_n} + \left(\frac{M_{\text{uy}}}{M_{\text{uy1}}}\right)^{\alpha_n} = 0.95$$

Design Example

$P = 2000$ kN, $M_x = 150$ kN-m, $M_y = 100$ kN-m, design a rectangular section (Higher bending moment acing along depth of the section)

$P = 2000$ kN, $M_x = 150$ kN-m, $M_y = 100$ kN-m
Grade of concrete M25; $f_{\text{ck}} = 25$ N/mm^2
Grade of steel Fe500; $f_y = 500$ N/mm^2
Partial safety factor for load, $\gamma_f = 1.5$
Let us initially consider only the effect of 2000 kN only,

$$A_{\text{sc}} = 1\% \text{ of } A_g = 0.01A_g, \ A_c = \left(A_g - A_{\text{sc}}\right) = 0.99A_g$$

$$\boldsymbol{P_u = 0.4 f_{ck} A_c + 0.67 f_y A_{sc} \Rightarrow 1.5 \times 2000 \times 10^3 = (0.4 \times 25 \times 0.99 A_g)}$$

$$\boldsymbol{+(0.67 \times 500 \times 0.01 A_g)}$$

$$\Rightarrow A_g = 2,26,415 \text{ mm}^2$$

Take B = 400 mm and D = 625 mm

$$\left(A_g\right)_{\text{prov.}} = B \times D = 400 \times 625 = 2,50,000 \text{ mm}^2$$

Let us now consider percentage of steel, $p = 2.05\%$, $p/f_{ck=}(2.05/25) = 0.082$

$$d' = 50 \text{ mm}, \ d' / D = 50 / 625 = 0.08 \approx 0.10$$

Considering, reinforcement equally distributed on all four sides

$$P_u / f_{ck} BD = (1.5 \times 2000 \times 1000) / (25 \times 400 \times 625) = 0.48$$

From chart 48 of SP: 16, $M_{ux1}/f_{ck}BD^2 = 0.093$

$$\Rightarrow M_{ux1} = 0.093 \times 25 \times 400 \times 625^2 = 363.28 \text{ kN-m}$$
$$M_{ux} = 1.5 \times M_x = 1.5 \times 150 = 225 \text{ kN-m} < M_{ux1}, \text{ So OK.}$$
$$b' / B = 50 / 400 = 0.125 \approx 0.15, \ P_u / f_{ck} BD = 0.48$$

From chart 49 of SP: 16, $M_{uy1}/f_{ck}DB^2 = 0.088$

$$\Rightarrow M_{uy1} = 0.088 \times 25 \times 625 \times 400^2 = 220 \text{ kN-m}$$
$$M_{uy} = 1.5 \times M_y = 1.5 \times 100 = 150 \text{ kN-m} < M_{uy1}, \text{ So Ok.}$$

From chart 63 of SP: 16, $P_{uz}/A_g = 18.75$, $P_{uz} = 18.75 \times 250{,}000$ N = 4687.5 kN

$$P_u / P_{uz} = (1.5 \times 2000) / 4687.5 = 0.64$$

$$\alpha_n = \left(\frac{2}{3}\right)\left(1 + \left(\frac{5}{2}\right)\left(\frac{P_u}{P_{uz}}\right)\right) = 1.733$$

$$\left(\frac{M_{ux}}{M_{ux1}}\right)^{\alpha_n} + \left(\frac{M_{uy}}{M_{uy1}}\right)^{\alpha_n} = 0.951, \text{ So Ok.}$$

Design Example

$P = 2000$ kN, $M_x = 150$ kN-m, $M_y = 100$ kN-m, Design a rectangular section (higher bending moment acing along width of the section)

$P = 2000$ kN, $M_x = 150$ kN-m, $M_y = 100$ kN-m
Grade of concrete M25; $f_{ck} = 25$ N/mm²
Grade of steel Fe500; $f_y = 500$ N/mm²
Partial safety factor for load, $\gamma_f = 1.5$
Initially, considering, only the effect of 2000 kN,

$$A_{sc} = 1\% \text{ of } A_g = 0.01A_g, A_c = \left(A_g - A_{sg}\right) = 0.99A_g$$

$$P_u = 0.4 f_{ck} A_c + 0.67 f_y A_{sc}$$

$$\Rightarrow 1.5 \times 2000 \times 10^3 = \left(0.4 \times 25 \times 0.99A_g\right) + \left(0.67 \times 500 \times 0.01A_g\right) \Rightarrow A_g$$

$$= 226{,}415 \text{ mm}^2$$

Considering $B = 400$ mm and $D = 625$ mm, $(A_g)_{prov.} = B \times D = 400 \times 625 = 2{,}50{,}000$ mm²

Considering percentage of steel, $p = 2.3825\%$, $p/f_{ck} = (2.3825/25) = 0.0953$

$$d' = 50 \text{ mm}, \; d' / D = 50 / 625 = 0.08 \approx 0.10$$

Considering reinforcement equally distributed on all four sides

$$P_u / f_{ck} BD = (1.5 \times 2000 \times 1000) / (25 \times 400 \times 625) = 0.48$$

From chart 48 of SP: 16, $M_{ux1}/f_{ck}B^2D = 0.108$

$$\Rightarrow M_{ux1} = 0.108 \times 25 \times 400^2 \times 625 = 270 \text{ kN-m}$$
$$M_{ux} = 1.5 \times M_x = 1.5 \times 150 = 225 \text{ kN-m} < M_{ux1}, \text{ So Ok.}$$
$$d' / B = 50 / 400 = 0.125 \approx 0.15, \; P_u / f_{ck} BD = 0.48$$

From chart 49 of SP: 16, $M_{uy1}/f_{ck}BD^2 = 0.0972$

$$\Rightarrow M_{uy1} = 0.0972 \times 25 \times 400 \times 625^2 = 379.68 \text{ kN-m}$$
$$M_{uy} = 1.5 \times M_y = 1.5 \times 100 = 150 \text{ kN-m} < M_{uy1}, \text{ So Ok.}$$

From chart 63 of SP: 16, $P_{uz}/A_g = 20$, $P_{uz} = 20 \times 250{,}000$ N $= 5000$ kN

$$P_u / P_{uz} = (1.5 \times 2000) / 5000 = 0.6$$

$$\alpha_n = \left(\frac{2}{3}\right)\left(1+\left(\frac{5}{2}\right)\left(\frac{P_u}{P_{uz}}\right)\right) = 1.667$$

$$\left(\frac{M_{ux}}{M_{ux1}}\right)^{\alpha_n} + \left(\frac{M_{uy}}{M_{uy1}}\right)^{\alpha_n} = 0.951, \text{ So Ok.}$$

Design Example

$P = 2000$ kN, $M_x = 150$ kN-m, $M_y = 100$ kN-m, Design a circular section

$P = 2000$ kN, $M_x = 150$ kN-m, $M_y = 100$ kN-m

Grade of concrete M25; $f_{ck} = 25$ N/mm^2

Grade of steel Fe500; $f_y = 500$ N/mm^2

Partial safety factor for load, $\gamma_f = 1.5$

Initially consider only the effect of 2000 kN,

$$A_{sc} = 1\% \text{ of } A_g = 0.01A_g,\ A_c = \left(A_g - A_{sc}\right) = 0.99A_g$$

$$P_u = 0.4 f_{ck} A_c + 0.67 f_y A_{sc}$$

$$\Rightarrow 1.5 \times 2000 \times 10^3 = \left(0.4 \times 25 \times 0.99A_g\right) + \left(0.67 \times 500 \times 0.01A_g\right) \Rightarrow A_g$$

$$= 226415 \text{ mm}^2$$

Considering circular section, $D = 565$ mm

$$\left(A_g\right)_{\text{prov.}} = \pi D^2 / 4 = 2{,}50{,}718 \approx 2{,}50{,}000 \text{ mm}^2$$

Considering percentage of steel, $p = 2\%$, $p/f_{ck} = (2.1/25) = 0.084$

$$d' = 50 \text{ mm},\ d'/D = 50/565 = 0.088 \approx 0.10$$

$$P_u / f_{ck} D^2 = (1.5 \times 2000 \times 1000)/(25 \times 565^2) = 0.376$$

From chart 60 of SP: 16, $M_{ux1}/f_{ck}D^3 = 0.062$

$$\Rightarrow M_{ux1} = 0.062 \times 25 \times 565^3 = 279.56 \text{ kN-m}$$

$$M_{ux} = 1.5 \times M_x = 1.5 \times 150 = 225 \text{ kN-m} < M_{ux1}, \text{ So Ok.}$$

$$d'/D = 50/565 = 0.088 \approx 0.10,\ P_u/f_{ck} D^2$$

$$= (1.5 \times 2000 \times 1000)/\left(25 \times 565^2\right) = 0.376$$

From chart 60 of SP: 16, $M_{uy1}/f_{ck}D^3 = 0.062$

$$\Rightarrow M_{uy1} = 0.062 \times 25 \times 565^3 = 279.56 \text{ kN-m}$$

$$M_{uy} = 1.5 \times M_y = 1.5 \times 100 = 150 \text{ kN-m} < M_{uy1}, \text{ So Ok.}$$

From chart 63 of SP: 16, $P_{uz}/A_g = 18.9$, $P_{uz} = 18.9 \times 250{,}718$ N = 4738.60 kN

$$P_u / P_{uz} = (1.5 \times 2000) / 4738.60 = 0.633$$

$$\alpha_n = \left(\frac{2}{3}\right)\left(1 + \left(\frac{5}{2}\right)\left(\frac{P_u}{P_{uz}}\right)\right) = 1.722$$

$$\left(\frac{M_{ux}}{M_{ux1}}\right)^{\alpha_n} + \left(\frac{M_{uy}}{M_{uy1}}\right)^{\alpha_n} = 0.950 \qquad 1.0304 \quad .$$

Design Example

$P = 2000$ kN, $M_x = 150$ kN-m, $M_y = 100$ kN-m

Grade of concrete M25; $f_{ck} = 25$ N/mm²

Grade of steel Fe500; $f_y = 500$ N/mm²

Partial safety factor for load, $\gamma_f = 1.5$

Initially considering only the effect of 2000 kN,

$$A_{sc} = 1\%\ of\ A_g = 0.01A_g, A_c = \left(A_g - A_{sc}\right) = 0.99A_g$$

$$P_u = 0.4 f_{ck} A_c + 0.67 f_y A_{sc}$$

$$\Rightarrow 1.5 \times 2000 \times 10^3 = \left(0.4 \times 25 \times 0.99A_g\right) + \left(0.67 \times 500 \times 0.01A_g\right)$$

$$A_g = 226415 \text{ mm}^2$$

$$\text{Maintaining}\left(\frac{\boldsymbol{B}}{\boldsymbol{D}} \approx \frac{\boldsymbol{M_y}}{\boldsymbol{M_x}}\right)$$

$\therefore B = 388$ mm and $D = 582$ mm

Let us provide a column section 425 mm × 625 mm

$$\left(A_g\right)_{\text{prov.}} = B \times D = 425 \times 625 = 265{,}625 \text{ mm}^2 \approx 250{,}000 \text{ mm}^2$$

Considering percentage of steel, $p = 2\%$, $p/f_{ck=}(2/25) = 0.08$

$$d' = 50 \text{ mm}, \ d'/D = 50/625 = 0.08 \approx 0.10$$

Considering reinforcement equally distributed on all four sides

$$P_u / f_{ck}BD = (1.5 \times 2000 \times 1000) / (25 \times 425 \times 625) = 0.45$$

From chart 48 of SP: 16, $M_{ux1}/f_{ck}BD^2 = 0.1$

$$\Rightarrow M_{ux1} = 0.1 \times 25 \times 425 \times 625^2 = 415 \text{ kN-m}$$
$$M_{ux} = 1.5 \times M_x = 1.5 \times 150 = 225 \text{ kN-m} < M_{ux1}, \text{ So Ok.}$$
$$b'/B = 50/425 = 0.12 \approx 0.15, P_u / f_{ck}BD = 0.45$$

From chart 49 of SP: 16, $M_{uy1}/f_{ck}DB^2 = 0.09$

$$\Rightarrow M_{uy1} = 0.09 \times 25 \times 625 \times 425^2 = 254 \text{ kN-m}$$
$$M_{uy} = 1.5 \times M_y = 1.5 \times 100 = 150 \text{ kN-m} < M_{uy1}, \text{ So Ok.}$$

From chart 63 of SP: 16, $P_{uz}/A_g = 18.72$, $P_{uz} = 18.72 \times 265{,}625$ N = 4972.5 kN

$$P_u / P_{uz} = (1.5 \times 2000) / 4972.5 = 0.603$$

$$\alpha_n = \left(\frac{2}{3}\right)\left(1 + \left(\frac{5}{2}\right)\left(\frac{P_u}{P_{uz}}\right)\right) = 1.672$$

$$\left(\frac{M_{ux}}{M_{ux1}}\right)^{\alpha_n} + \left(\frac{M_{uy}}{M_{uy1}}\right)^{\alpha_n} = 0.77 < 1, \text{ So Ok}$$

Design Example

$P = 2000$ kN, $M_x = 150$ kN-m, $M_y = 100$ kN-m

Grade of concrete M25; $f_{ck} = 25$ N/mm²

Grade of steel Fe500; $f_y = 500$ N/mm²

Partial safety factor for load, $\gamma_f = 1.5$ Initially considering only the effect of 2000 kN,

$$A_{sc} = 1\% \ of \ A_g = 0.01A_g, A_c = (A_g - A_{sc}) = 0.99A_g$$

$$P_u = 0.4 f_{ck} A_c + 0.67 f_y A_{sc}$$

$$\Rightarrow 1.5 \times 2000 \times 10^3 = (0.4 \times 25 \times 0.99A_g) + (0.67 \times 500 \times 0.01A_g), A_g$$

$$= 2{,}26{,}415 \text{ mm}^2$$

Considering $\frac{D_x}{D_y} \approx \frac{M_x}{M_y}$, $a = b = 300$ mm

$\therefore D_x = 680$ mm, $D_y = 455$ mm

From Figure 10.4 $A1 = 91{,}800$ mm² and $A_2 = 159{,}000$ mm²

$$(A_g)_{prov.} = D_x \times D_y = 680 \times 455 = 2{,}50{,}000 \text{ mm}^2$$

Considering service load,

$$P = P_x + P_y \tag{10.2}$$

and

$$\frac{P_x}{P_y} = \frac{A_2}{A_1} \tag{10.3}$$

Solving Eqs. 10.2 and 10.3, $P_x - 1268$ kN and $P_y - 732$ kN

$$d' = 50 \text{ mm}, \; d' / D_x = 50 / 680 = 0.07 \approx 0.10$$

$$P_{\text{ux}} / f_{\text{cks}} \; a \; D_x = (1.5 \times 1268 \times 1000) / (25 \times 300 \times 680) = 0.373$$

$$M_{\text{ux}} / f_{\text{ck}} \; a \; D_x^2 = (1.5 \times 150 \times 10^6) / (25 \times 300 \times 680^2) = 0.065$$

Considering reinforcement equally distributed on all four sides
From chart 48 of SP: 16, $p_x/f_{\text{ck}} = 0.037 \Rightarrow p = 0.037{\times}25 = 0.925\%$

$$d' / D_y = 50 / 456 = 0.11 \approx 0.15, \; P_{\text{uy}} / f_{\text{ck}} \; a \; D_y$$

$$= (1.5 \times 732 \times 1000) / (25 \times 300 \times 456) = 0.373$$

$$M_{\text{uy}} / f_{\text{ck}} BD_y^2 = (1.5 \times 100 \times 10^6) / (25 \times 300 \times 456^2) = 0.096$$

Considering reinforcement equally distributed on all four sides
From chart 49 of SP: 16, $p_y/f_{\text{ck}} = 0.074 \Rightarrow p = 0.074 \times 25 = 1.85\%$

$$A_{\text{scx}} = p_x BD_x /100 = 1887 \text{ mm}^2, \; A_{\text{scy}} = p_y BD_y /100 = 2531 \text{ mm}^2$$

$$A_{\text{sc}} = \left(A_{\text{scx}} + A_{\text{scy}}\right) = 4418 \text{ mm}^2$$

$$p = \frac{A_{\text{sc}}}{A} \times 100 = (4418 \times 100) / 250{,}800 = 1.76\% \Rightarrow p / f_{\text{ck}} = 0.0704$$

Design of Short Columns under Axial Compression and Biaxial Bending Moments

Generally, reinforced concrete columns are subjected to an axial force P_u, and biaxial moments M_{ux} and M_{uy} about x and y axes respectively. Such a section may be considered to be acted upon by an axial load P_u at eccentricities $e_x = M_{\text{ux}}/P_u$ and $e_y = M_{\text{uy}}/P_u$. If P_u, e_x, and e_y are given, one can calculate $M_{\text{ux}} = e_x \times P_u$ and $M_{\text{uy}} = e_y \times P_u$. The method provided in IS 456: 2000 (clause 38.6) may be applied as follows:

$$\left(\frac{\mathbf{M}_{\mathbf{ux}}}{\mathbf{M}_{\mathbf{ux1}}}\right)^{\alpha_n} + \left(\frac{\mathbf{M}_{\mathbf{uy}}}{\mathbf{M}_{\mathbf{uy1}}}\right)^{\alpha_n} \leq \mathbf{1.0} \tag{10.4}$$

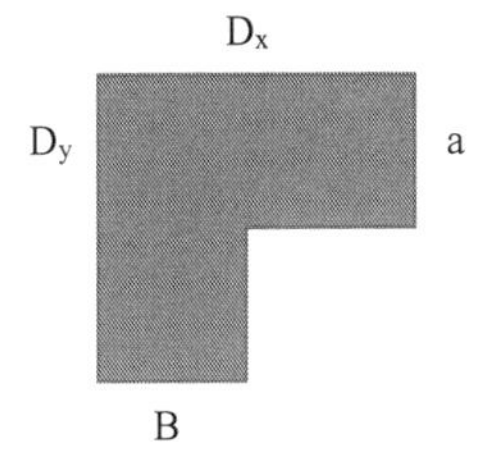

FIGURE 10.5 L-shaped column.

where M_{ux} and M_{uy} are the moments about x and y axes, respectively, due to the design loads. M_{ux1} and M_{uy1} are the ultimate uniaxial moment capacities about x and y axes, respectively, under an axial load P_u (Figure 10.5).

α_n is an exponent which depends on the ratio on the P_u/P_{uz}, where $\mathbf{P_{uz} = 0.45 f_{ck} A_c + 0.75 f_y A_{sc}}$

P_u/P_{uz}	α_n
≤0.2	1.0
≤0.8	2.0

Linear interpolation may be done. Alternatively, the value of α_n can be determined by the following equation as follows

$$\alpha_n = \left(\frac{2}{3}\right)\left(1 + \left(\frac{5}{2}\right)\left(\frac{P_u}{P_{uz}}\right)\right)$$

Load P_{uz} may be evaluated from SP16 handbook also. Interaction curve between P_u and M_u are given in SP16, for different conditions in SP16.

Design Example 1

Design of column under axial compression and moments.

$$P_u = \gamma_f \cdot P = 1400 \text{ kN}$$

$$M_u = \gamma_f \cdot M = 280 \text{ kN-m}$$

Let us consider as a short column i.e. $\frac{le}{b} < 12$ and only effect of axial compression (Figure 10.6).

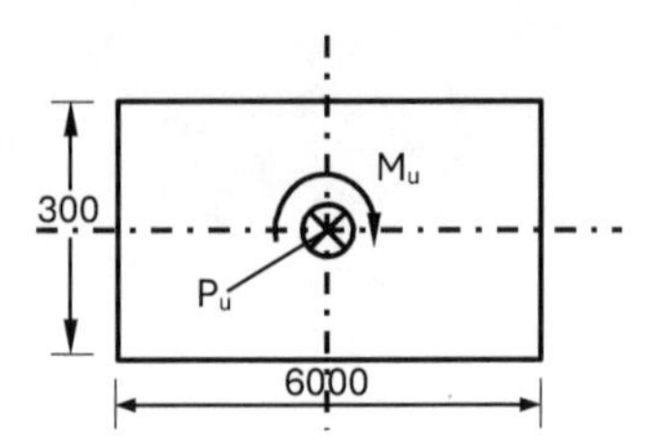

FIGURE 10.6 Column under axial compressive load and uniaxial bending moment.

Therefore, $P_u = 0.4 f_{ck} A_c + 0.67 f_y A_{sc}$

Let us consider M20 concrete & Fe415 steel.

Let us consider $A_{sc} = 1\% A_g = 0.01A_g$, $A_c = A_g - A_{sc} = 0.99A_g$

$$\therefore\ 1400 \times 10^3 = 0.4 \times 20 \times (A_g - 0.01) + 0.67 \times 415 \times (0.01A_g)$$

$$A_g = 129{,}864.114 \text{ mm}^2$$

Let $B = 300$ mm Then $D = \dfrac{129{,}864.114}{300} = 433$ mm

Considering effect of moment, let us consider the column section 300×600 as shown above.

Therefore, $b = 300$ mm & $D = 600$ mm

Let us consider that reinforcement will equally distributed on all four sides. i.e. $= S_1 \approx S_2$

Let, effective cover = 60 mm i.e., $d' = 60$ mm

$$\frac{d'}{D} = \frac{60}{600} = 0.10$$

$$\frac{P_u}{f_{ck} bD} = \frac{1400 \times 10^3}{20 \times 300 \times 600} = 0.39,$$

$$\frac{M_u}{f_{ck} bD^2} = \frac{280 \times 10^6}{20 \times 300 \times 600^2} = 0.13$$

From chart 44 of SP16, $\dfrac{p}{f_{ck}} = 0.11$ i.e., $p = 0.11 \times 20 = 2.2\%$

$$A_{sc} = \frac{2.2}{100} \times 300 \times 600 = 3960 \text{ mm}^2$$

Let us provide eight nos. 25Ø bars. → $A\phi = 3928\text{ mm}^2$ slightly less than the requirement, and it is quite acceptable (Figure 10.7).

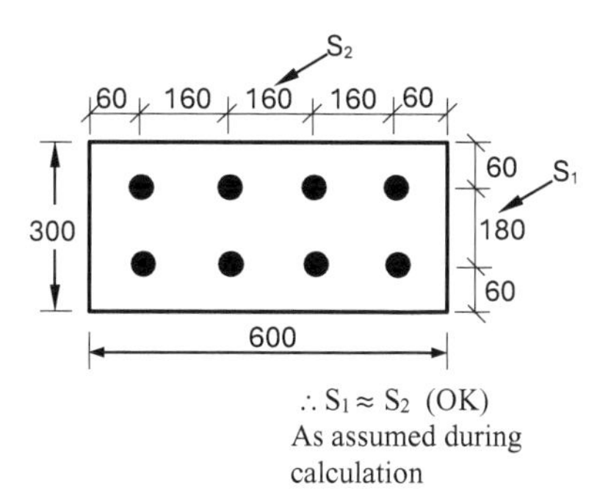

FIGURE 10.7 Column reinforcement.

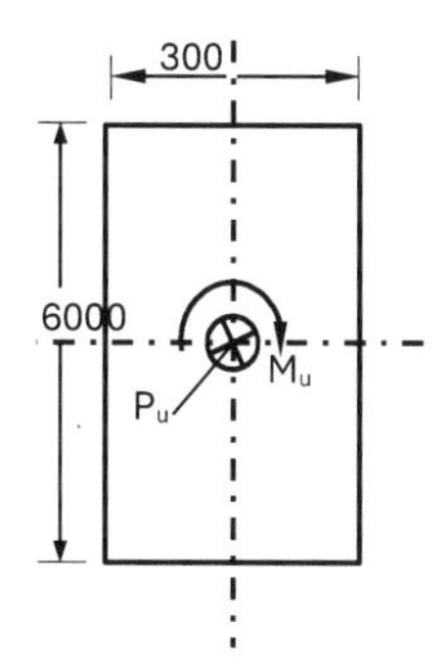

FIGURE 10.8 Column under axial compressive load and uniaxial bending moment.

Design Example 2

Design of column under axial compression and moments (Figure 10.8).

$$P_u = 1400 \text{ kN}, \ M_u = 200 \text{ kN-m}$$

Solution:

Initial choice of the section

Let $A_{sc} = 1\%$ of A_g and considering a short column,

$$\therefore \ 1400 \times 10^3 = 0.4 \times 20 \times (0.99A_g) + 0.67 \times 415 \times (0.01A_g)$$

$$A_g = 130{,}841 \text{ mm}^2$$

Let us consider, 300 × 600 column section as example 1

$$d' = 60 \text{ mm} \qquad \frac{d'}{b} = \frac{60}{300} = 0.20$$

Let us consider, reinforcement distributed equally on two opposite sides.
Referring chart 34 of SP16

$$\frac{P_u}{f_{ck}bD} = \frac{1400 \times 10^3}{20 \times 600 \times 300} = 0.39, \qquad \frac{M_u}{f_{ck}bD^2} = \frac{200 \times 10^6}{20 \times 600 \times 300^2} = 0.185$$

From chart 34 of SP16, $\frac{p}{f_{ck}} = 0.175$ i.e., $p = 0.175 \times 20 = 3.5\%$

$$A_{sc} = \frac{3.5}{100} \times 300 \times 600 = 6300 \text{ mm}^2$$

Let us provide 32Ø bars. $\rightarrow A\phi = \frac{\pi \times 32^2}{4} = 804 \text{ mm}^2$

$$\text{No of bars required} = \frac{6300}{804} = 7.38 \approx 8 \text{ nos.}$$

Reinforcement is equally distributed on two sides along the moment as assumed during calculation (Figure 10.9).

Design Example 3

Design of column under axial compression and biaxial moments.

$$P_u = 1300 \text{ kN}, M_{ux} = 190 \text{ kN-m}, M_{uy} = 110 \text{ kN-m, M25 \& Fe415}$$

Solution:

Let us assume it is a short column.

$$P_u = 0.4 f_{ck} A_c + 0.67 f_y A_{sc}$$

Let us consider $A_{sc} = 1\% A_g = 0.01 A_g$, $A_c = A_g - A_{sc} = 0.99 A_g$

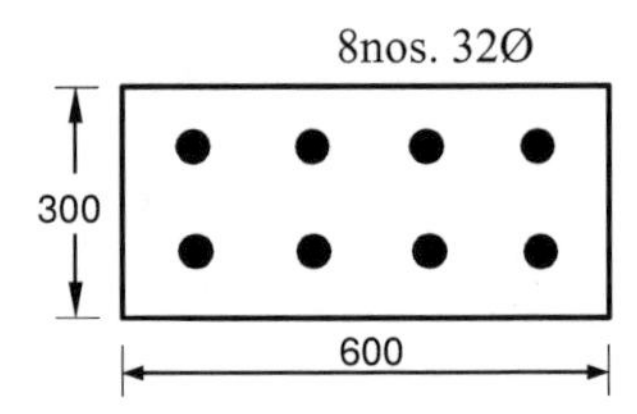

FIGURE 10.9 Column reinforcement.

$$\therefore 1300 \times 10^3 = 0.4 \times 250 \times (0.99A_g) + 0.67 \times 415 \times (0.01A_g)$$
$$= 9.9\ A_g + 2.78A_g = 12.68\ A_g$$

$$A_g = \frac{1300 \times 10^3}{12.68} = 102{,}526\ \text{mm}^2$$

Let us provide a square column $b = D = \sqrt{102{,}526} = 320$ mm

Let us provide 400 mm × 400 mm column section (i.e. $b = D = 400$ mm)

$$\frac{P_u}{f_{ck}bD} = \frac{1300 \times 10^3}{25 \times 400 \times 400} = 0.325, \text{ Let } d' = 60 \text{ mm.}$$

$$\frac{d'}{D} = \frac{60}{400} = 0.15$$

Let us consider, reinforcement distributed equally on all four sides.

Let us provide $p = 3.7\%$, i.e., $\frac{p}{f_{ck}} = \frac{3.7}{25} = 0.15$

Chart 45- sp16 to be used as $\frac{d'}{D} = 0.15$

$$\frac{M_{ux1}}{f_{ck}bD^2} = 0.165 \Rightarrow M_{ux1} = 0.165 \times 25 \times 400 \times 400^2 = 264 \times 10^6 \text{N} - \text{mm}$$

$$= 264 \text{ kN} - \text{mm} > M_{ux} = 190 \text{ kN} - \text{mm}$$
$$M_{uy} = 110 \text{ kN} - \text{mm}$$

So, OK

Biaxial check

Should be $\left(\frac{M_{ux}}{M_{ux1}}\right)^{\alpha_n} + \left(\frac{M_{uy}}{M_{uy1}}\right)^{\alpha_n} \leq 1$

$$A_{sc} = \frac{3.7}{100} \times 400 \times 400 = 5920 \text{ mm}^2$$

$$P_{uz} = 0.45 f_{ck} A_c + 0.75 f_y A_{sc} = 0.45 \times 25 \times (400 \times 400 - 5920) + 0.75 \times 415 \times 5920$$
$$= 1{,}733{,}400 + 1{,}842{,}600 = 3{,}576{,}000 \text{ N} = 3576 \text{ kN.}$$

$$\frac{P_u}{P_{uz}} = \frac{1300}{3576} = 0.36, \text{ therefore } \alpha_n = 1 + \frac{0.36 - 0.2}{0.8 - 0.2} \times (2 - 1) = 1.27$$

$$\left(\frac{M_{ux}}{M_{ux1}}\right)^{\alpha_n}+\left(\frac{M_{uy}}{M_{uy1}}\right)^{\alpha_n}=\left(\frac{190}{264}\right)^{1.27}+\left(\frac{110}{264}\right)^{1.27}=0.986<1 \text{ So, OK}$$

Therefore, $A_{sc} = 3.7\%$ of A_g can be provided.

$$A_{sc} = \frac{3.7}{25}\times 400\times 400 = 5920 \text{ mm}^2$$

Design Example 4

Design of column under axial compression and biaxial moments.

$$P_u = 1300 \text{ kN}, M_{ux} = 190 \text{ kN-m}, M_{uy} = 110 \text{ kN-m}, \text{M25 \& Fe415}$$

Solution:

Let us assume as a short column.

$$P_u = 0.4 f_{ck} A_c + 0.67 f_y A_{sc}$$

Let us consider $A_{sc} = 1\% A_g = 0.01 A_g$, $A_g = 102{,}526 \text{ mm}^2$

Let us provide a rectangular column

$$\frac{D}{b}=\frac{M_{ux}}{M_{uy}}=\frac{190}{110}\Rightarrow D=\frac{190}{110}b=1.73\,b$$

$$A_g = b\times(1.73b)=102{,}526\Rightarrow b=\sqrt{\left(\frac{102{,}526}{1.73}\right)}=243 \text{ mm}$$

$$D = 1.73\times 243 = 420 \text{ mm}$$

Let us provide a column section 300 mm × 525 mm,

$A_g = 300\times 525 = 15{,}750 \text{ mm}^2$ more or less same area of 400 × 400 column.

$$\frac{P_u}{f_{ck}bD}=\frac{1300\times 10^3}{25\times 300\times 525}=0.33, \text{ Let } d' = 50 \text{ mm.}$$

$$\frac{d'}{D}=\frac{50}{525}\approx 0.1$$

Let us provide reinforcement distributed equally on all four sides (Figure 10.10).

Let us provide $p = 2.5\%$, i.e., $\frac{p}{f_{ck}}=\frac{2.5}{25}=0.10$

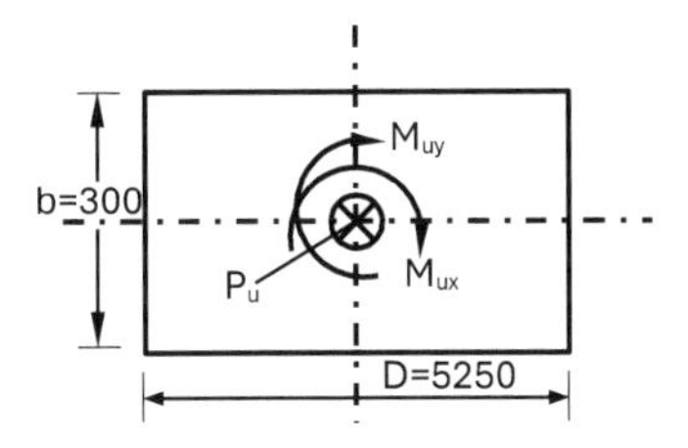

FIGURE 10.10 Column under axial compressive load and biaxial moment.

Chart 44 sp16

$$\frac{M_{\text{ux1}}}{f_{\text{ck}}bD^2} = 0.136 \quad \Rightarrow M_{\text{ux1}} = 0.136 \times 25 \times 300 \times 525^2 = 281{,}137{,}500 \text{ N-mm}$$

$$\frac{d'}{b} = \frac{50}{300} = 0.166 \approx 0.20$$

Chart 46 sp16

$$\frac{M_{\text{uy1}}}{f_{\text{ck}}Db^2} = 0.105 \Rightarrow M_{\text{uy1}} = 0.105 \times 25 \times 525 \times 300^2 = 124{,}031{,}250 \text{ N-mm}$$

$$P_{uz} = 0.45 f_{\text{ck}} A_c + 0.75 f_y A_{\text{sc}}$$

where $A_{\text{sc}} = 2.5\%$ of $(300 \times 525) = 3937.5\,\text{mm}^2, f_y = 415\ \text{N/mm}^2$,

$$f_{\text{ck}} = 25 \text{ N/mm}^2,\ A_c = (300 \times 525 - A_{\text{sc}}) = 153{,}562.5 \text{ mm}^2$$

$$P_{\text{uz}} = 0.45 \times 25 \times 153{,}562.5 + 0.75 \times 415 \times 3937.5 = 2{,}953{,}125 \text{ N-mm}$$

$$\frac{P_u}{P_{\text{uz}}} = \frac{1300 \times 1000}{2953125} = 0.44, \text{ therefore } \alpha_n = 1 + \frac{0.44 - 0.2}{0.8 - 0.2} \times (2 - 1) = 1.4$$

$$\left(\frac{M_{\text{ux}}}{M_{\text{ux1}}}\right)^{\alpha_n} + \left(\frac{M_{\text{uy}}}{M_{\text{uy1}}}\right)^{\alpha_n} = \left(\frac{190}{281}\right)^{1.4} + \left(\frac{110}{124}\right)^{1.4} = 1.42 < 1 \text{ So, OK}$$

$$\text{Therefore, } A_{\text{sc}} = 2.5\% \text{ of } A_g = \frac{2.5}{110} \times 300 \times 525 = 3580 \text{ mm}$$

Less than provided for 400×400
Therefore, it is more economic than 400×400 column section.

DESIGN OF SLENDER COLUMNS USING IS 456

Behavior of Slender Columns

Due to the applied eccentricity e, 'primary moments' $M_{pr} = Pe$ are developed not only at the end sections of the column, but all along the height [Fig. 13.29(b)]. The bending of the column causes it to deflect laterally, thereby introducing additional displacement (load)-dependent eccentricities. If the lateral deflection of the longitudinal axis is denoted as Δ, then the total eccentricity is $e + \Delta$, and the total moment M at any section is given by

$$M = P(e + \Delta)$$

It should be noted that the lateral deflection Δ_{max} is not only due to the curvature produced by the primary moment M_{pr} but also due to the $P-\Delta$ moment. Hence, the variation of M_{max} with P is nonlinear, with M_{max} increasing at a faster rate as P increases. The axial thrust P effectively reduces the flexural stiffness of the column ('beam column'), and, in the case of a very slender column, it may so happen that the flexural stiffness is effectively reduced to zero, resulting in an instability (buckling) failure. On the other hand, in the case of a very short column, the flexural stiffness is so high that the lateral deflection Δ is negligibly small; consequently, the $P-\Delta$ moment is negligible, and the primary moment M_{pr} alone is of significance in such a case.

IS Code Procedures for Design of Slender Columns

In routine design practice, only first-order structural analysis (based on the linear elastic theory and undeflected frame geometry) is performed, as second-order analysis is computationally difficult and laborious. In recognition of this, the Code recommends highly simplified procedures for the design of slender columns, which either attempt to predict the increase in moments (over primary moments) or, equivalently, the reduction in strength, due to, slenderness effects. An alternative method called the 'moment magnification method' is adopted by the ACI and Canadian codes.

The primary moments should not be less than those corresponding to the minimum eccentricities specified by the Code.

Strength Reduction Coefficient Method (Elastic approach)

This is a highly simplified procedure, which is given in the Code for the working stress method of design. According to this procedure (B-3.3 of the Code), the permissible stresses in concrete and steel are reduced by multiplication with a strength reduction coefficient C_r, given by

$$C_r = 1.25 - l_{eff} / 48b$$

where d is the least lateral dimension of the column (or diameter of the core in a spiral column). Alternatively, for more exact calculations,

$$C_r = 1.25 - l_{\text{eff}} \ / 160 \ i_{\text{min}}$$

It is recommended in the Explanatory Handbook to the Code that instead of applying the strength reduction factor C_r to the 'permissible stresses', this factor may be directly applied to the load-carrying capacity estimated for a corresponding short column. Furthermore, it may be noted that although this method has been prescribed for WSM, it can be extended to the limit state method (LSM) for the case of axial loading (without primary bending moments).

Limit State Approach to Design Slender Column

The method prescribed by the Code (Cl. 39.7.1) for slender column design by the limit state method is the 'additional moment method', which is based on Ref. 13.20, 13.21. According to this method, every slender column should be designed for biaxial eccentricities which include the $P-\Delta$ moment ("additional moment") components

$$e_{\text{ax}} \equiv M_{\text{ax}} \ / \ P_u \text{ and } e_{\text{ay}} \equiv M_{\text{ay}} / P_u \ :$$

$$\sim M_{\text{ux}} = P_u \left(e_x + e_{\text{ax}} \right) = M_{\text{ux}} + M_{\text{ax}}$$

$$\sim M_{\text{uy}} = P_u \left(e_y + e_{\text{ay}} \right) = M_{\text{ay}} + M_{\text{ay}}$$

Here, M_{ux} and denote the total design moments; $M \sim M\text{uy}_{\text{ux}}$, M_{uy} denote the primary factored moments (obtained from first-order structural analyses); and M_{ax}, M_{ay} denote the additional moments with reference to bending about the major and minor axes respectively

Accordingly, the following expressions for additional moments M_{ax}, M_{ay} are obtained, as given in the Code (Cl. 39.7.1):

$$M_{\text{ax}} = P_u e_{\text{ax}} = \left[P_u . D \ / \ 2000 \right] \left(l_{ex} / D_x \right)^2$$

$$M_{\text{ay}} = P_u e_{\text{ay}} = \left[P_u . D \ / \ 2000 \right] \left(l_{\text{ey}} / D_y \right)^2$$

where l_{ex} and l_{ey} denote the effective lengths, and D_x and D_y denote the depths of the rectangular column section with respect to bending about the major axis and minor axis, respectively.

Design Example

Determine the reinforcement required for a column which is restrained against sway, with the following data:

Size of column: 450 × 350 mm, Concrete Grade: M 25

Characteristic strength of reinforcement = 415 N/mm^2

Effective length for bending parallel to larger dimension, l_{ex}: 6.5 m

Effective length for bending parallel to shorter dimension, l_{ey}: 5.5 m

Unsupported length: 7.2 m

Factored load: 1800 kN

Factored moment in the direction of larger dimension: 45 kNm at top and 25 kNm at bottom

Factored moment in the direction of shorter dimension: 35 kNm at top and 22 kNm at bottom

The column is bent in double curvature. Reinforcement will be distributed equally on four sides

$$\frac{l_{ex}}{D} = 6500 / 450 = 14.4 > 12$$

$$\frac{l_{ey}}{b} = 5500 / 350 = 15.7 > 12$$

Therefore, the column is slender about both the axes.

From Table-I, SP-16

$$\text{For } \frac{l_{ex}}{D} = 14.4, \frac{e_x}{D} = 0.104$$

$$\text{For } \frac{l_{ey}}{b} = 15.7, \frac{e_y}{b} = 0.124$$

Additional moments:

$$M_{ax} = P_u e_x = 1800 \times 0.104 \times 0.45 = 84.24 \text{ kN-m}$$

$$M_{ay} = P_u e_y = 1800 \times 0.124 \times 0.35 = 78.12 \text{ kN-m}$$

The above moments will have to be reduced in accordance with 39.7.1.1 of the Code; but multiplication factors can be evaluated only if the reinforcement is known.

Trial 1:

Let's assume $p = 3$ (with reinforcement equally on all four sides)

$$A_g = 450 \times 350 = 157{,}500 \text{ mm}^2$$

From chart 63, SP-16; $P_{uz}/A_g = 20.2$ N/mm^2

Therefore $P_{uz} = 20.2 \times 157{,}500/1000 = 3181.5$ kN

Calculation of P_b:
Assuming 25 mm dia bars with 45 mm cover

$$d' / D \text{ (about } xx\text{-axis)} = 57.5 / 450 = 0.13$$

Chart or Table for $d'/d = 0.15$ will be used.

$$d' / D \text{ (about } yy\text{-axis)} = 57.5 / 350 = 0.164$$

Chart or Table for $d'/d = 0.2$ will be used.
From Table 60, SP-16:

$$P_b \text{ (about } xx\text{-axis)} = \left(k_1 + k_2 \frac{p}{f_{ck}}\right) f_{ck} bD = \left(0.196 + 0.203 \times \frac{3}{25}\right) 25$$

$$\times 350 \times 450/1000 = 867.67 \text{ kN}$$

$$P_b \text{ (about } yy\text{-axis)} = \left(k_1 + k_2 \frac{p}{f_{ck}}\right) f_{ck} bD = \left(0.184 + 0.028 \times \frac{3}{25}\right) 25 \times 350$$

$$\times 450 / 1000 = 737.731 \text{ kN}$$

$$k_x = \frac{P_{uz} - P_u}{P_{uz} - P_{bx}} = \frac{3181.5 - 1800}{3181.5 - 867.67} = 0.597$$

$$k_y = \frac{P_{uz} - P_u}{P_{uz} - P_{by}} = \frac{3181.5 - 1800}{3181.5 - 737.73} = 0.565$$

The additional moments calculated earlier, will now be multiplied by the above values of k.

$$M_{ax} = 84.24 \times 0.597 = 50.29 \text{ kN-m}$$

$$M_{ay} = 78.12 \times 0.565 = 44.14 \text{ kN-m}$$

The additional moments due to slenderness effects should be added to the initial moments after modifying the initial moments as follows (see Note 1 under 39.7.1 of the Code):

$$M_{ux} = 0.6 \times 45 - 0.4 \times 25 = 17 \text{ kNm}$$

$$M_{uy} = 0.6 \times 35 - 0.4 \times 22 = 12.2 \text{ kNm}$$

The above actual moments should be compared with those calculated from minimum eccentricity consideration (see 25.4 of the Code) and a greater value is to be taken as the initial moment for adding the additional moments.

$$e_x = 7200 / 500 + 450 / 30 = 29.4 \text{ mm}$$

$$e_y = 7200 / 500 + 350 / 30 = 26.1 \text{ mm}$$

Both e_x and e_y are greater than 20 mm

Moments due to minimum eccentricity:

$$M_{\text{ux}} = 1800 \times 29.4 / 1000 = 53 \text{ kNm} > 17 \text{ kNm}$$

$$M_{\text{uy}} = 1800 \times 26.1 / 1000 = 47 \text{ kNm} > 12.2 \text{ kNm}$$

Total moments for which the column is to be designed are:

$$M_{\text{ux}} = 53 + 50.29 = 103.29 \text{ kNm}$$

$$M_{\text{uy}} = 47 + 44.14 = 91.14 \text{ kNm}$$

The section is to be checked for biaxial bending.

$$P_u / f_{\text{ck}}\, bD = 1800 \times 1000 / (25 \times 350 \times 450) = 0.46$$

$$\frac{p}{f_{\text{ck}}} = \frac{3}{25} = 0.12$$

Referring to chart 45 ($d'/D = 0.15$)

$$M_u / f_{\text{ck}} bD^2 = 0.11$$

$$M_{\text{ux1}} = 0.11 \times 25 \times 350 \times 450^2 / 10^6 = 194.91 \text{ kNm}$$

Referring to chart 46 ($d'/D = 0.2$)

$$M_u / f_{\text{ck}} Db^2 = 0.1$$

$$M_{\text{uy1}} = 0.1 \times 25 \times 450 \times 350^2 / 10^6 = 137.81 \text{ kNm}$$

$$M_{\text{ux}} / M_{\text{ux1}} = 103.29 / 194.91 = 0.53$$

$$M_{\text{uy}} / M_{\text{uy1}} = 91.14 / 137.81 = 0.66$$

$$P_u / P_{uz} = 1800 / 3181.5 = 0.57$$

Referring to chart 64, the maximum allowable value of M_{ux}/M_{ux1} corresponding to the above values of M_{uy}/M_{uy1} and P_u/P_{uz} is 0.64 which is higher than the actual value of 0.53. The assumed reinforcement of 3% is therefore satisfactory.

$$A_s = pbD / 100 = 4725 \text{ mm}^2$$

Concluding Remarks

Practically, all the columns carry axial compressive loads and biaxial moments. Generally, reinforcement is equally distributed on all four sides in practice to avoid any human mistakes though compromising lever arm advantage. Circular columns are inefficient in carrying biaxial moments in comparison to square columns of the same area of cross section. Architectural constraints regarding orientation of columns should not be in the way of economic design of a rectangular section. *L*-shaped has shown the best results and can be considered at corner or any other places avoiding architectural constraints regarding orientation of columns though in many situations square section is preferred but leads to increase in cost (Figure 10.11).

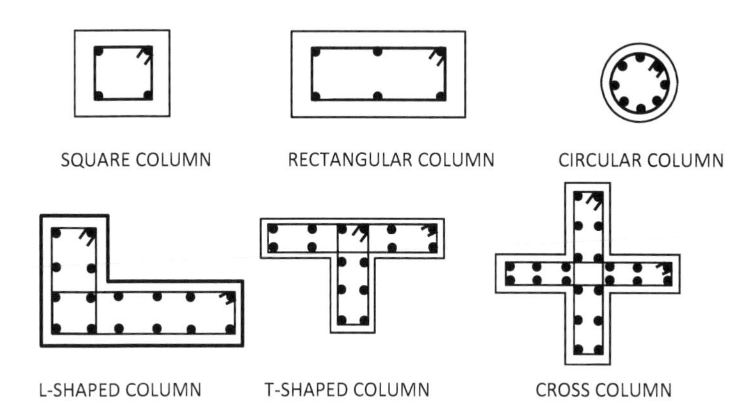

FIGURE 10.11 Different shape of RCC column showing reinforcements.

11 Design of Foundation

PREAMBLE

Foundation may be categorized as **Shallow foundations and Deep foundations.** Shallow foundations may be isolated footing, combined footing, strip, raft, etc. are considered when soil is sufficiently strong within a relatively short depth below the ground surface. Rafts support multiple columns on a large plan area. **Substructure or foundation** effectively supports the superstructure by transmitting the applied load, i.e., vertical and horizontal forces and moments, to the soil below, without exceeding the **safe bearing capacity** of the soil and ensuring that the **settlement** of the structure is within permissible limits. Foundation should provide safety against overturning or sliding and/or possible pullout. The choice of the type of foundation depends on vertical and horizontal forces and moments and the nature of the soil strata at founding level. It is necessary to avoid differential settlements. The isolated footings can have different shapes in plan. Some of the popular shapes of footings are: (i) square, (ii) rectangular, and (iii) circular. The isolated footings essentially consist of bottom slab. These bottom slabs can be flat, stepped or sloping in nature. The bottom of the slab is reinforced with steel mesh to resist the two internal forces namely bending moment and shear force. Combined/strip footing are footings that support the loads from two or more columns, closely spaced. The term 'pedestal' is also used to refer to that portion of a column below ground level where the cross-sectional dimensions are enlarged. The provision of a pedestal is optional, depending on the choice of design engineers. Pedestals are also used to support structural steel columns, and the load transfer between the steel column and the concrete pedestal is achieved generally through gusseted steel base plates with holding down bolts. The plan area of a footing base slab is decided so as to limit the maximum soil bearing pressure induced below the footing to keep within a safe limit. This safe limit to the soil pressure is in geotechnical report. The main considerations in determining the allowable soil pressure, as well as fixing the depth of foundation, are (i) that the soil does not fail under shear and (ii) that the settlements, both overall and differential, are within permissible limit depending on the type of shallow foundation. If there is an overlapping problem considering the demand of depending on column load and moments, if isolated footings are attempted, the pressure bulb below footing may also overlap which leads to huge settlement and tilting. In these cases, it is better to provide combined/strip footing. Generally, combined footing is used for two columns, and the term 'strip footing' is used, if there are more than two columns 'Raft or Mat foundation' is used when there are multiple columns and overlapping problems exist. The geometry of the footing should preferably be so selected as to ensure that the centroid of the footing area coincides with the resultant of the column loads (including consideration of moments if any, at the

DOI: 10.1201/9781003208204-11

column bases). This will result in a uniform distribution of soil pressure, which is desirable in order to avoid possible tilting of the footing. The footing may be rectangular or trapezoidal in shape, depending on the relative magnitudes of loads on the two columns. The base slab of the isolated footing is subjected to two-way bending. It should be checked against one-way and two-way shear (punching). The flexural reinforcement in both the direction is to be designed and provided. One-way shear is critical at a distance, equal to the effective depth of the footing slab, from column/pedestal. One-way shear in footing is considered similar to that of slabs. Considering the footing as a wide beam, the critical section is taken along a vertical plane extending the full width of the footing, located at a distance equal to the effective depth of footing (i.e., considering a dispersion angle of 45°) from the face of the column, pedestal, or wall. Shear forces are also induced in columns, which may result in significant horizontal forces at column bases, under lateral loads. These are resisted by friction between the underside of the footing and the soil below and also by passive resistance of the soil adjoining the sides of the footing, and in some cases, by 'keys' cast integrally with the footing. Deep foundations are provided when adequate safe bearing capacity is available at larger depth below GL.

There are different types of deep foundations, such as pile foundation, pier foundation, and well foundation.

When lateral loads act on a structure, adequate stability of the structure as a whole should be ensured at the foundation level – against the possibilities of ***overturning*** **and** ***sliding*****.** Instability due to overturning may also occur due to eccentric loads, in footings for columns that support cantilevered beams/slabs. The Code (Cl. 20) recommends a factor of safety of *not less than* 1.4 against both sliding and overturning† under the most adverse combination of the applied *characteristic* loads. In cases where dead loads contribute to improved safety, i.e., increased frictional resistance against sliding or increased restoring moment against overturning moment, only 0.9 times the characteristic dead load should be considered. It may be noted that problems of overturning and sliding are relatively rare in reinforced concrete buildings, but are commonly encountered in such structures such as retaining walls, chimneys, industrial sheds, etc. The resistance against sliding is obtained by friction between the concrete footing base and the soil below as well as the passive resistance of the soil in contact with the vertical faces of the footing. Improved resistance against sliding can be obtained by providing a local 'shear key' at the base of the footing, as is sometimes done in foundations for retaining walls. The restoring moment, counterbalancing the overturning moment due to lateral/eccentric loads, is generally derived from the weight of the footing plus backfill. In some cases, this may call for footings with large base area and large depths of foundation. However, in cases where the overturning moment (not due to wind or earthquake) is not reversible, the problem can be more economically solved by suitably making the column/wall eccentric to the center of the footing. Another possibility, relatively rare in practice, is the case of pullout of a foundation supporting a tension member. Such a situation is encountered, for example, in an overhead tank (or silo) structure (supported on multiple columns), subjected to a very severe lateral wind load. Under minimal gravity load conditions (tank empty), the windward columns are likely to be under tension, with the result that the forces acting on these column foundations will tend to pull out the column-footing from the soil. The counteracting forces, comprising the self-weight of the footing and the weight of the overburden, should be sufficiently large to prevent such a 'pullout'. If the tensile forces are excessive, it may be necessary to resort to tension piles for proper anchorage (Figures 11.1–11.13).

DESIGN OF ISOLATED FOOTING

Design Example 1

$$\text{Axial load from column, } P = 1000 \text{ kN}$$

$$\text{Safe Bearing Capacity of soil } = 120 \text{ kN/m}^2$$

$$\therefore \text{ Area of footing required } = \frac{1.05 \times 1000}{120} \text{ m}^2$$

$$= 8.75 \text{ m}^2 \text{ [Considering } 5\% \text{ extra on account of}$$

$$\text{self-weight of footing]}$$

Let us provide a square footing.
Therefore, $B = \sqrt{8.75} \text{ m} = 2.96 \text{ m}$.
Let us provide a square isolated footing of size **3000 mm × 3000 mm**.

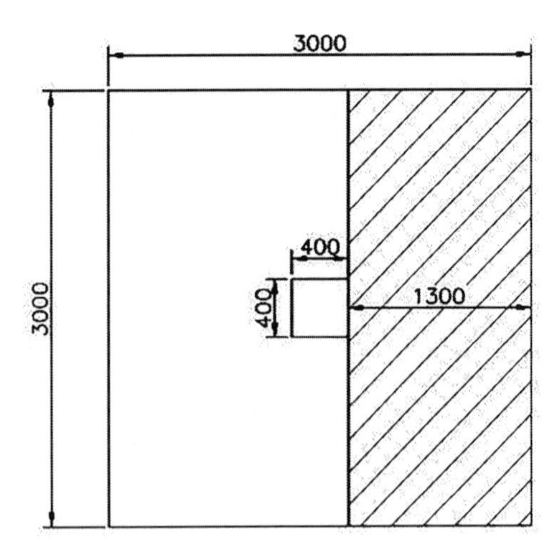

FIGURE 11.1 Plan of isolated footing.

$$\therefore \text{ Net upward soil reaction, } p = \frac{1000}{(3\times 3)} \text{ kN / m}^2 = 111.11 \text{ kN / m}^2$$

$$\text{Bending Moment about the face of the column } = \left(3\times 1.3\times 111.11\times \frac{1.3}{2}\right)\text{kN-m}$$

$$= 281.66 \text{ kN-m}$$

$$\therefore \text{ Bending Moment per metre width, } M = \frac{281.66}{3} \text{ kN-m} = 93.89 \text{ kN-m}$$

$\therefore$ Design Bending Moment per metre width, $M_u = 1.5\times \text{M} = (1.5\times 93.89)$ kN-m $= 140.84$ kN-m

As per Cl. G.1.1.c), Annex-G, Page 96, IS 456:2000, for M 25 grade concrete and Fe 500 grade steel.

$$\text{Moment of resistance factor, } Q_{\text{ub}} = 0.36 f_{\text{ck}} \frac{X_{\text{umax}}}{d}\left(1-0.42\frac{X_{\text{umax}}}{d}\right)$$

$$= 0.36\times 25\times 0.46\times (1-0.42\times 0.46) = 3.34 \text{ N / mm}^2$$

$$\left[\frac{X_{\text{umax}}}{d} = 0.46 \text{ for Fe 500 grade steel, as per Cl. 38.1, Page -70, IS 456:2000}\right]$$

Applying "Limit State Method of design" as per IS 456, we get

$$M_u = Q_{\text{ub}} \times b^2$$

or

$$140.84\times 10^6 = 3.34\times 1000\times d^2$$

or

$$d = 205 \text{ mm.}$$

$$\therefore \text{ Overall depth required, } D_{\text{res}} = \left(205+50+\frac{16}{2}\right)\text{mm}$$

$$= 263 \text{ mm. [Considering 50 mm cover and 16 mm dia. bar]}$$

Let us provide $D = 450$ mm.

$$\therefore \text{ Effective depth, } d = \left(450-50-\frac{16}{2}\right)\text{mm} \doteq 392 \text{ mm.}$$

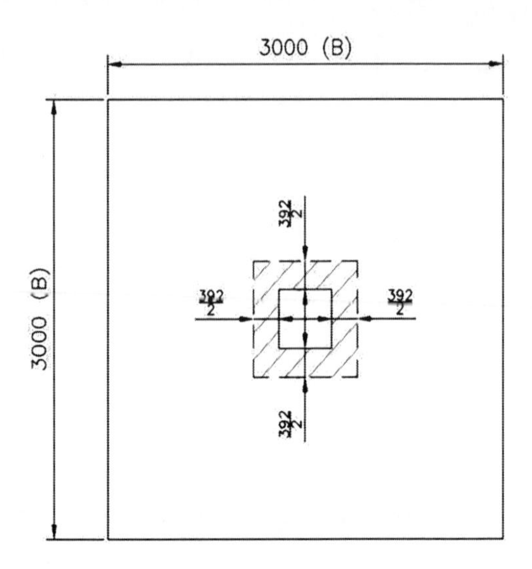

FIGURE 11.2 Critical zone of two-way shear stress.

Check against Two-Way Shear Stress or Punching Shear Stress

$$\text{Width of punching shear zone, } B_p = \left(400 + \frac{392}{2} + \frac{392}{2}\right)\text{mm} = 792 \text{ mm.}$$

$$\text{Two way shear at critical section } \left(\frac{d}{2} \text{ from the face of the column}\right), V_p$$

$$= \left(B \times B - B_p \times B_p\right).p = \{(3 \times 3 - 0.792 \times 0.792) \times 111.11\}\text{kN} = 930.29\text{kN}$$

$$\therefore \text{ Design Two way shear, } V_{\text{pu}} = (1.5 \times 930.29) \text{ kN} = 1395.44 \text{ kN}$$

$$\text{Punching Surface, } S_p = \text{ Perimeter of Punching zone } \times d$$

$$= (4 \times 792 \times 392) \text{ mm}^2$$

$$= 1241856 \text{ mm}^2$$

$$\therefore \text{ Punching Shear Stress developed, } \tau_{\text{vu}} = \frac{V_{\text{pu}}}{Sp} = \frac{1395.44 \times 1000}{1{,}241{,}856} = 1.12 \text{ N/mm}^2$$

As per Cl. 31.6.3.1, Page 58, IS 456:2000,

$$\text{Allowablepunchingshearstress,} \tau_{\text{va}} = K_s.\tau_c\text{,where } K_s = 0.5 + \beta_c = 0.5 + \frac{400}{400} = 1.5$$

But, $K_s \ngtr 1$, So $K_s = 1$

$$\tau_c = 0.25\sqrt{f_{\text{ck}}} = (0.25 \times \sqrt{25})\ \text{N/mm}^2 = 1.25\ \text{N / mm}^2$$

$\therefore \tau_{\text{xa}} = K_s.\tau_c = (1 \times 1.25)\ \text{N/mm}^2 = 1.25\ \text{N/mm}^2 > \tau_{\text{xu}} = 1.12\ \text{N/mm}^2$, **Hence ok**.

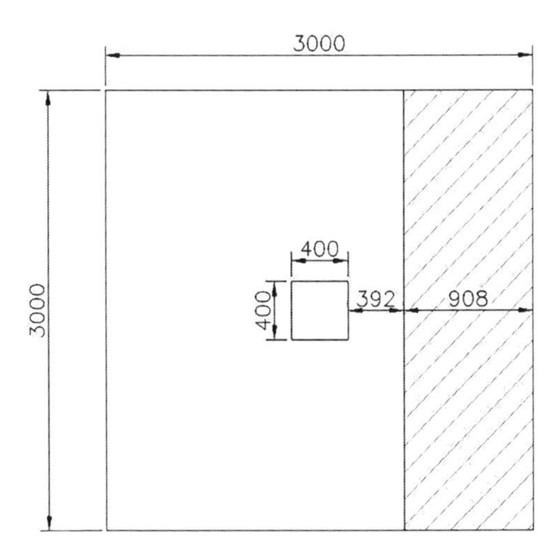

FIGURE 11.3 Critical zone of one-way shear stress.

Check against One-Way Shear Stress

One way Shear force, $V = (0.908 \times 3 \times 111.11)$ kN

$= 302.66$ kN acting on a width of 3.0 m.

$\therefore$ One way Shear Stress developed, $\tau = \dfrac{1.5 \times 302.66 \times 1000}{3000 \times 392}\ \text{N/mm}^2 = 0.39\text{N/mm}^2$

$$\frac{M_u}{bd.d} = \frac{140.84 \times 1{,}000{,}000}{1000 \times 392 \times 392}\ \text{N/mm}^2 = 0.92\,\text{N/mm}^2$$

From Table 3, Page 49, SP 16:1980, for $\dfrac{M_u}{bd.d} = 0.92\,\text{N/mm}^2$ and Fe 500 grade steel,

$$p_t = 0.221\%$$

From Table 19, Page 73, IS456:2000, for $p_t = 0.221\%$ and M 25 grade concrete,

$\tau_c = 0.34\ \text{N/mm}^2 < \tau = 0.39\ \text{N/mm}^2$, **Hence, it is not ok**.

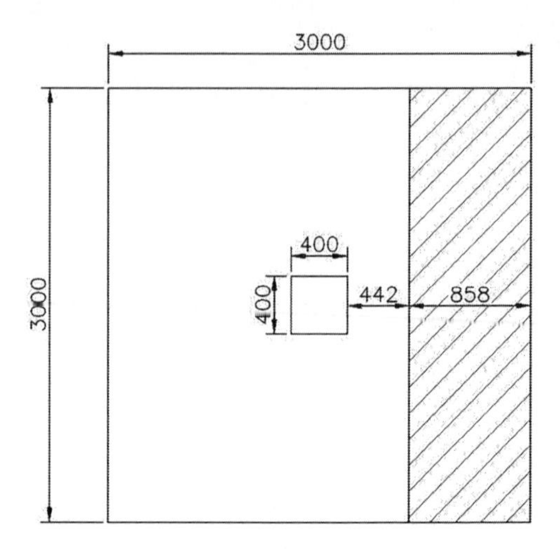

FIGURE 11.4 Load transfer from critical zone of one-way shear stress.

Let us provide overall depth, $D = 500\,\text{mm}$.

$$\therefore \text{ Effective depth, } d = \left(500 - 50 - \frac{16}{2}\right)\text{mm} = 442 \text{ mm.}$$

$$\text{One way Shear force, } V = (0.858 \times 3 \times 111.11)\text{kN}$$

$$= 285.997 \text{ kN acting on a width of 3.0 m.}$$

$$\therefore \text{ One way Shear Stress developed, } \tau = \frac{1.5 \times 285.997 \times 1000}{3000 \times 442} \text{N/mm}^2$$

$$= 0.32 \text{ N/mm}^2$$

As $\tau = 0.32$ N/mm$^2 < \tau_c = 0.34$ N/mm^2, **Hence ok**.

Calculation of Reinforcement

A_{st} required as per $p_t = 0.221\%$ is given by $A_{st} = \frac{p_t.b.d}{100} = \frac{0.221 \times 1000 \times 442}{100} \text{mm}^2$ $= 976.82 \text{ mm}^2$

As per Cl. 26.5.2.1, Page 48, IS 456:2000, Min. reinforcement required = 0.12% of bD

$$= \left(\frac{0.12}{100} \times 1000 \times 500\right)\text{mm}^2 = 600 \text{ mm}^2$$

Let us provide 12 T bars, $\therefore$ Spacing of bars, $S = \left(\dfrac{\dfrac{\pi}{4} \times 12 \times 12}{976.82} \times 1000 \right)$ mm = 115.78 mm.

Let us provide **12 T bars @ 100 mm c/c both ways**.

Design Example 2

Isolated Footing with Pedestal

$$\text{Axial load from column, } P = 1000 \text{ kN}$$

$$\text{Safe Bearing Capacity of soil } = 120 \text{ kN / m}^2$$

$$\therefore \text{ Area of footing required } = \frac{1.05 \times 1000}{120} \text{m}^2 = 8.75 \text{ m}^2$$

[Considering 5% extra on account of self-weight of footing]

Let us provide a square footing.

Therefore, $B = \sqrt{8.75}$ m = 2.96 m

Let us provide a square isolated footing of size **3000 mm × 3000 mm** with a pedestal of size **800 mm × 800 mm × 600 mm**.

$$\therefore \text{ Net upward soil reaction, } p = \frac{1000}{(3 \times 3)} \text{kN / m}^2 = 111.11 \text{ kN / m}^2$$

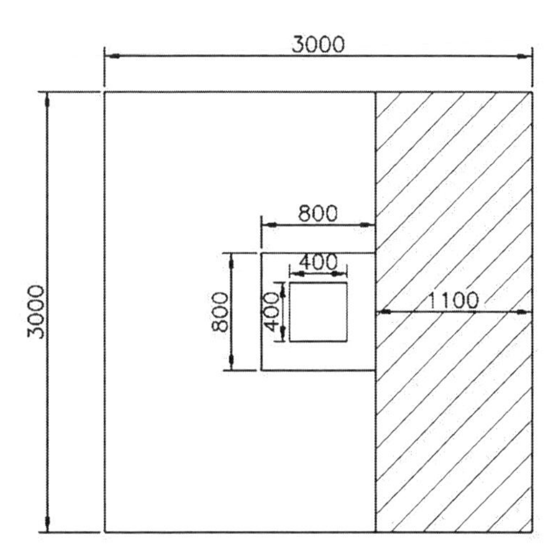

FIGURE 11.5 Load transfer at the face of pedestal.

Bending Moment about the face of the pedestal

$$= \left(3 \times 1.1 \times 111.11 \times \frac{1.1}{2}\right) \text{kN-m} = 201.66 \text{ kN-m}$$

$$\therefore \text{ Bending Moment per metre width, } M = \frac{201.66}{3} \text{ kN-m} = 67.22 \text{ kN-m}$$

$$\therefore \text{ Design Bending Moment per metre width, } M_u = 1.5 \times M$$

$$= (1.5 \times 67.22) \text{ kN-m} = 100.83 \text{ kN-m}$$

As per Cl. G.1.1.c), Annex-G, Page 96, IS 456:2000, for M 25 grade concrete and Fe 500 grade steel.

$$\text{Moment of resistance factor, } Q_{\text{ub}} = 0.36 f_{\text{ck}} \frac{X_{\text{umax}}}{d}\left(1 - 0.42 \frac{X_{\text{umax}}}{d}\right)$$

$$= 0.36 \times 25 \times 0.46 \times (1 - 0.42 \times 0.46) = 3.34 \text{N/mm}^2$$

$$\left[\frac{X_{\text{umax}}}{d} = 0.46 \text{ for Fe 500 grade steel, as per Cl. 38.1, Page-70, IS 456:2000}\right]$$

Applying “Limit State Method of design” as per IS 456, we get

$$M_u = Q_{\text{ub}} \times bd^2$$

or

$$100.83 \times 10^6 = 3.34 \times 1000 \times d^2 \text{ or, } \quad d = 174 \text{ mm}$$

$$\therefore \text{ Overall depth required, } D_{\text{req}} = \left(174 + 50 + \frac{16}{7}\right) \text{mm} = 232 \text{mm.}$$

[Considering 50 mm cover and 16 mm dia. bar]

Let us provide $D = 350$ mm.

$$\therefore \text{ Effective depth, } d = \left(350 - 50 - \frac{16}{2}\right) \text{mm} = 292 \text{ mm.}$$

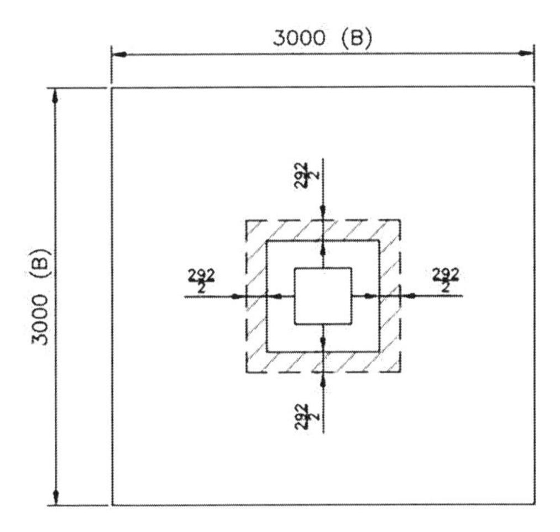

FIGURE 11.6 Critical zone of two-way shear stress with pedestal.

Check against Two-Way Shear Stress or Punching Shear Stress

Width of punching shear zone, $B_p = \left(800 + \frac{292}{2} + \frac{292}{2}\right)$ mm = 1092 mm.

Two way shear at critical section $\left(\frac{d}{2} \text{ from the face of the pedestal }\right), V_D$

$$=\left(B \times B - B_p \times B_p\right) \times p = \{(3 \times 3 - 1.092 \times 1.092) \times 111.11\} \text{ kN} = 867.50 \text{ kN}$$

$\therefore$ Design Two way shear, $V_{\text{pu}} = (1.5 \times 867.50)\text{kN} = 1301.25 \text{ kN}$

Punching Surface, $S_p =$ Perimeter of Punching zone $\times d$

$= (4 \times 1092 \times 292) \text{ mm}^2 = 1{,}275{,}456 \text{ mm}^2$

$\therefore$ Punching Shear Stress developed, $\tau_{\text{vu}} = \frac{V_{\text{pu}}}{S_p} = \frac{1301.25 \times 1000}{1{,}275{,}456} \text{ N/mm}^2$

$= 1.02 \text{ N/mm}^2$

As per Cl. 31.6.3.1, Page 58, IS 456:2000,

Allowable punching shear stress, $\tau_{\text{va}} = K_s.\tau_c$, where $K_s = 0.5 + \beta_c = 0.5 + \frac{400}{400} = 1.5$

But, $K_s \not> 1$, So $K_s = 1$

$$\tau_c = 0.25\sqrt{f_{\text{ck}}} = (0.25 \times \sqrt{25}) \text{ N/mm}^2 = 1.25 \text{ N/mm}^2$$

$\therefore \tau_{\text{va}} = K_s.\tau_c = (1 \times 1.25) \text{ N/mm}^2 = 1.25 \text{ N / mm}^2 > \tau_{\text{vu}} = 1.02 \text{ N/mm}^2$, **Hence ok**.

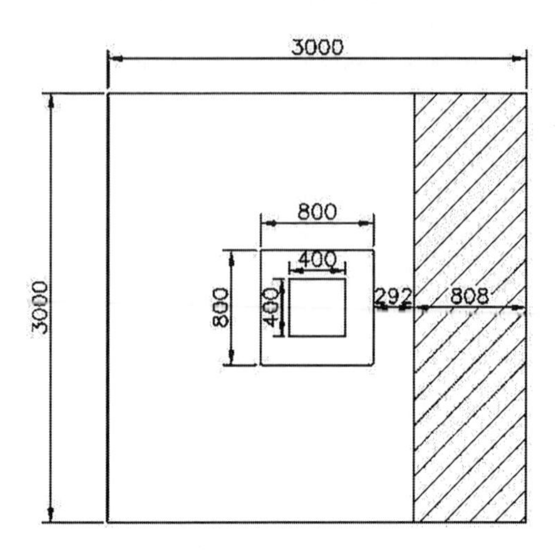

FIGURE 11.7 Load transfer from critical zone of one-way shear stress with pedestal.

Check against One-Way Shear Stress

One way Shear force, $V = (0.808 \times 3 \times 111.11)$ kN

= 269.33 KN acting on a width of 3.0 m.

$$\therefore \text{ One way Shear Stress developed, } \tau = \frac{1.5 \times 269.33 \times 1000}{3000 \times 292} \text{ N/mm}^2 = 0.46 \text{ N/mm}^2$$

$$\frac{M_u}{bd.d} = \frac{100.83 \times 1{,}000{,}000}{1000 \times 292 \times 292} \text{N/mm}^2 = 1.18 \text{ N/mm}^2$$

From Table 3, Page 49, SP 16:1980, for $\frac{M_u}{bd.d} = 1.18 \text{ N/mm}^2$ and Fe 500 grade steel,

$$p_t = 0.288\%$$

From Table 19, Page 73, IS 456:2000, for $p_t = 0.288\%$ and M 25 grade concrete,

$$\tau_c = 0.38 \text{ N / mm}^2 < \tau = 0.46 \text{ N / mm}^2, \textbf{Hence it is not ok}.$$

For $p_t = 0.450\%$, $\tau_c = 0.46$ N/mm² $= \tau = 0.46$ N/mm², **Hence ok**.

Calculation of Reinforcement

A_{st} required as per $p_t = 0.450$ is given by $A_{st} = \frac{p_t.b.d}{100} = \frac{0.450 \times 1000 \times 292}{100} \text{mm}^2$ $= 1314.00 \text{ mm}^2$

As per Cl. 26.5.2.1, Page 48, IS 456:2000, Min. reinforcement required = 0.12% of bD

$$= \left(\frac{0.12}{100} \times 1000 \times 350 \right) \text{mm}^2 = 420 \text{ mm}^2$$

Let us provide 16 T bars.

$$\therefore \text{ Spacing of bars, } S = \left(\frac{\frac{\pi}{4} \times 16 \times 16}{1314.00} \times 1000 \right) \text{mm} = 153 \text{ mm.}$$

Let us provide **16 T bars @ 150 mm C/C both ways**.

DESIGN OF COMBINED FOOTING

Safe Bearing Capacity of soil $= 120 \text{ kN/m}^2$

Total axial load from columns $= (P_1 + P_2) = (1000 + 1250) \text{ kN} = 2250 \text{ kN}$

$$\therefore \text{ Area of footing required } = \frac{1.05 \times 2250}{120} \text{ m}^2 = 19.69 \text{ m}^2$$

[Considering 5% extra on account

of self-weight of footing]

Spacing between columns, $S_c = 5000$ mm.

$$\therefore \text{ The C.G. distance from centre of edge column, } n = \frac{P_2 . S_c}{P_1 + P_2}$$

$$= \frac{1250 \times 5000}{(1000 + 1250)} \text{mm}$$

$= 2778 \text{ mm} = 2.78 \text{ m.}$

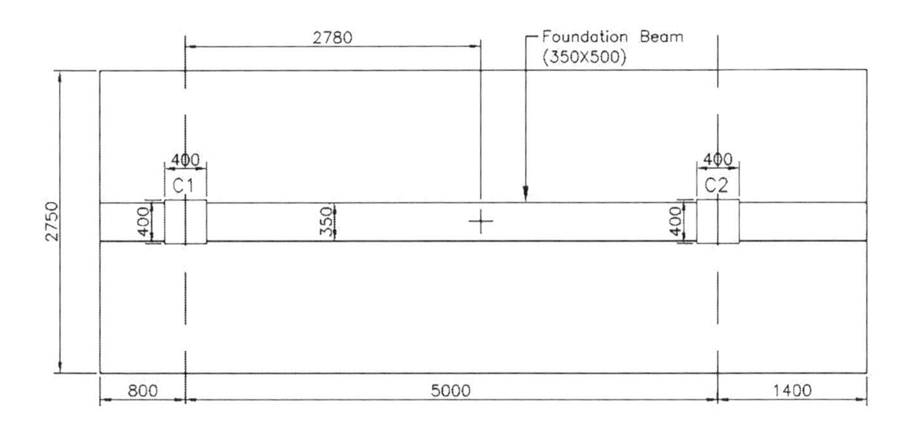

FIGURE 11.8 Plan of combine footing.

Let us assume the edge of the footing is 800 mm away from the center of the edge column.

$$\therefore \text{ Length of the footing, } L = 2(0.8 + 2.78)\text{ m} = 7.16\text{ m} \simeq 7.20\text{ m}$$

$$\therefore \text{ Required width of footing, } B = \frac{19.69}{7.20}\text{ m} = 2.73\text{ m} \simeq 2.75\text{ m}$$

Hence provide a combined footing of size **7.20 m × 2.75 m**.

Bending Moment and Shear Force in the Longitudinal Direction

$$\text{Ultimate loads are : } P_{u1} = (1000 \times 1.5)\text{ kN} = 1500\text{ kN}$$

$$P_{u2} = (1200 \times 1.5)\text{ kN} = 1875\text{ kN}$$

Now, treating the footing as a wide beam of width = 2.75 m, design upward soil pressure $= \frac{(1500 + 1875)}{7.2}$ kN/m = 468.75 kN/m

The bending moment and shear force are calculated as follows:

Shear Force Calculation

$$\text{Shear force at point } A = 0$$

$$\text{Shear force at left of point } B = (468.75 \times 0.8)\text{ KN} = 375\text{ kN}$$

$$\text{Shear force at right of point } B = (468.75 \times 0.8 - 1500)\text{ KN} = -1125\text{ kN}$$

$$\text{Shear force at point } D = 0$$

$$\text{Shear force at right of point } C = -(468.75 \times 1.4)\text{ KN} = -656.25\text{ kN}$$

$$\text{Shear force at left of point } C = (1875 - 468.75 \times 1.4)\text{ KN} = 1218.75\text{ kN}$$

Bending Moment Calculation

$$\text{Bending moment at point } A = 0$$

$$\text{Bending moment at point } B = \left\{468.75 \times \frac{(0.8 \times 0.8)}{2}\right\} = 150\text{ kN-m } \textbf{(Sagging)}$$

$$\text{Bending moment at point } D = 0$$

$$\text{Bending moment at point } C = \left\{468.75 \times \frac{(1.4 \times 1.4)}{2}\right\} = 459.38\text{ kN-m } \textbf{(Sagging)}$$

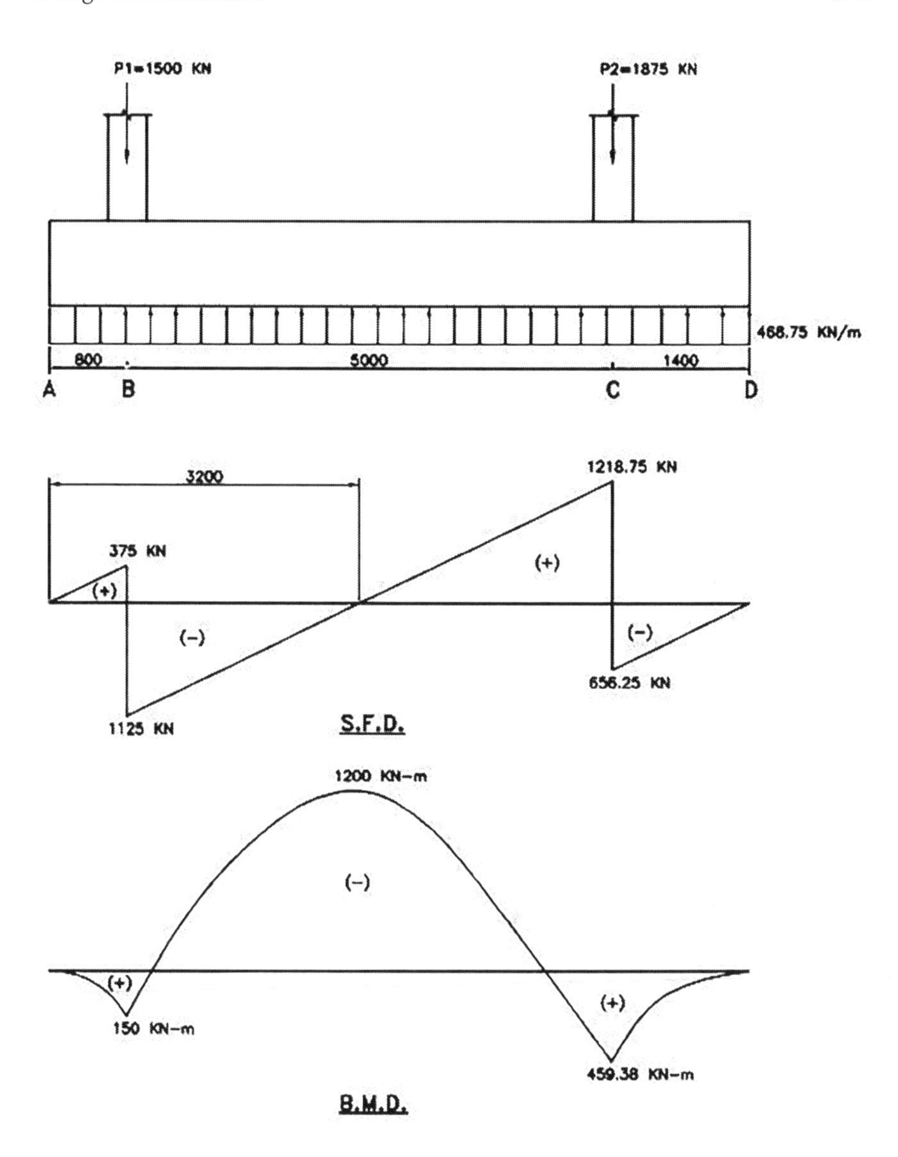

FIGURE 11.9 Bending moment and shear force diagram of combined footing.

Let us assume that the point of zero shear lies at a distance 'x' from end A.

$$\therefore 468.75x - 1500 = 0$$

or

$$x = \frac{1500}{468.75} = 3.2 \text{ m from end } A.$$

$$\therefore \text{ Maximum bending moment } = \left\{468.75 \times \frac{(3.2 \times 3.2)}{2} - 1500 \times (3.2 - 0.8)\right\} \text{kN-m}$$

$$= -1200 \text{ kN-m } \textbf{(Hogging)}$$

Calculation of Main Reinforcement and Shear Reinforcement of Foundation Beam

Let us consider the section of the Foundation beam be 350 mm × 500 mm.

As per Cl. G.1.1.c), Annex-G, Page 96, IS 456:2000, for M 25 grade concrete and Fe 500 grade steel.

$$\text{Moment of resistance factor, } Q_{\text{ub}} = 0.36 f_{\text{cks}} \frac{X_{\text{umax}}}{d}\left(1 - 0.42 \frac{X_{\text{umax}}}{d}\right)$$

$$= 0.36 \times 25 \times 0.46 \times (1 - 0.42 \times 0.46) = 3.34 \text{N/mm}^2$$

$$\left[\frac{X_{\text{umax}}}{d} = 0.46 = \text{ 0.46 for Fe 500 grade steel, as per Cl. 38.1, Page-70, IS 456:2000}\right]$$

Applying "Limit State Method of design" as per IS 456, we get

$$M_u = Q_{\text{ub}} \times bd^2$$

or

$$1200 \times 10^6 = 3.34 \times 2750 \times d^2 \text{ [As the moment is acting over a width of 2.75 m]}$$

or

$$d = 362 \text{ mm.}$$

$\therefore$ Overall depth required, $D_{\text{req}} = \left(362 + 50 + 8 + \frac{20}{2}\right)\text{mm} = 430 \text{ mm} < 500 \text{ mm}$,

Hence ok. [Considering 50 mm cover and 20 mm dia. Bar]

$$\therefore \text{ Provided effective depth, } d = \left(500 - 50 - 8 - \frac{20}{2}\right)\text{mm} = 432 \text{ mm.}$$

$$\frac{M_u}{bd.d} = \frac{1200 \times 1{,}000{,}000}{2750 \times 432 \times 432} \text{N/mm}^2 = 2.34 \text{ N/mm}^2$$

From Table 3, Page 49, SP 16:1980, for $\frac{M_u}{bd.d} = 2.34 \text{ N/mm}^2$ and Fe 500 grade steel,

$$p_t = 0.614\%$$

$\therefore A_{st}$ required for foundation beam as per $p_t = 0.614\%$ is given by $A_{st} = \dfrac{p_t \cdot b \cdot d}{100}$

$$= \frac{0.614 \times 350 \times 432}{100}\text{mm}^2 = 928.37\text{mm}^2$$

Providing, **3 nos. 20 T bar (All through)**.

$$\therefore \text{ Ast provided } = \left(3 \times \frac{\pi}{4} \times 20^2\right)\text{mm}^2 = 942.48\text{ mm}^2 > A_{st} \text{ required}$$

$= 928.37\text{ mm}^2$, **Hence ok**.

As per Cl. 26.5.1.2, Page 47, IS 456:2000,

Maximum $A_{st} = 0.04\ bD$

$$= (0.04 \times 350 \times 500)\text{ mm}^2 = 7000\text{ mm}^2$$

$\therefore$ Maximum $A_{st} = 7000\text{ mm}^2 > A_{st}$ provided $= 942.48\text{ mm}^2$, **Hence ok**.

As per Cl. 26.5.1.1.a), Page 47, IS 456:2000, Minimum A_{st} can be obtained as:

$$\frac{A_s}{bd} = \frac{0.85}{f_y}$$

or

$$\frac{100\ A_s}{bd} = \frac{85}{f_y} = \frac{85}{500} = 0.17\%$$

$$\therefore p_t \text{ provided } = \frac{100 A_s}{bd} = \frac{100 \times 942.48}{350 \times 432} = 0.623\% > \text{ Minimum } p_t$$

$= 0.17\%$, **Hence ok**.

Similarly, for support moment at B, $M_u = 150$ kN-m (over a width of 2.75 m)

$$\frac{M_u}{bd.d} = \frac{150 \times 1{,}000{,}000}{2750 \times 432 \times 432}\text{ N/mm}^2 = 0.29\text{ N/mm}^2$$

From Table 3, Page 49, SP 16:1980, for $\dfrac{M_u}{bd.d} = 0.29\text{ N/mm}^2$ and Fe 500 grade steel,

$p_t = 0.070\% <$ Minimum $p_t = 0.17\%$, **Hence minimum A_{st} is to be provided**.

$$\therefore \text{ Minimum } A_{st} \text{ to be provided } = \frac{p_t.b.d}{100} = \frac{0.17 \times 350 \times 432}{100}\text{mm}^2 = 257.04\text{ mm}^2$$

Providing, 3 **nos. 12 T bar (All through)**.

$$\therefore A_{st} \text{ provided } = \left(3\times\frac{\pi}{4}\times 12^2\right)\text{mm}^2 = 339.29\text{mm}^2 > \text{Minimum } A_{st}$$

$= 257.04\ \text{mm}^2$, **Hence ok**.

Similarly, for support moment at C, $M_u = 459.38$ kN-m (over a width of 2.75 m)

$$\frac{M_u}{bd.d} = \frac{459.38 \times 1{,}000{,}000}{2750 \times 432 \times 432}\text{N/mm}^2 = 0.90\ \text{N/mm}^2$$

From Table 3, Page 49, SP 16:1980, for $\frac{M_u}{bd.d} = 0.90\ \text{N/mm}^2$ and Fe 500 grade steel,

$$p_t = 0.216\% > \text{Minimum } p_t = 0.17\%, \textbf{Hence ok}.$$

We can provide **3 nos. 12 T bar (All through)**.

$$\therefore p_t \text{ provided } = \frac{100As}{bd} = \frac{100\times 339.29}{350\times 432} = 0.224\% > \text{Minimum } p_t$$

$= 0.17\%$, **Hence ok**.

Design against Shear

Maximum design shear force at support $C = 1218.75$ kN per 2.75 m width.

$$\therefore V_u(\text{ For } 350 \text{ mm width }) = \frac{1218.75\times 0.35}{2.75}\text{kN} = 155.11\text{ kN}$$

From Table 61, Page 178, SP 16:1980, for $p_t = 0.224\%$, $\tau_c' = 0.344$ N/mm^2

$$\therefore \text{ Design shear strength developed } = \tau_c'.bd = \frac{0.344\times 350\times 432}{1000}\text{kN} = 52.01\text{ kN}$$

As per Table 20, Page 73, IS 456:2000, maximum shear stress for M 25 grade of concrete, $\tau_{c\,max} = 3.1$ N/mm^2

$$\text{Nominal shear stress developed in the section, } \tau_s = \frac{V_u}{bd} = \frac{155.11\times 1000}{350\times 432}\text{N/mm}^2$$

$= 1.03\ \text{N/mm}^2$

As $\tau_c = 1.03$ N/mm^2 $< \tau c_{max} = 3.1$ N/mm^2, the section is sufficient and need not to be re-designed.

Shear to be carried by shear reinforcement, $V_{us} = V_u - \tau_c' \cdot bd = (155.11 - 52.01)$ kN $= 103.10$ kN

Let us provide 8T – 2L stirrups.

As per Clause no. 40.4, Page 73, IS 456:2000,

$$V_{us} = \frac{0.87\ f_y\ A_{sv}\ d}{S_v}$$

or

$$103.10 \times 10^3 = \frac{0.87 \times 500 \times \frac{\pi}{4} \times 8 \times 8 \times 2 \times 432}{S_v} \quad \text{or, } S_v = 183.24 \text{ mm}$$

Let us provide **8T–2L Stirrups @ 175 mm C/C**.

$$\therefore \frac{x_1}{52.01} = \frac{2.4}{143.18} \quad \text{and} \quad \frac{x_2}{52.01} = \frac{2.6}{155.11}$$

$$\text{or, } x_1 = 0.87\text{m} \qquad \text{or, } x_2 = 0.87 \text{ m}$$

Therefore, provide 8T–2L stirrups @ 175 mm C/C from support A to 1.53 m toward mid-span and from support C to 1.73 m toward mid-span.

Minimum shear reinforcement is to be provided at mid-span.

As per Clause no. 26.5.1.6, Page 48, IS 456:2000,

$$\frac{A_{sv}}{b\ S_v} \geq \frac{0.4}{0.87\ f_y}$$

$$\text{or, } \frac{2 \times \frac{\pi}{4} \times 8 \times 8}{350 \times Sv} \geq \frac{0.4}{0.87 \times 500} \quad \text{or, } S_v \leq 312 \text{ mm}$$

But as per Clause no. 26.5.1.4, Page 47, IS 456:2000, maximum spacing of shear reinforcement should be least of the following:

a. $0.75d = (0.75 \times 432)$ mm = 324 mm
b. 300 mm.

Let us provide **8T–2L Stirrups @ 250 mm C/C** as nominal shear reinforcement.

Design of Footing Slab

Width of footing slab on either side of the foundation beam

$$= \frac{(2750 - 350)}{2} = 1200 \text{ mm.}$$

Now, design upward soil pressure $= \frac{(1500+1875)}{(7.2\times 2.75)}\text{kN/m}^2 = 170.45\text{ kN/m}^2$.

Consideration of Bending Moment

Bending Moment about the face of the column $= \left(7.2\times 1.2\times 170.45\times \frac{1.2}{2}\right)$kN-m

$= 883.61$ kN-m

$\therefore$ Design Bending Moment per metre width, $M_u = \frac{883.61}{7.2}$ kN-m $= 122.72$ kN-m

As per Cl. G.1.1.c), Annex-G, Page 96, IS 456:2000, for M 25 grade concrete and Fe 500 grade steel.

$$\text{Moment of resistance factor, } Q_{\text{ub}} = 0.36 f_{\text{ck}} \frac{X_{\text{umax}}}{d}\left(1 - 0.42\frac{X_{\text{umax}}}{d}\right)$$

$$= 0.36 \times 25 \times 0.46 \times (1 - 0.42 \times 0.46)$$

$$= 3.34\text{ N/mm}^2$$

$\left[\frac{X_{\text{umax}}}{d} = 0.46 \text{ for Fe 500 grade steel, as per Cl. 38.1, Page-70, IS 456:2000}\right]$

Applying "Limit State Method of design" as per IS 456, we get

$$M_u = Q_{\text{ub}} \times bd^2$$

$$\text{or, } 122.72\times 10^6 = 3.34\times 1000\times d^2$$

$$\text{or, } d = 192\text{ mm.}$$

$$\therefore \text{ Overall depth required, } D_{\text{req}} = \left(192 + 50 + \frac{16}{2}\right)\text{mm} = 250\text{ mm.}$$

[Considering 50 mm cover and 16 mm dia. bar]

Let us provide, $D = 350$ mm.

$$\therefore \text{ Effective depth, } d = \left(350 - 50 - \frac{16}{2}\right)\text{ mm} = 292\text{ mm.}$$

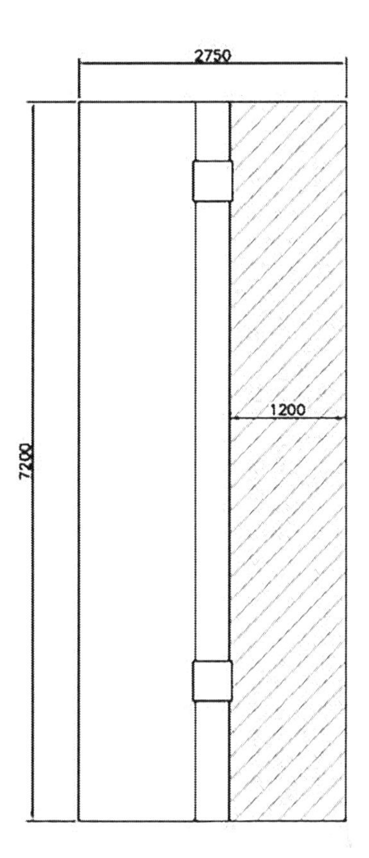

FIGURE 11.10 Load transfer from footing slab to edge of foundation beam.

Check against One-Way Shear Stress

One way Shear force, $V = (0.908 \times 7.2 \times 170.45)$ kN

$= 1114.33$ kN acting on a width of 7.2 m.

$$\therefore \text{ One way Shear Stress developed, } \tau = \frac{1114.33 \times 1000}{7200 \times 292} \text{ N/mm}^2 = 0.53 \text{ N/mm}^2$$

$$\frac{M_u}{bd.d} = \frac{122.72 \times 1{,}000{,}000}{1000 \times 292 \times 292} \text{N/mm}^2 = 1.44 \text{ N/mm}^2$$

From Table 3, Page 49, SP 16:1980, for $\frac{M_u}{bd \cdot d} = 1.44$ N/mm^2 and Fe 500 grade steel,

$$p_t = 0.356\%$$

From Table 61, Page 178, SP 16:1980, for $p_t = 0.356\%$ and M 25 grade concrete,

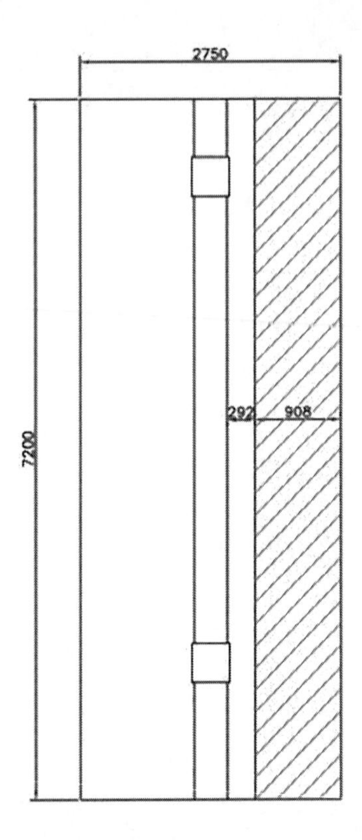

FIGURE 11.11 Load transfer to the critical section of one-way shear.

$$\tau_c' = 0.42 \text{ N/mm}^2 < \tau = 0.53 \text{ N/mm}^2, \textbf{Hence it is not ok}.$$

For $p_t = 0.633\%$, $\tau_c' = 0.54 \text{ N/mm}^2 > \tau = 0.53 \text{ N/mm}^2$, **Hence ok**.

Calculation of Reinforcement

A_{st} required as per $p_t = 0.633$ is given by $A_{st} = \dfrac{p_t.b.d}{100} = \dfrac{0.633 \times 1000 \times 292}{100} \text{ mm}^2$ $= 1848.36 \text{ mm}^2$

As per Cl. 26.5.2.1, Page 48, IS 456:2000, Min. reinforcement required = 0.12% of bD

$$= \left(\frac{0.12}{100} \times 1000 \times 350\right) \text{mm}^2 = 420 \text{ mm}^2$$

Let us provide 16 T bars.

$$\therefore \text{ Spacing of bars, } S = \left(\frac{\frac{\pi}{4} \times 16 \times 16}{1848.36} \times 1000\right) \text{mm} = 108 \text{ mm}.$$

Hence provide **16 T bars @ 100 mm C/C as main reinforcement**.

Let us use 10 T bars as distribution reinforcement.

$$\therefore \text{ Spacing of bars, } S = \left(\frac{\frac{\pi}{4} \times 10 \times 10}{420} \times 1000 \right) \text{mm} = 186 \text{ mm.}$$

Hence provide **10 T bars @ 175 mm C/C as distribution reinforcement**.

Reinforcement Detail

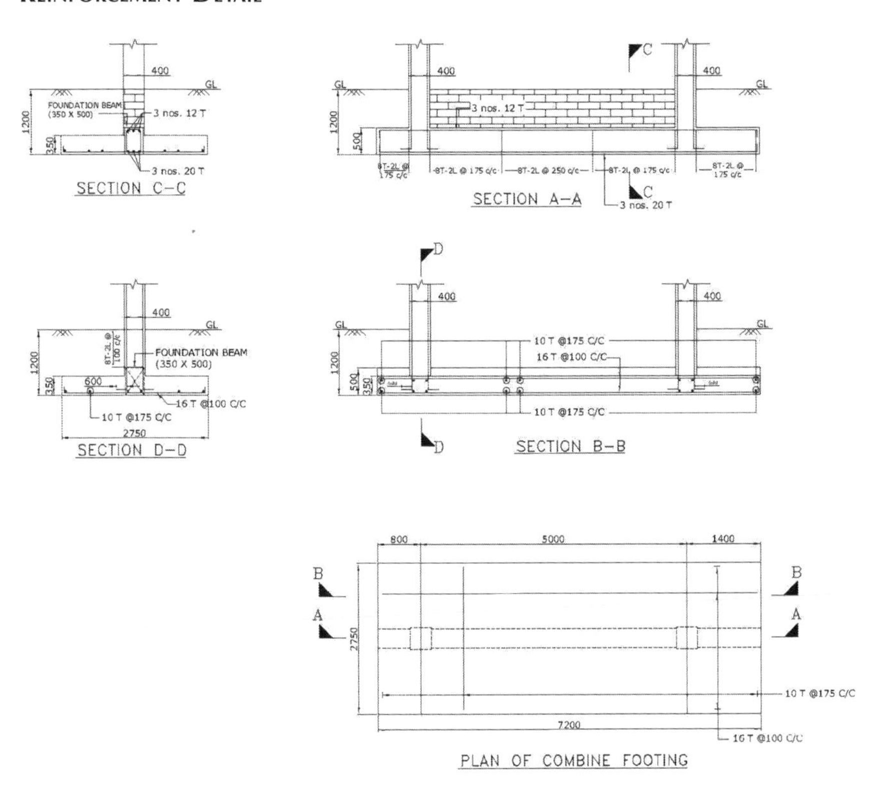

FIGURE 11.12 Reinforcement detaining of combined footing.

Design of a Strip Footing

A single strip footing slab or slab with a foundation beam is provided for more than two columns. It has to be designed as a continuous member. Column may be subjected to axial load together with uniaxial moments. Under these circumstances, generally analysis and design are done using commercial software, under different combination of loads, using limit state approach as recommended in IS456 and IS13920. In this case, four spans are considered.

Beam Marked-1

Beam Length	: 3225 mm
Breadth (B)	: 350 mm
Depth (D)	: 750 mm
Effective Depth (d)	: 695 mm
Design Code	: IS 456 + IS 13920-2016
Beam Type	: Ductile Beam
Grade of Concrete (F_{ck})	: M25 N/sqmm
Grade of Steel	: Fe415 N/sqmm
Clear Cover (C_{min})	: 25 mm
E_s	: 2×10^5 N/sqmm

Flexure Design of Beam *B*1

	Beam Bottom			Beam Top		
	Left	Mid	Right	Left	Mid	Right
M_u (kNm)	121.07	-	283.95	111.75	165.66	30.31
M_{ud} (kNm)	121.07	0	283.95	111.75	165.66	30.31
M_{uLim} (kNm)	584	584	584	584	584	584
R	0.716	0	1.68	0.661	0.98	0.179
P_{tmin} (%)	0.289	0.289	0.289	0.289	0.289	0.289
$P_t$clc (%)	0.289	0.289	0.508	0.289	0.289	0.289
$P_c$clc (%)	0	0	0	0	0	0
P_tPrv (%)	0.413	0.413	0.553	0.413	0.413	0.413
A_{st}Calc (sqmm)	703.37	703.37	1236.44	703.37	703.37	703.37
A_{st}Prv (sqmm)	1005.3	1005.3	1344.6	1005.3	1005.3	1005.3
Reinforcement Provided	5-T16	5-T16	5-T16 3-T12	5-T16	5-T16	5-T16

Shear Design of Beam *B*1

	Left	Mid	Right
PtPrv (%)	0.413	0.553	0.553
V_u (kN)	401.97	–	503.01
V_{ut} (kN)	401.97	–	503.01
M_h (kNm)	234.87		234.87
M_s (kNm)	234.87		306.29
V_{ud} (kN)	401.97	–	503.01
T_v (N/sqmm)	1.65	–	2.07
T_c (N/sqmm)	0	–	0

(Continued)

V_c (kN)	0	–	0
$V_{us} = V_{ud} - V_c$ (kN)	401.97	–	503.01
Legs	4	–	4
Stirrup Rebar	8	–	8
A_{sv} Reqd (sqmm/m)	1601.91	–	2004.57
S_vCalc (mm)	95	–	75
S_vPrv (mm)	95	–	75
A_{sv} Total Prv (sqmm)	2116.63	–	2681.07

Beam Marked-2

Beam Length	: 3400 mm
Breadth (B)	: 350 mm
Depth (D)	: 750 mm
Effective Depth (d)	: 695 mm
Design Code	: IS 456 + IS 13920-2016
Beam Type	: Ductile Beam
Grade of Concrete (F_{ck})	: M25 N/sqmm
Grade of Steel	: Fe415 N/sqmm
Clear Cover (C_{min})	: 25 mm
E_s	: 2×10^5 N/sqmm

Flexure Design Beam $B2$

	Beam Bottom			Beam Top		
	Left	Mid	Right	Left	Mid	Right
M_u (kNm)	276.08	–	264.64	30.82	135.02	36.53
M_{ud} (kNm)	276.08	0	264.64	30.81	135.02	36.53
M_{uLim} (kNm)	584	584	584	584	584	584
R	1.633	0	1.565	0.182	0.799	0.216
P_{tmin} (%)	0.289	0.289	0.289	0.289	0.289	0.289
$P_t$clc (%)	0.493	0.289	0.47	0.289	0.289	0.289
$P_c$clc (%)	0	0	0	0	0	0
P_tPrv (%)	0.506	0.413	0.506	0.413	0.413	0.413
A_{stCalc} (sqmm)	1198.77	703.37	1144.49	703.37	703.37	703.37
A_{st}Prv (sqmm)	1231.5	1005.3	1231.5	1005.3	1005.3	1005.3
Reinforcement Provided	5-T16 2-T12	5-T16	5-T16 2-T12	5-T16	5-T16	5-T16

Shear Design of Beam B_2

	Left	Mid	Right
P_tPrv (%)	0.506	0.506	0.506
V_u (kN)	480.4	–	473.68
T_u (kNm)	0	–	0
V_{Tu} (kN)	0	–	0
V_{Ut} (kN)	480.4	–	473.68
M_h (kNm)	234.87	–	234.87
M_s (kNm)	282.92	–	282.92
V_{ud} (kN)	480.4	–	473.68
T_v (N/sqmm)	1.97	–	1.95
T_c (N/sqmm)	0	–	0
V_c (kN)	0	–	0
$V_{us} = V_{ud} - V_c$ (kN)	480.4	–	473.68
Legs	4	–	4
Stirrup Rebar	8	–	8
A_{sv} Reqd (sqmm/m)	1914.49	–	1887.69
S_{vCalc} (mm)	75	–	75
S_{vPrv} (mm)	75	–	75
A_{sv} Total *Prv* (sqmm)	2681.07	–	2681.07

SFR Design

Beam Width	= 350 mm
Beam Depth	= 750 mm
Web Depth	= 750 <= 750 mm Side Face Reinforcement not required.

Beam B_3

Beam Length	3401 mm
Breadth (*B*)	: 350 mm
Depth (*D*)	: 750 mm
Effective Depth (*d*)	: 695 mm
Design Code	: IS 456 + IS 13920-2016
Beam Type	: Ductile Beam
Grade of Concrete (*F*ck)	: M25 N/sqmm
Grade of Steel	: Fe415 N/sqmm
Clear Cover (C_{min})	: 25 mm
E_s	: 2×10^5 N/sqmm

Flexure Design of B_3

	Beam Bottom			Beam Top		
	Left	Mid	Right	Left	Mid	Right
M_u (kNm)	263.04	–	297.92	32.45	125.13	15.01
M_{ud} (kNm)	263.04	0	297.92	32.45	125.13	15.01
M_{uLim} (kNm)	584	584	584	584	584	584
R	1.556	0	1.762	0.192	0.74	0.089
P_{tmin} (%)	0.289	0.289	0.289	0.289	0.289	0.289
P_{tclc} (%)	0.467	0.289	0.536	0.289	0.289	0.289
P_{cclc} (%)	0	0	0	0	0	0
P_{tPrv} (%)	0.506	0.413	0.553	0.413	0.413	0.413
A_{st}Calc (sqmm)	1136.92	703.37	1303.88	703.37	703.37	703.37
A_{st}Prv (sqmm)	1231.5	1005.3	1344.6	1005.3	1005.3	1005.3
Reinforcement Provided	5-T16 2-T12	5-T16	5-T16 3-T12	5-T16	5-T16	5-T16

Shear Design of B_3

	Left	Mid	Right
P_tPrv (%)	0.506	0.553	0.553
V_u (kN)	466.92	–	487.44
V_{Tu} (kN)			
V_{ut} (kN)	466.92	–	487.44
M_h (kNm)	234.87	–	234.87
M_s (kNm)	282.92	–	306.29
V_{ud} (kN)	466.92	–	487.44
T_v (N/sqmm)	1.92	–	2
T_c (N/sqmm)	0	–	0
V_c (kN)	0	–	0
$V_{us} = V_{ud} - V_c$ (kN)	466.92	–	487.44
Legs	4	–	4
Stirrup Rebar	8	–	8
Asv Reqd (sqmm/m)	1860.77	–	1942.54
S_vCalc (mm)	75	–	75
S_vPrv (mm)	75	–	75
A_{sv} Total Prv (sqmm)	2681.07	–	2681.07

Beam B_4	**B_4**
Beam Length	: 3600 mm
Breadth (B)	: 350 mm
Depth (D)	: 750 mm
Effective Depth (d)	: 695 mm

(*Continued*)

Design Code	: IS 456 + IS 13920-2016
Beam Type	: Ductile Beam
Grade of Concrete (Fck)	: M25 N/sqmm
Grade of Steel	: Fe415 N/sqmm
Clear Cover (Cmin)	: 25 mm
E_s	: 2×10^5 N/sqmm

Flexure Design of B_4

	Beam Bottom			Beam Top		
	Left	**Mid**	**Right**	**Left**	**Mid**	**Right**
M_u (kNm)	337.46	-	157.52	48.37	209.35	138.34
T_u (kNm)	0	0	0	0	0	0
M_{Tu} (kNm)	0	0	0	0	0	0
M_{ud} (kNm)	337.46	0	157.52	48.37	209.35	138.34
M_{uLim} (kNm)	584	584	584	584	584	584
R	1.996	0	0.932	0.286	1.238	0.818
P_{tmin} (%)	0.289	0.289	0.289	0.289	0.289	0.289
$P_t$clc (%)	0.616	0.289	0.289	0.289	0.365	0.289
$P_c$clc (%)	0	0	0	0	0	0
P_tPrv (%)	0.646	0.413	0.413	0.413	0.413	0.413
AstCalc (sqmm)	1498.96	703.37	703.37	703.37	888.42	703.37
AstPrv (sqmm)	1570.8	1005.3	1005.3	1005.3	1005.3	1005.3
Reinforcement Provided	5-T16 5-T12	5-T16	5-T16	5-T16	5-T16	5-T16

Shear Design of Beam $B4$

	Left	**Mid**	**Right**
P_tPrv (%)	0.646	0.413	0.413
V_u (kN)	555.1	–	455.1
V_{ut} (kN)	555.1	–	455.11
M_h (kNm)	234.87	–	234.87
M_s (kNm)	351.7	–	234.87
V_{ud} (kN)	555.1	–	455.11
T_v (N/sqmm)	2.28	–	1.87
T_c (N/sqmm)	0	–	0
V_c (kN)	0	–	0
$V_{us} = V_{ud} - V_c$ (kN)	555.1	–	455.11
Legs	4	–	4
Stirrup Rebar	8	–	8
A_{sv} Reqd (sqmm/m)	2212.17	–	1813.68
S_vCalc (mm)	75	–	95
S_vPrv (mm)	75	–	95
A_{sv} Total Prv (sqmm)	2681.07	–	2116.63

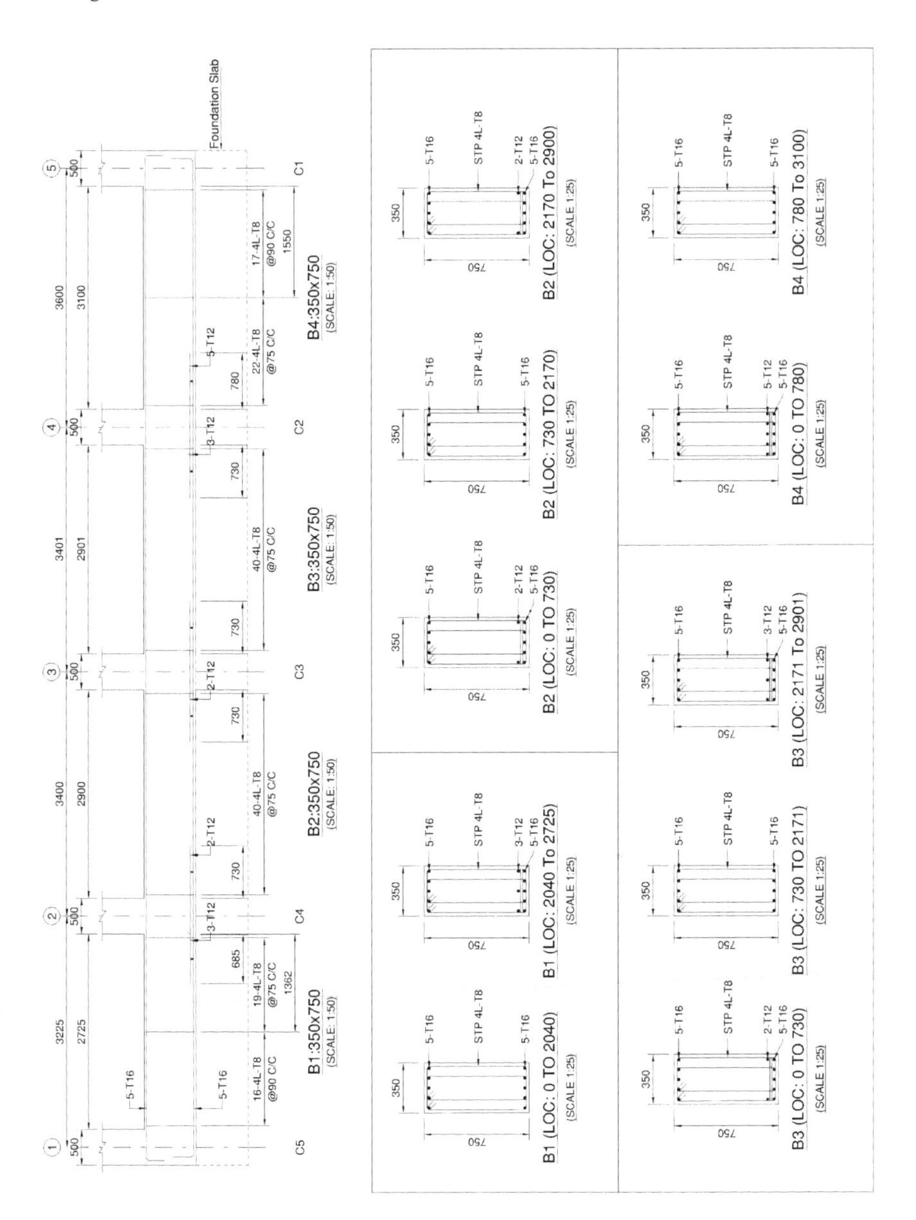

FIGURE 11.13 Reinforcement detailing of strip footings.

Design of Raft Foundation

A slab or slab with a foundation beam is provided for a huge number columns. It has to be designed as a plate/stiffed plate. Column may be subjected to axial load together with biaxial moments. Under these circumstances, generally analysis and design are done using commercial software, under different combination of loads and finally RCC design is made using limit state approach as recommended in IS456 and IS13920 (Figure 11.14).

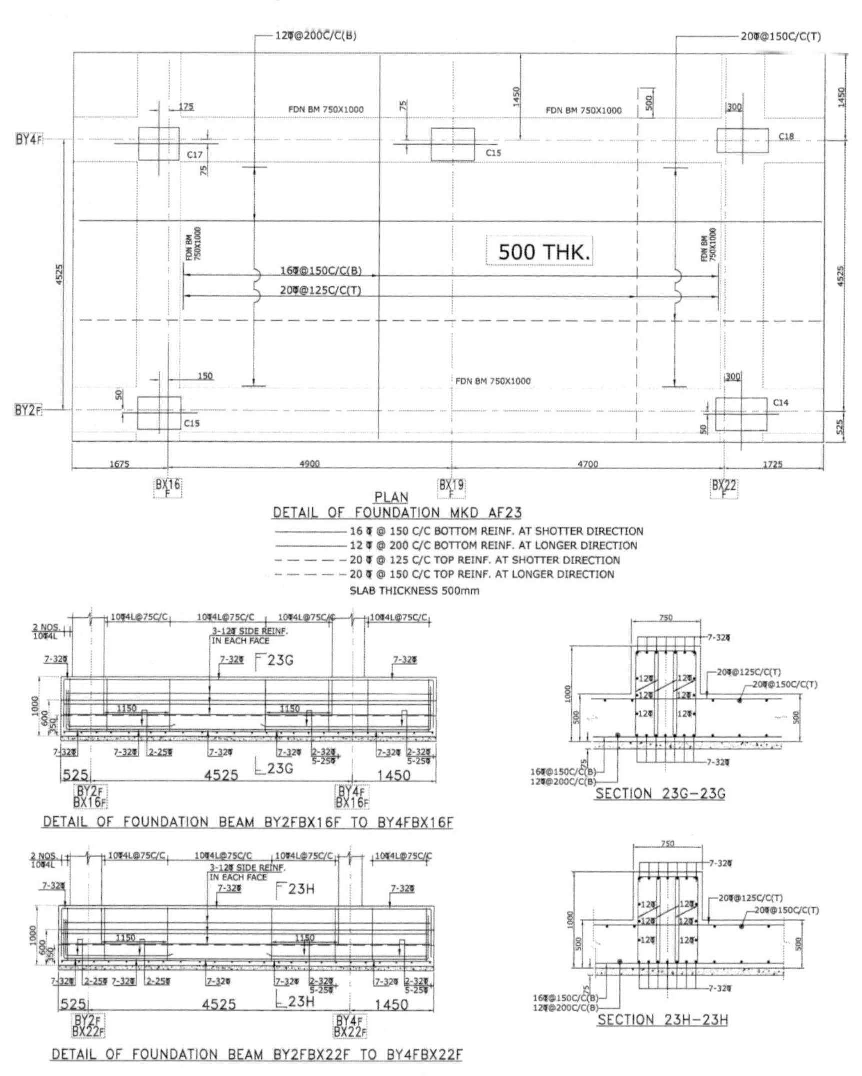

FIGURE 11.14 (a) Reinforcement detailing of a raft foundation. (b) Reinforcement detailing of raft foundation.

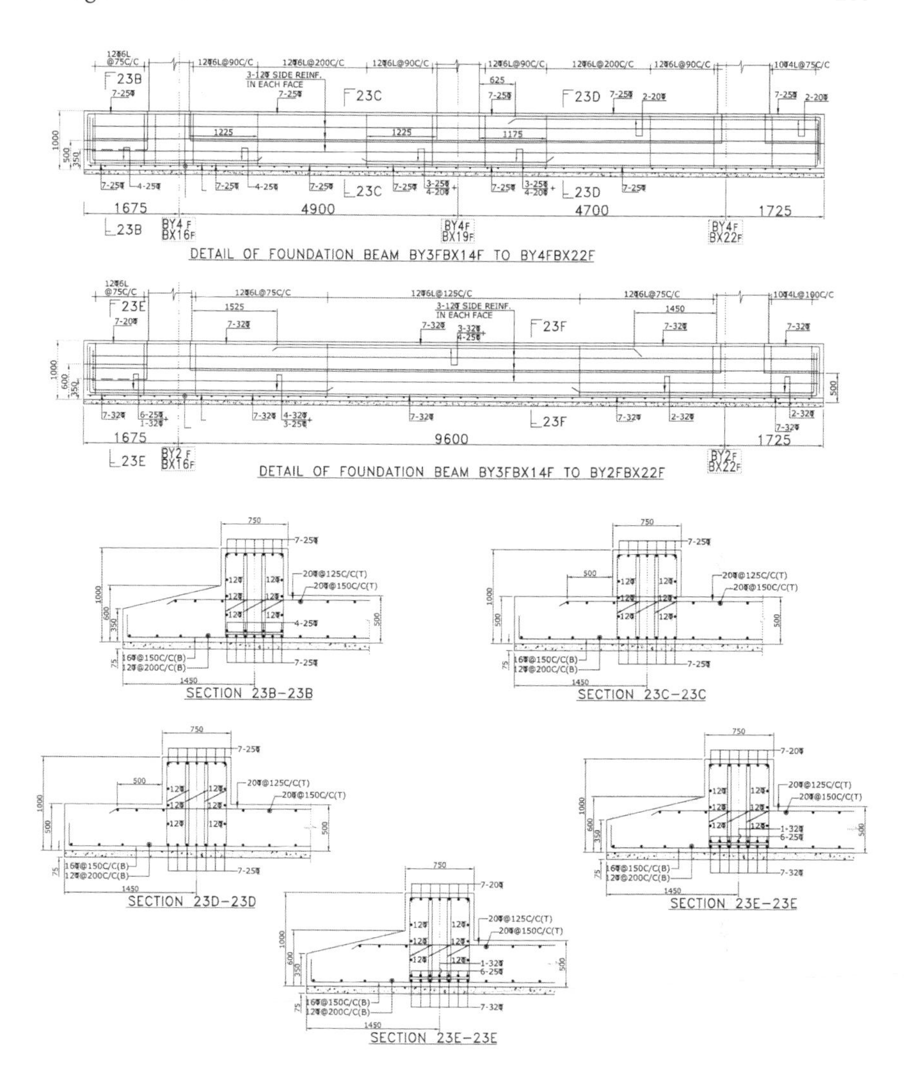

FIGURE 11.14 (a) Reinforcement detailing of a raft foundation. (b) Reinforcement detailing of raft foundation.

DESIGN OF PILE FOUNDATION

A pile cap has been designed under a huge number of load combination. Columns are subjected to axial load together with biaxial moments. Under these circumstances, generally analysis and design are done using commercial software, under different combination of loads and finally RCC design is made using limit state approach as recommended in IS456 and IS13920. An example is presented below (Figure 11.15).

Code References: IS 456

1. P_{tmax}	clause 26.5.1.1
2. P_{tmin}	clause 26.5.2.1
3. P_t	clause 38
4. T_c	clause 40.2.1
5. T_{cmax}	clause 40.2.3
6. A_{sv}Req	clause 40.4
7. Min Shear Reinf	clause 26.5.1.6
8. Max Stirrup Spacing	clause 26.5.1.5
9. One-Way Shear Criteria	clause 34.2.4.2
10. Load transfer	clause 34.4
11. Ptnominal	clause 34.5.2
12. crack width calculation	Annex-F
13. cracking	clause 35.3.2

S 2911 Part 1 Section 2

1. Pile capacity check	: clause 5.10
Design Code	: IS 456 + IS 13920-2016
Column Size	: 300 mm × 500 mm
Concrete Grade	: M25
Steel Grade	: Fe415
Clear Cover	: 50 mm
Top of pile cap below ground	: 3.4 m
Density of Soil	= 18 kN/cum
Founding Depth	= 4 m
Pile Capacity in Compression	= 360 kN
Pile Capacity in Tension	= 250 kN
Pile Capacity in Shear	= 60 kN
No. of Piles	= 4
Pile Diameter	= 450 mm
Pile cap Offset	= 150 mm
Pile Spacing	= 3 times diameter of pile
Pile cap Size	= 2100 × 2100 mm
Pile cap Depth	= 600 mm
Pile cap weight	= 66.15 Kn

Maximum Load on Pile

Critical Load Combination	: DL + LL + EQ
P_{comb}	= 478.96 kN
P_{total}	= 545.11 kN
M_x	= 9.87 kNm
M_y	= 12.61 kNm
P	= 136.28 kNm
P_{mx}	= 25.879 kN
P_{my}	= 4.67 kN
Maximum load on pile	= 166.83 kN
Allowable load on pile	= 360 × 1.25 kN

Maximum Load on Pile Group

Critical Load Combination	: DL + LL
P_{comb}	= 358.39 kN
P_{total}	= P_{comb} + Pile cap Wt. = 424.54 kN
M_x	= 0.13 kNm
M_y	= −3.84 kNm
Maximum load on pile group	= 424.54 kN
Allowable load on pile group	= 4 × 360 + 0.4 × 360 = 1584 kN

Maximum Shear on Pile Group

Critical Load Combination	: DL − EQ
P_{comb}	= 208.42 kN
P_{total}	= P_{comb} + Pile cap Wt. = 274.57 kN
M_x	= −69.43 kNm
M_y	= −20.23 kNm
V_x	= −14.12 kNm
V_y	= −31.02 kNm
Maximum shear on pile group	= sqrt(-14.12^2 + -31.02^2) = 34.09 kN
Shear capacity of pile group	= 4 × 60 × 1 × 1.25 = 300 kN

Check for Uplift Load on Pile

No uplift in any pile

Design for Bending

Bottom Reinforcement along Column-D

Critical Load Combination	: 1.5 DL +1.5 LL
P_{comb}	= 537.59 kN
P_{total}	= P_{comb} + Pile cap Wt. = 636.815 kN
M_x	= 0.19 kNm
M_y	= −5.76 kNm
P_{pile}	= Max Load on pile = 161.41 kN
D_{eff}	= 540 mm
B_{eff}	= 750 mm
D_fCol	= 0.43 m
BM_{ux}	= P_{pile} × DfCol = 68.6 kNm
P_tReq	= 0.24%
A_{st}Req (*BM*)	= 1327 mm²/m
A_{st}Prv	= T12 @ 100 C/C = 1131 mm²/m

Top Reinforcement along Column-D

$$D = 600 \text{ mm}$$

$$A_{st}\text{Req (BM)} = \text{Min } P_t \text{ for Top Reinforcement} \times D \times 1000$$

$$= 0.06\% \times 600 \times 1000 = 360 \text{ mm}^2/\text{m}$$

$$A_{st}Prv = \text{T10 @ 215 C/C} = 365.3 \text{ mm}^2/\text{m}$$

Bottom Reinforcement along Column-B

Critical Load Combination	: 1.5 DL + 1.5 LL
P_{comb}	= 537.59 kN
P_{total}	= P_{comb} + Pile cap Wt.= 636.815 kN
Mx	= 0.19 kNm
M_y	= −5.76 kNm
P_{pile}	= Max Load on pile = 161.41 kN
D_{eff}	= 520 mm
B_{eff}	= 750 mm
D_fCol	= 0.52 m
BM_{uy}	= *P*pile × D_fCol = 84.74 kNm
P_tReq	= 0.2 %
A_{st}Req (BM)	= 1065.06 mm²/m
A_{st}Prv	= T12 @ 105 C/C= 1077.14 mm²/m

Top Reinforcement along Column-B

$$D = 600 \text{ mm}$$

$$A_{st}\text{Req (BM)} = \text{Min } P_t \text{ for Top Reinforcement} \times D \times 1000$$

$$= 0.06\% \times 600 \times 1000 = 360 \text{ mm}^2/\text{m}$$

$$A_{st}\text{Prv} = \text{T10 @ 215 C/C} = 365.3 \text{ mm}^2/\text{m}$$

Design for One-Way Shear: Along Column-*D*

Critical Load Combination	: 1.5 DL + 1.5 EQ
P_{comb}	= 674.33 kN
P_{total}	= P_{comb} + Pile cap Wt. = 773.56 kN
M_x	= 105.09 kNm
M_y	= 19 kNm
P_{pile}	= Max Load on pile = 239.3466 kN

Location of critical section is at $d/2$ from face of column

$$\text{Section Location from column center} = 520 \text{ mm}$$

Data for Piles

Pile No	Load (kN)	Shear (kN)
P_1	147.43	124.5
P_2	225.27	190.23
P_3	161.5	136.38
P_4	239.35	202.11

Design Shear Force (V_u)	= Max. of (Shear due to $P_1 + P_3$, $P_2 + P_4$) = 392.35 kN
D_{eff}	= 540 mm
B_{eff}	= 2100 mm
T_v	= $V_u/(B_{eff} \times D_{eff})$ = 0.35 N/sqmm
T_c	= 0.33 N/sqmm
A_v	= 155 mm
Shear enhancement factor	=2 × *Deff*/*Av* = 6.97
Enhanced shear strength (T_{ce})	= 2.33 N/sqmm
T_v	<Tce

Hence Shear Reinforcement is not required

Along Column-B

Critical Load Combination	: 1.5 DL + 1.5 EQ
P_{comb}	= 674.33 kN
P_{total}	= P_{comb} + Pile cap Wt. = 773.56 kN
M_x	= 105.09 kNm
M_y	= 19 kNm
P_{pile}	= Max Load on pile = 239.3466 kN

Location of critical section is at $d/2$ from face of column

Section Location from column center = 410 mm

Data for Piles

Pile No	Load (kN)	Shear (kN)
P_1	147.43	147.43
P_2	225.27	225.27
P_3	161.5	161.5
P_4	239.35	239.35

Design Shear Force (Vu)	= Max. of (Shear due to $P_1 + P_2$, $P_3 + P_4$) = 400.85 kN
D_{eff}	= 520 mm
B_{eff}	= 2100 mm
T_v	= $V_u/(B_{eff} \times D_{eff})$ = 0.37 N/sqmm
T_c	= 0.33 N/sqmm
Shear enhancement factor	= $2 \times D_{eff}/A_v$ = 3.92
Enhanced shear strength (T_{ce})	= 1.31 N/sqmm

$T_v < T_{ce}$, Hence Shear Reinforcement is not required

Design of Face Reinforcement

$$\text{AsfrReq} = \text{SFR}\ \% \times D \times B_{\text{eff}}\text{sfr} = 0.05 \times 600 \times 500 / 100 = 150 \text{ sqmm}$$

$$\text{Asfr Prv} = \text{2-T10}$$

$$\text{Spacing} = 157 \text{ mm}$$

Check for Load Transfer from Column to Pile Cap

Critical Load Combination	: 1.5 DL + 1.5 EQ
P	= 674.33 kN
A_2	= 0.15 sqm
A_1	= 4.41 sqm
Base Area	= 4.41 sqm
A_1	< Base Area
Modification Factor	= SquareRoot (A_1/A_2) < = 2
SquareRoot ($A1/A2$)	= 5.42
Thus, Modification Factor	= 2
Concrete Bearing Capacity	= 0.45 × F_{ck} × Modification Factor × Column Area
	= 3375 kN
Concrete Bearing Capacity	>P, Hence Safe.

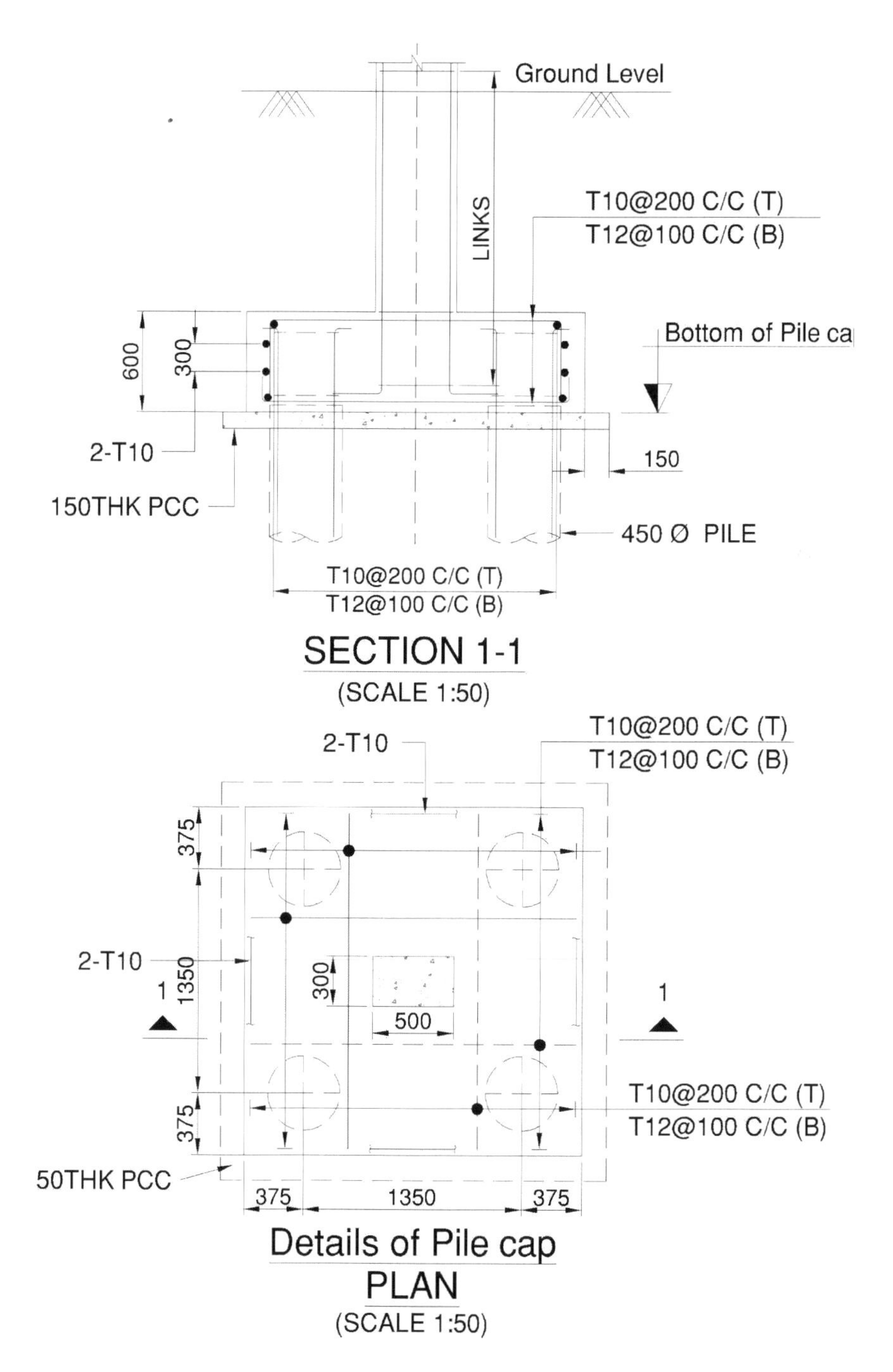

FIGURE 11.15 Reinforcement detailing of pile foundation.

12 Design of Retaining Walls

PREAMBLE

A retaining wall is a structure that serves to prevent, usually earth or rocks on a slope, from falling or collapsing. Some construction projects require removing soil or cutting away segments of a mountain, leaving voids or very vertical faces that may crumble and collapse. This mostly applies to landscapes featuring small hills where these walls act as a necessary barrier to prevent the soil from sliding forward in a landslide. Gravity retaining walls, cantilever retaining walls, counterfort retaining walls, etc. are generally adopted. This wall is connected to the foundation and rests on the slab foundation. These walls may be RCC, prestressed concrete, or precast concrete. RCC retaining wall is used generally up to 10 m height. Stem and base slab are the two main components of the cantilever retaining wall. The concrete units are facing with monolithic perpendicular stems, forming the shape of a "T". The stems internally stabilize the wall, providing pullout resistance against the lateral earth pressure exerted on the back of the facing. Generally, if the height is more than 10 m, counterfort retaining wall is suggested. Counterforts are provided to support base slab and stem to arrive at economical solution (Figures 12.1–12.3).

DIFFERENT TYPES OF RETAINING WALLS

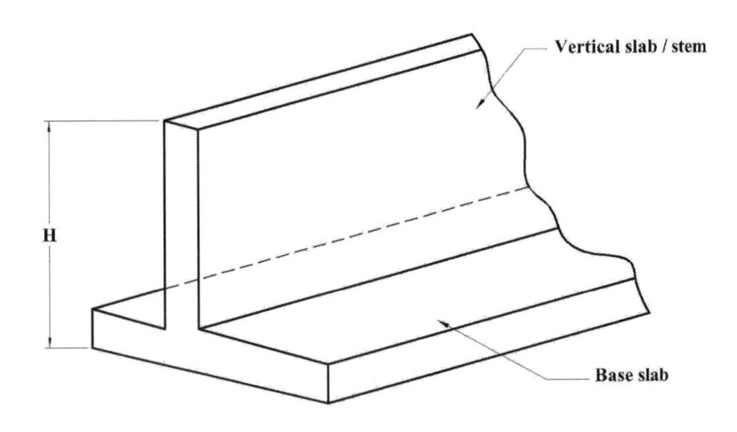

FIGURE 12.1 Cantilever or invert T-type.

DOI: 10.1201/9781003208204-12

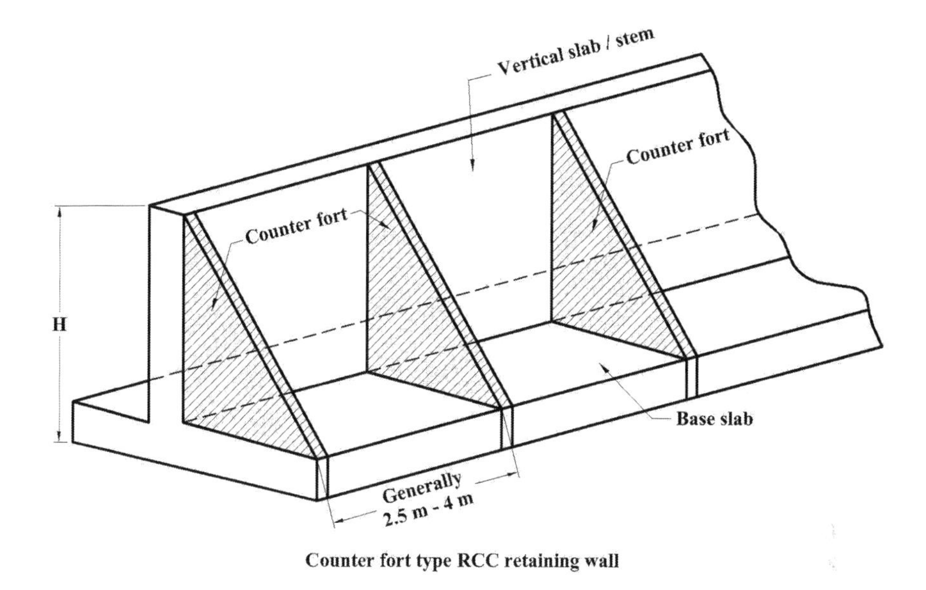

FIGURE 12.2 Counterfort type RCC.

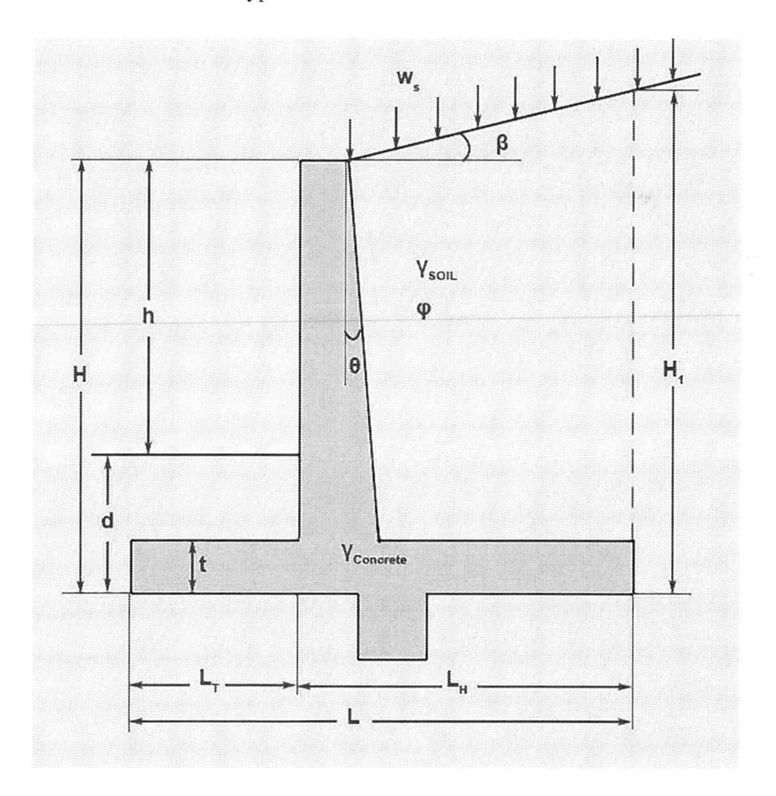

FIGURE 12.3 Cantilever type retaining wall with shear key.

EARTH PRESSURES AND STABILITY REQUIREMENT

For the design of the RCC retaining wall, the following three stability aspects have to be considered (Figure 12.4).

1. **Stability against 1.Overturning Moment**

   Stabilizing moment due to vertical loads ($\mathbf{M_s}$) should be greater than Overturning Moment ($\mathbf{M_o}$) due to horizontal loads and a factor of safety as recommended by code is to be ensured. i.e., $\mathbf{0.9M_s/M_o > 1.4}$ **(Factor of safety against overturning as recommended by code)**
2. **Stability against Sliding**

   Frictional resistance **(μW)** between bottom of base slab and soil should be greater than horizontal load ($\mathbf{P_H}$) (where **W** = Total vertical load and **μ** is the coefficient of friction between soil and concrete) and a factor of safety as recommended by code is to be ensured. The value of **μ** is available in the handbook, i.e., $\mathbf{0.9\mu W / P_H > 1.4}$ **(Factor of safety against sliding as recommended by code)**
3. **Stability against Soil Bearing Capacity**

   Max soil pressure at the base level (P_{max}) < Safe bearing pressure of soil. And also minimum soil pressure at base level (P_{min}) ≥ 0, in order to ensure full contact with soil.

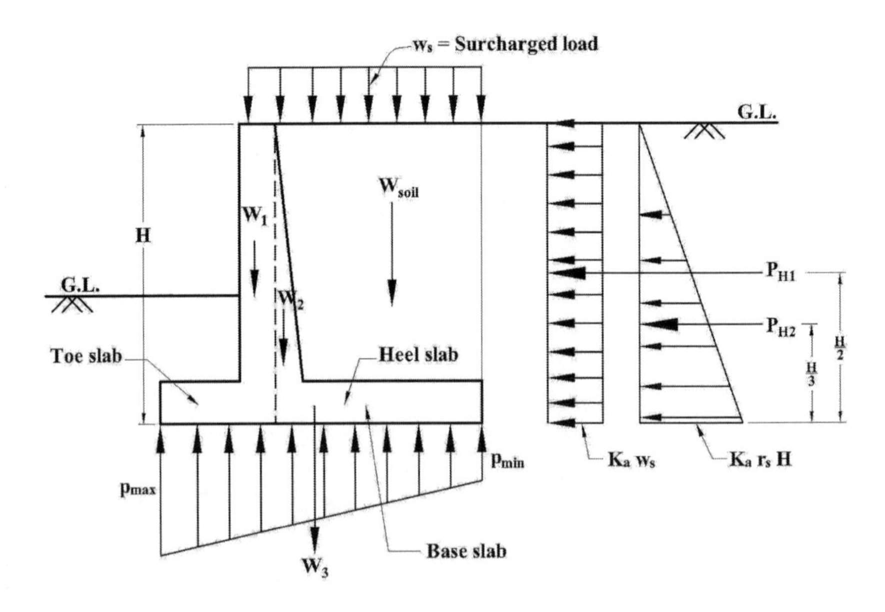

FIGURE 12.4 Earth pressure and soil reaction on retaining wall.

DESIGN OF A CANTILEVER TYPE RETAINING WALL

Basic Steps to Check Stability of a Cantilever Type Retaining Wall

Step I – Selection of Thickness of Stem/Vertical Slab:

Considering a cantilever type retaining wall

Considering, 1 m length of the retaining wall.

Bending moment about the junction of stem and base slab,

$$M = (K_a W_s)(H_1)\left(\frac{H_1}{2}\right) + \frac{1}{2}(K_a \gamma_s H_1)(H_1)\left(\frac{H_1}{3}\right)$$

where H is the total height of the retaining wall, $H = H_1 + D_b$

where H_1 = Height of stem, D_b = Depth of Base slab

Applying limit state design methodology as per IS456

Factored bending moment, $M_u = \gamma_f M$, where γ_f = partial safety factor against load

It is known,

$$M_u = Q_u bd^2 \_\_\_\_ d = \sqrt{\left(\frac{M_u}{Q_u b}\right)}$$

Minimum thickness of the stem (D) should be 250 mm. Stem slab may be tapered.

For counterfort type retaining wall, design moments have to be obtained from plate analysis and stem/vertical slab should be uniform, otherwise procedure of stability check is same as cantilever type retaining wall. However, some geometry of counterforts are to be decided in the beginning (Figures 12.5–12.16).

Step II – Check against Overturning:

$$\text{Factor of safety against overturning} = \frac{0.9M_s}{M_o} > 1.4$$

where M_s = Stabilizing moment $=\sum W X$

M_o = Overturning moment due to soil pressurc $=\sum W Y$

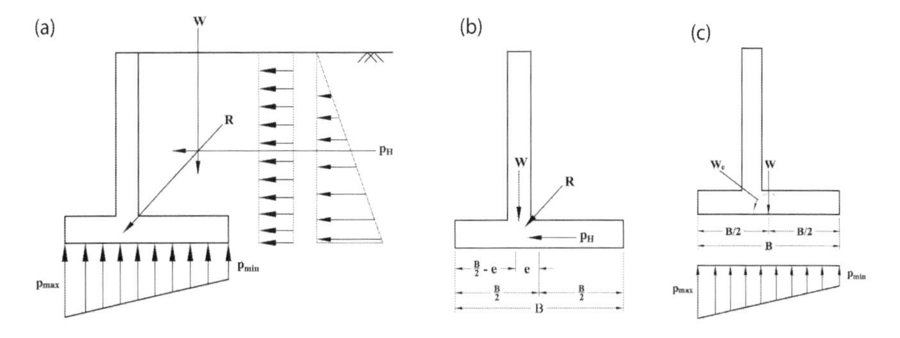

FIGURE 12.5 Earth pressure and soil reaction on retaining wall.

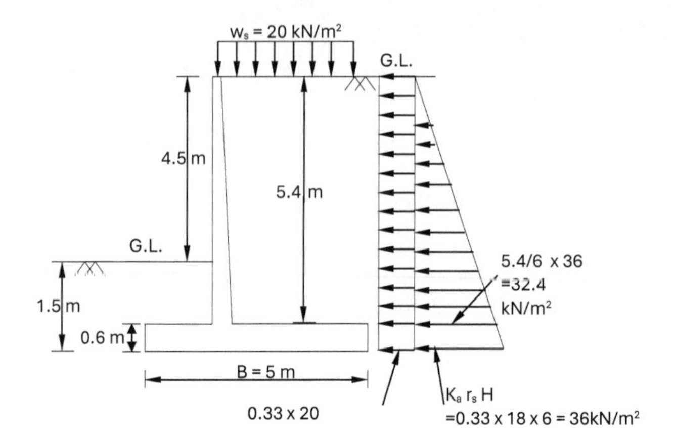

FIGURE 12.6 Earth pressure on retaining wall.

Step III – Check against Sliding:

$$\text{Factor of safety against sliding} = \frac{0.9(\mu W)}{P_H} > 1.4$$

where, $W = \Sigma\, W$ and $P_H = \Sigma\, P_H$

Step IV – Check against Bearing Capacity:

$$\text{Upward Soil reaction } p = \frac{W}{A} \pm \frac{M}{Z},$$

W = Total Vertical load, A = Area of base slab, M = Net moment acting on the system and

Z = Section modulus of base slab

Considering base slab width B and 1 m width design strip

$$p = \frac{W}{(B \times 1)} \pm \frac{M}{\left(B^2 / 6\right)} \quad \text{i.e., } p = \frac{W}{B}\left(1 + \frac{6e}{B}\right), \text{ where } e = M / W,\ e \le \frac{B}{6}$$

Therefore,

Maximum and minimum upward soil reaction

$$p_{\max} = \frac{W}{B}\left(1 + \frac{6e}{B}\right) \text{and } p_{\min} = \frac{W}{B}\left(1 - \frac{6e}{B}\right)$$

It is to be noted that $p_{\max <}$ Safe bearing capacity of soil (SBC) and $p_{\min \ge}$ 0 to ensure the full contact of the base slab with soil

$$\text{i.e } \frac{W}{B}\left(1 - \frac{6e}{B}\right) \ge 0$$

Equating moment about edge of toe slab

$$M = \text{Net moment} = M_S - M_O = W\left(\frac{B}{2} - e\right)$$

Therefore, $e = \dfrac{M_O - M_S}{W} + \dfrac{B}{2}$ which should be $\leq \dfrac{B}{6}$

DESIGN EXAMPLE OF A CANTILEVER TYPE RETAINING WALL

H = Height of the retaining wall (Bottom of base slab to GL) = 6 m,

W_s = Surcharge load on heel slab = 20 kN/m^2 at GL

Unit weight of soil, γ_s = 18 kN/m^3, μ = 0.48, Concrete M20, Steel Fe415,

Angle of friction of cohesion less soil, $\varphi = 30°$, Safe bearing capacity of soil = 150 kN/m^2

Assuming a thickness of the base slab (D_b) = H/10= 6000/10 = 600 mm

Assuming, Width of base (B) = 0.8H to 0.9H = 4.8 m to 5.4 m, Considering, B = 5 m

Angle of friction of cohesion less soil, $\varphi = 30°$,

Hence,

$$K_a = \text{Active earth pressure coefficient } = \frac{1-\sin\varnothing}{1+\sin\varnothing} = \frac{1-\sin 30}{1+\sin 30} = 0.33$$

Consider 1 m length of the retaining wall.

M = Bending moment about the junction of the base slab and vertical slab

$$M = \left(6.67 \times 5.4 \times 1 \times \frac{5.4}{2} + \frac{1}{2} \times 32.4 \times 5.4 \times 1 \times \frac{5.4}{3}\right) \text{kN-m /}$$

metre length of retaining wall

$$= 61.23 + 157.46 = 218.69 \text{ kN-m / meter length}$$

Applying limit state method of design as per IS456, Q_u = 2.76 N/mm^2 for M20 and Fe415 steel. Effective depth,

$$d = \sqrt{\frac{1.5 \times 218.69 \times 10^6}{2.76 \times 1000}} = 345 \text{ mm}$$

D = Overall depth = 345 + 50 + 10 = 405 mm, considering clear cover to tension reinforcement = 50 mm and diameter of tension reinforcement = 20 mm

Providing, D = 450 mm i.e., d = 450 – 50 – 10 = 390 mm

$$\frac{M_u}{bd^2} = \frac{1.5 \times 218.69 \times 10^6}{1000 \times 390^2} = 2.15 \text{ N/mm}^2$$

From SP-16, Table 2, percentage reinforcement required for M20 and Fe415 steel, P_t = 0.6040% ie.

$$A_{st} = \frac{0.604 \times 1000 \times 390}{100} = 2355 \text{ mm}^2/\text{m}$$

Spacing 20 ϕ HYSD Fe 415 bars,

$$S_p = \frac{1000 \times A_{\varnothing}}{A_{st}} = \frac{1000 \times 314}{2355} = 133 \text{ mm}$$

Providing, 20ϕ @125 mm c/c as main tension reinforcement (Table 12.1).

Check against Overturning

$$\text{Factor of safety against overturning} = \frac{0.9M_s}{M_0} = \frac{0.9 \times (1539.37)}{336} = 4.12 > 1.4. \text{ OK}$$

Check against Sliding

$$\text{Factor of Safety against sliding} = \frac{0.9(\mu W)}{P_H} = \frac{0.9 \times (0.48 \times 498.43)}{148}$$

$$= 1.45 > 1.4. \text{OK}$$

TABLE 12.1
Loads and Moments Related to Stability Check of Retaining Wall

Vertical Load	Magnitude (kN)	Distance from 'O' (m)	Moment about 'O' (kN-m)
w_s	$20 \times 3.5 \times 1 = 70$	$\frac{3.5}{2} + 1.5 = 3.25$	$70 \times 3.25 = 227.5$
W_{soil1}	$3.05 \times 5.4 \times 18 = 296.46$	$\frac{3.05}{2} + 0.45 + 1.5 = 3.475$	$296.46 \times 3.475 = 1030.2$
W_{soil2}	$0.5 \times 5.4 \times 0.2 \times 18 = 9.72$	$\frac{0.2}{3} + 0.25 + 1.5 = 1.88$	$9.72 \times 1.88 = 18.3$
W_1	$0.25 \times 5.4 \times 25 = 33.75$	$\frac{0.25}{2} + 1.5 = 1.625$	$33.75 + 1.625 = 54.84$
W_2	$0.5 \times 5.4 \times 0.2 \times 25 = 13.5$	$\frac{0.25}{3} + 1.5 = 1.58$	$13.5 \times 1.58 = 21.33$
W_3	$0.6 \times 5 \times 25 = 75$	$\frac{5}{2} = 2.5$	$75 \times 2.5 = 187.5$
Total	$W = 498.43$ kN		$M_s = 1539.67$ kN-m

Vertical Load	Magnitude (kN)	Distance from 'O' (m)	Moment about 'O' (kN-m)
P_{H1}	$6.67 \times 6 \times 1 = 40$	$\frac{6}{2} = 3$	$40 \times 3 = 120$
P_{H2}	$\frac{1}{2} \times 36 \times 6 \times 1 = 108$	$\frac{6}{2} = 3$	$108 \times 3 = 216$
	$P_H = 148$ kN		$M_0 = 336$ kN-m

Check against Stability

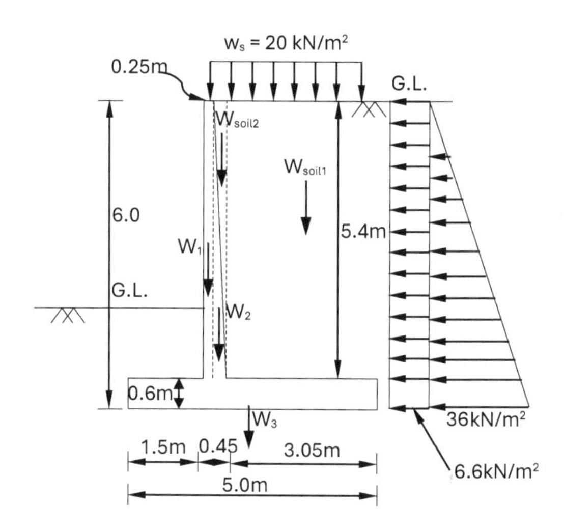

FIGURE 12.7 Earth pressure on retaining wall.

Check against Bearing Capacity

We have

$$\frac{B}{2} - e = \frac{M_s - M_o}{W}$$

i.e.

$$\frac{5}{2} - e = \frac{1539.37 - 336}{498.43}$$

or

$$\frac{5}{2} - e = 2.41 \text{ or,} \quad e = 2.5 - 2.41, \quad e = 0.09 \text{ m}$$

Again,

$$\frac{B}{6} = \frac{5}{6} = 0.83\,\text{m} > e = 0.09\,\text{m, OK}$$

Therefore, it is ensured that the base slab will be in full contact with soil.

$$p_{\max} = \frac{W}{B}\left(1+\frac{6e}{B}\right) = \frac{498.43}{5}\left(1+\frac{6\times 0.09}{5}\right) = 110 \text{ kN/m}^2 < 150 \text{ kN/m}^2$$

$$p_{\min} = \frac{W}{B}\left(1-\frac{6e}{B}\right) = \frac{498.43}{5}\left(1-\frac{6\times 0.09}{5}\right) = 89 \text{ kN/m}^2$$

Design of Heel Slab and Toe Slab

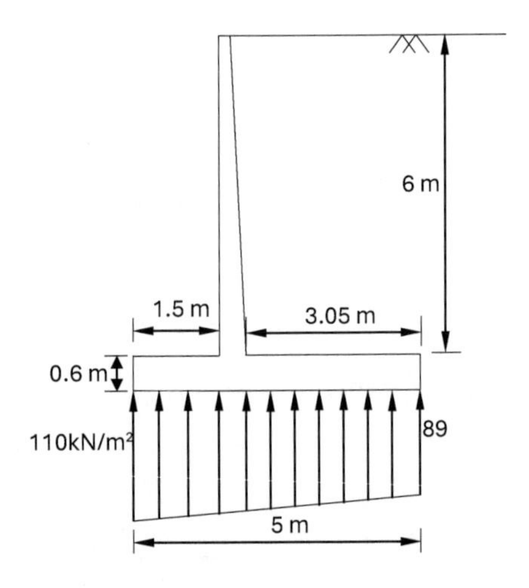

FIGURE 12.8 Soil reaction at the base level.

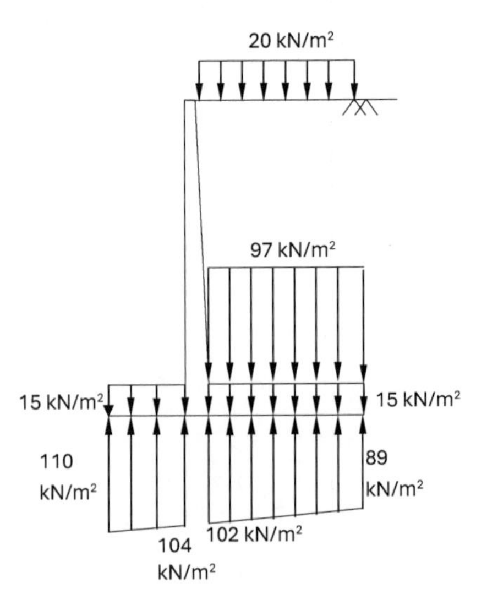

FIGURE 12.9 Load on base slab.

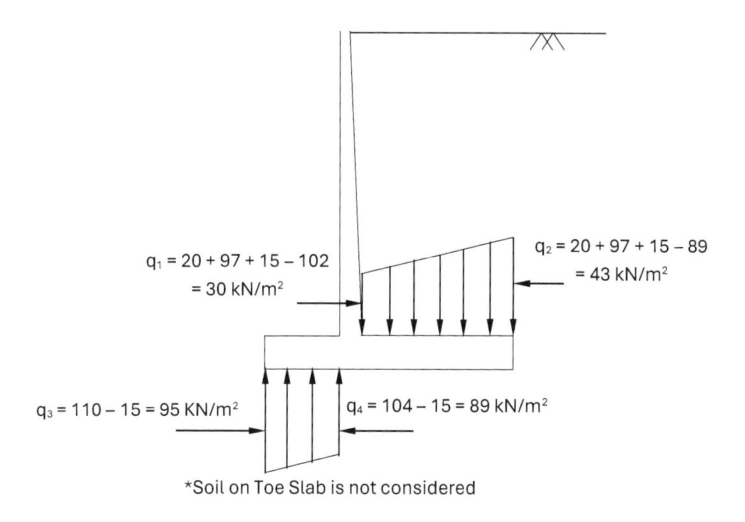

FIGURE 12.10 Net soil pressure on heel slab and toe slab.

Surcharge load on heel slab = 20 kN/m²

Self-weight of soil on heel slab = 5.4 × 18 = 97 kN/m²

Self-weight of heel slab = 0.6 × 25 = 15 kN/m²

$$p_1 = 89 + \frac{3050}{5000} \times (110 - 89) = 102\,\text{kN / m}^2$$

$$p_2 = 89 + \frac{3500}{5000} \times (110 - 89) = 104\,\text{kN / m}^2$$

Design of Heel Slab

Consider 1 m length of the base slab.

Bending moment about the junction of heel slab and vertical slab (Stem)

$$= 30 \times 3.05 \times \frac{3.05}{2} + \frac{1}{2} \times (43 - 30) \times 3.05 \times \left(\frac{2}{3} \times 3.05\right)$$

$$= 139.5 + 40.31 = 179.81\,\text{kN-m / meter length}$$

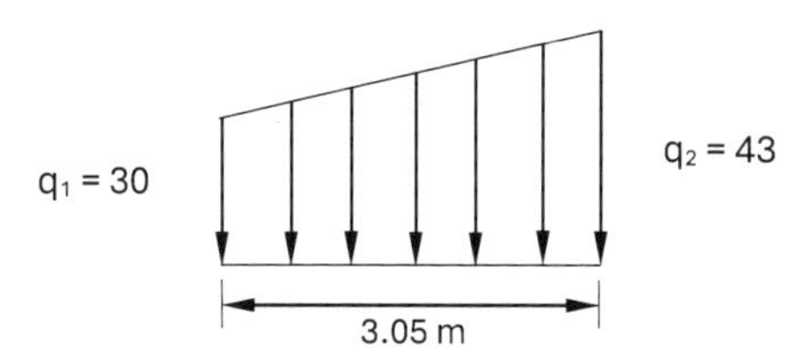

FIGURE 12.11 Net pressure on heel slab.

We know

$$d = \sqrt{\frac{M_u}{Q_{nb} b}} = \sqrt{\frac{1.5 \times 179.81 \times 10^6}{2.76 \times 1000}} = 313\,\text{mm}$$

$$D = 313 + 50 + 10 = 373\,\text{mm}$$

Providing $D = 400\,\text{mm}$ instead of 600 mm assumed initially. $d = 400 - 60 = 340\,\text{mm}$

$$\frac{M_u}{b \cdot d^2} = \frac{1.5 \times 179.84 \times 10^6}{1000 \times 340^2} = 2.33\,\text{kN / mm}^2$$

$p_t = 0.769\%$ from Table 2 of SP 16, $A_{st} = \frac{0.769}{100} \times 1000 \times 340 = 2614\,\text{mm}^2/\text{m}$

$$S = \frac{314 \times 1000}{2614} = 120\,\text{mm c / c},$$

Providing 20 ф @ 100 mm c/c as main tension reinforcement in heel slab.

Check for One-Way Shear of Heel Slab

One-way shear is critical at a distance $d = 340\,\text{mm}$ from the face of the stem

V = Shear force at a distance d = 340 mm from the face of the vertical slab (Stem)

$$= \frac{1}{2} \times (28.5 + 43) \times 3.05 = 109\,\text{kN / meter length}$$

We have, $p_t = 0.769\%$, from Table 19 of IS 456, $\tau_c = 0.56\ \text{N/mm}^2$

$$\tau = \frac{V_u}{bd} = \frac{1.5 \times 109 \times 10^3}{1000 \times 340} = 0.45\,\text{N / mm}^2 < 0.56\,\text{N / mm}^2, \text{OK}$$

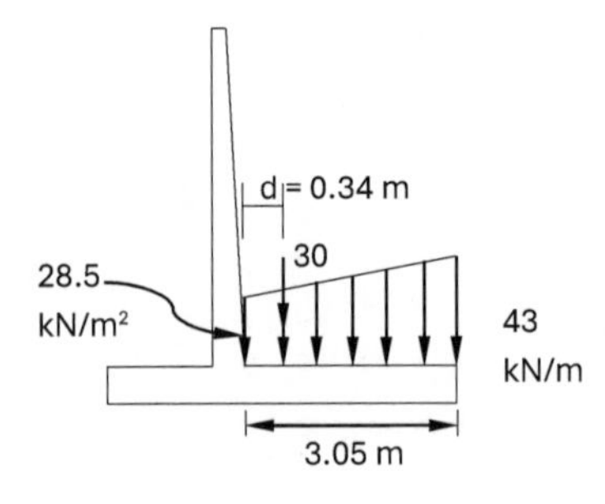

FIGURE 12.12 Soil reaction on heel slab.

Design of Toe Slab

Consider 1 m length of the base slab.

Bending moment about the face of the vertical slab (Stem)

$$= 89 \times 1.5 \times \frac{1.5}{2} + \frac{1}{2} \times (95 - 89) \times 1.5 \times \left(\frac{2}{3} \times 1.5\right)$$

$$= 100 + 4.5 = 104.5\,\text{kN-m / meter length}$$

Already, $D = 400\,\text{mm}$ provided for base slab and toe slab is a part of base slab,

$$d = 400 - 60 = 340\,\text{mm} \qquad \frac{M_u}{b \cdot d^2} = \frac{1.5 \times 104.5 \times 10^6}{1000 \times 340^2} = 1.36\,\text{kN / mm}^2$$

$$p_t = 0.412\% \text{ from Table 2 of SP 16, } A_{st} = \frac{0.412}{100} \times 1000 \times 340 = 1400\,\text{mm}^2\text{ / m}$$

$$\text{Spacing of reinforcement } (S) = \frac{201 \times 1000}{1400} = 143\,\text{mm c / c}$$

Providing 16 ϕ @ 100 mm c/c as main tension reinforcement in toe slab.

Finally, providing 20 ϕ @ 100 mm c/c at top face of the heel slab and 16 ϕ @ 100 mm c/c at the bottom face of the toe slab otherwise minimum reinforcement may be provided. Therefore, 16 ϕ @ 200 mm c/c may be provided at bottom of heel slab and 20 ϕ @ 200 mm c/c may be provided at top of toe slab as indicated in detail of reinforcement shown in Figure 12.15 and 12.16

Distribution Steel

$$\text{For vertical slab stem } 0.12\% \text{ of} \left(\frac{250 + 450}{2}\right) \times 1000 = 420\,\text{mm}^2\text{ / m}$$

Providing 12 ϕ @ 250 mm c/c both way at the outer face, i.e., compression face in both ways and inner face of the vertical slab (stem) along length of the retaining wall.

For base slab, 0.12% of $400 \times 1000 = 480\,\text{mm}^2$ / m.

Providing 12 ϕ @ 200 mm c/c at top and bottom.

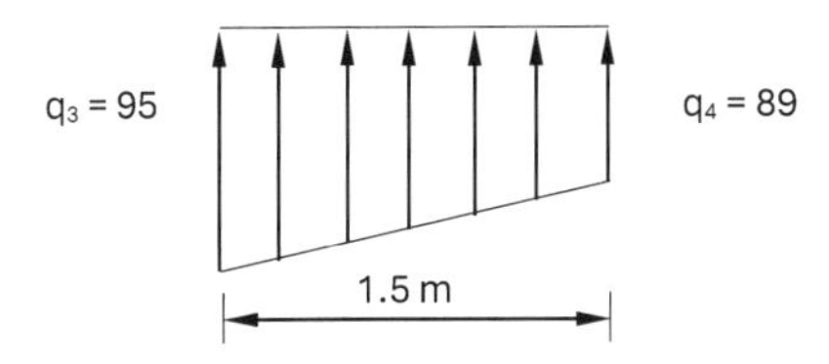

FIGURE 12.13 Net pressure on toe slab.

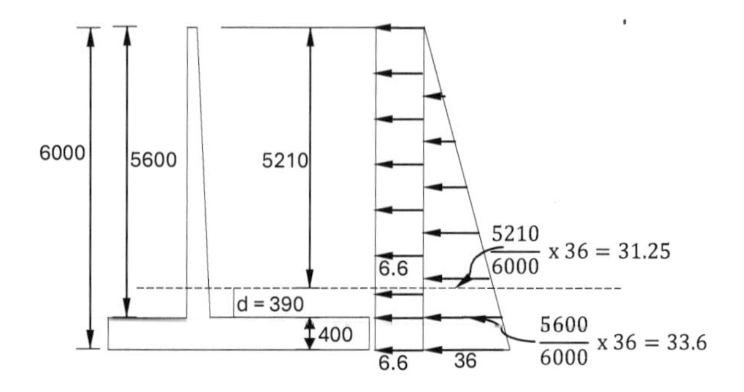

FIGURE 12.14 Pressure diagram for stem/vertical slab.

Check for One-Way Shear (Vertical Slab/Stem)

At a distance $d = 390\,\text{mm}$ from the top face of the base slab
$M =$ Bending moment about the face of the base slab

$$= 6.67 \times 5.6 \times \frac{5.6}{2} + \frac{1}{2} \times 33.6 \times 5.6 \times \frac{5.6}{3} = 104.6 + 175.6$$

$$= 280.2\,\text{kN-m / metre length}$$

$$\frac{M_u}{b \cdot d^2} = \frac{1.5 \times 280.2 \times 10^6}{1000 \times 390^2} = 2.76\,\text{kN / mm}^2 \; p_t = 0.955\% \text{ from Table 2 of SP 16}$$

$$\text{V} = \text{ Design Shear force } = 6.67 \times 5.210 + \frac{1}{2} \times 31.25 \times 5.210$$

$$= 34.75 + 81.4 = 116\,\text{kN / meter length}$$

From Table 19 of IS 456, $\tau_c = 0.61$ N/mm^2

$$\tau = \frac{v_u}{bd} = \frac{1.5 \times 116 \times 10^3}{1000 \times 390} = 0.446\,\text{N / mm}^2 < 0.61\,\text{N / mm}^2, \text{OK}$$

Curtailment of Reinforcement in Vertical Slab (Stem)

Bending moment at a distance 3 m from top of stem slab has been obtained and the tension steel requirement was reduced more than 50%. Therefore, the spacing of tension reinforcement may double beyond 2 m from the top of the stem/vertical slab, considering the development length of around 1 m as indicated in reinforcement drawing (Table 12.2).

TABLE 12.2
Calculation for Cattlemen of Reinforcement in Vertical Slab

Distance from Top (m)	D (mm)	$d=(D-d')$ (mm)	M (kN/m)	$M_u = 1.5M$ (kN/m)	$\frac{M_u}{b \cdot d^2}$	p_t (%)	$\frac{A_{st}}{m}$ (mm²/m)	S (mm)
5.6	450	390	280.2	420.3	2.76	0.955	3725	25ϕ@ 125 c/c
4	391	331	117.36	176.04	1.61	0.498	1649	20ϕ@ 125 c/c
3	356	296	57.0	85.50	0.97	0.286	847	16ϕ@ 125 c/c
2	320	260	21.34	32.01	0.47	0.2	520	16ϕ@ 125 c/c
0	250	–	–	–	–	0.2	520	16ϕ@ 125 c/c

$d' = 60$ mm.

At a particular distance from top spacing requirement will be double i.e., 25 ϕ@ 250 mm c/c will be sufficient.

[a] According to minimum tension steel criteria as per IS 456 for Fe415 HYSD bars = 0.2%.

Reinforcement Detailing

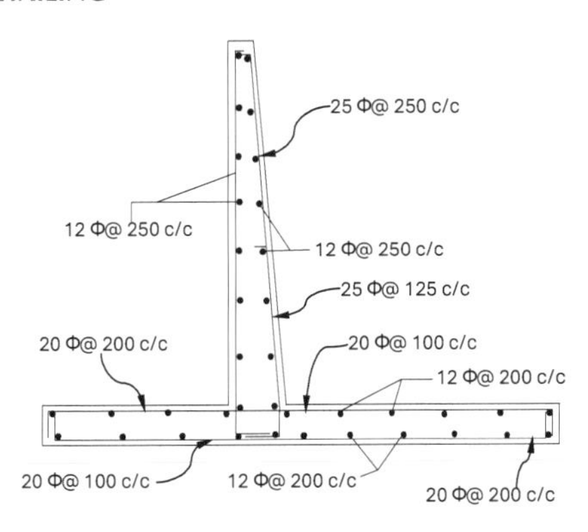

FIGURE 12.15 Reinforcement details of retaining wall.

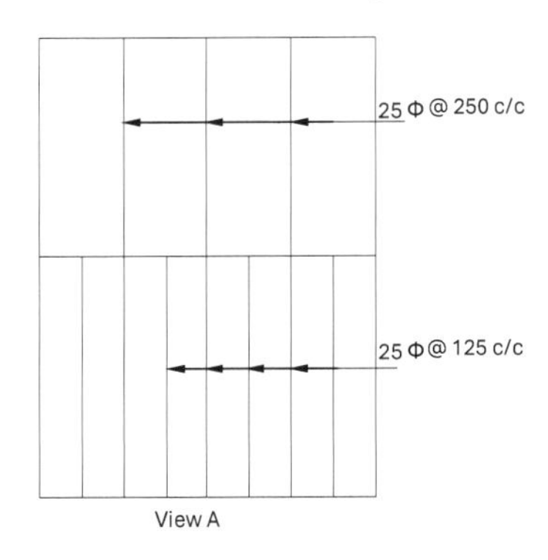

FIGURE 12.16 Vertical reinforcement of stem.

DESIGN EXAMPLE OF COUNTERFORT RETAINING WALLS

Example 1

The top earth retained is horizontal, and soil weights 18 kN/m³ with angle of internal friction $\phi = 30°$. Coefficient of friction between concrete and soil may be taken as 0.48, height of counterfort retaining wall = 8 m. Grade of concrete M20 & Grade of steel Fe415. $Q_u = 2.76$ N/mm² (Figures 12.17 and 12.18).

Solution:

Given data -
SBC = 225 kN/m²;
Unit weight of concrete = 25 kN/m³;
Unit weight of soil = 18 kN/m³.
$f_{ck} = 20$ N/mm²; $f_y = 415$ N/mm²; $\phi = 30°$; $\mu = 0.48$
Height of retaining wall = $H = 8$ m.

$$K_a = \frac{1-1/2}{1+1/2} = 0.333; \quad K_p = \frac{1}{K_a} = 3.$$

$Q_u = 2.76$ for Fe 415 & M20
Let thickness of base slab and steam slab = 0.55 m
Length of the heel slab = 2.3 m
Length of toe slab = 1.4 m
$\therefore H_1 = 8 - 0.55 = 7.45$ m
Let clear spacing of counter fort = 2.5 m & thickness of counterfort = 0.5 m
Hence effective spacing of counterforts = 3.0 m

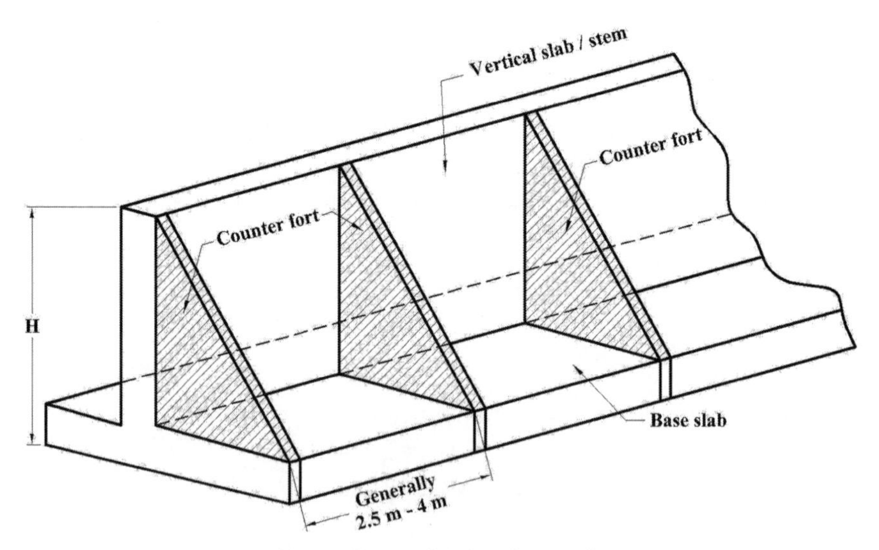

FIGURE 12.17 Counterfort retaining wall.

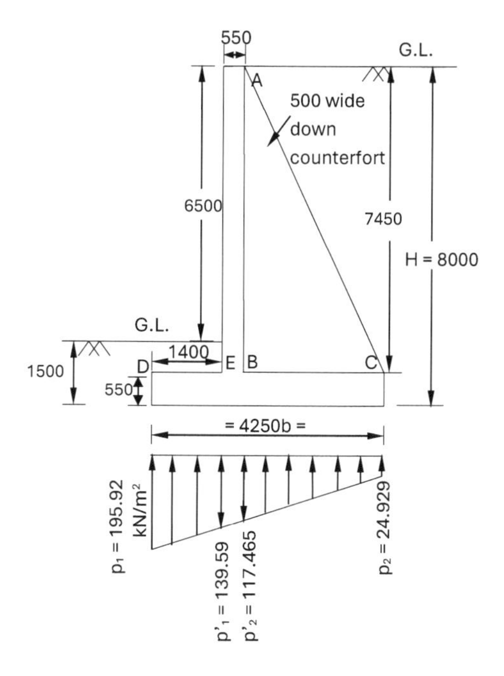

FIGURE 12.18 Soil reaction on base slab.

Stbility of Wall

The preliminary dimensions of wall are marked in figure 12.3

Let,

W_1 = weight of stem, per meter length.
W_2 = weight of base slab.
W_3 = weight of soil on heel slab.

The calculations are arranged in Table 12.3.

$$\therefore \quad \text{Resissting moment } M_R = 1251.896 \text{ kN-m.} \qquad (12.1)$$

TABLE 12.3
Loads and Moments Related to Stability Check of Counterfort Retaining Wall

Sl. No.	Designation	Force (kN)	L.A. (m)	Moment about Toe (kN-m)
1.	W_1	$0.55 \times 7.45 \times 1 \times 25 = 102.4375$	1.675	171.5828
2.	W_2	$0.55 \times 4.25 \times 1 \times 25 = 58.4375$	2.125	124.1797
3.	W_3	$2.3 \times 7.45 \times 1 \times 18 = 308.43$	3.1	956.133
		ΣW = 469.305		**M_R = 1251.896**

$$\text{Horizontal earth presure } P_H = K_a\gamma\frac{H}{2} = \frac{1}{3}\times\frac{1}{2}\times 18\times 8^2 = 192\,\text{kN}. \quad (12.2)$$

$$\text{Overturning moment } M_o = P_H\cdot\frac{H}{2} = 192\times\frac{8}{3} = 512\ \text{kN-m}. \quad (12.3)$$

$$\text{F.S. against overturning } = \frac{0.9\times M_R}{M_o} > 1.4 \quad\Rightarrow\quad \frac{0.9\times 1251.896}{512}$$

$$= 2.2 > 1.4\ (\text{Hence safe})$$

$$\text{F.S. against overturning } = \frac{0.9\times\mu\times w}{P_H} = \frac{0.9\times 0.48\times 469.305}{192} = 1.056 < 1.4$$

Hence, the wall is safe against overturning but unsafe against sliding. A shear key will have to be provided under the base.

Pressure Distribution

$$\text{Net moment } \Sigma M = M_R - M_0 = 1251.896 - 512 = 739.896\ \text{kN-m}.$$

Distance x from toe, of the point of application of the resultant is given by

$$x = \frac{\Sigma M}{\Sigma W} = \frac{739.896}{469.305} = 1.5766$$

$$\therefore\quad e = \frac{b}{2} - x = 2.125 - 1.5766 = 0.5484\,\text{m} < \frac{b}{6} = 0.7083$$

$$p_1 = \frac{\Sigma W}{b}\left(1+\frac{6e}{b}\right) = \frac{469.305}{4.25}\left(1+\frac{6\times 0.5484}{4.25}\right)$$

$$= 195.92\ \text{kN/m}^2 < 225\ \text{kN/m}^2.\ (\text{Hence safe}).$$

Pressure p_2 under the heel is given by

$$P_2 = \frac{\Sigma W}{b}\left(1-\frac{6e}{b}\right) = \frac{469.305}{4.25}\left(1-\frac{6\times 0.5484}{4.25}\right) = 24.929\,\text{kN/m}^2$$

The pressure intensity p_1' under E is

$$p_1' = 195.92 - \frac{195.92 - 24.929}{4.25}\times 1.4 = 139.59\,\text{kN / m}^2$$

The pressure intensity p_2' under B is

$$p_2' = 195.92 - \frac{195.92 - 24.929}{4.25}\times 1.95 = 117.465\,\text{kN / m}^2$$

Design of Heel Slab

Effective spacing between counterforts = 3 m. The pressure distribution below the heel slab is shown in figure 12.18. Consider a strip, 1 m wide, near the outer edge *C*. The upward pressure intensity is 24.929 kN/m², which is minimum at *C*.

Downward load due to weight of earth = 7.45 × 18 = 134.1 kN/m²,
and that due to self-weight of heel slab = 0.55 × 25 = 13.75 kN/m²,
Hence net downward intensity of load *p* is given by

$$p = 134.1 + 13.75 - 24.929 = 122.921\,\text{kN} / \text{m}^2,$$

∴ Maximum negative bending moment in heel slab, at counterforts is

$$M_1 = \frac{pl^2}{10} = \frac{122.921 \times 3^2}{10} = 110.6289\,\text{kN-m} = 110.6289 \times 10^6\ \text{N-mm}$$

$$M_u = \gamma_f \cdot M_1 = 1.5 M_1 = 1.5 \times 110.6289 = 165.9433\ \text{kN-m}$$

$M_u = Q_u \cdot bD^2$, where, $b = 1000\,\text{mm}$, and for Fe 415 & M20 $Q_u = 2.76\,\text{N/mm}^2$.

or,

$$d = \sqrt{\frac{M_u}{Q_u b}} = \sqrt{\frac{165.9433 \times 10^6}{2.76 \times 1000}} = 245.2028\,\text{mm}$$

Adopt $D = 0.65$ m
Hence, $d = 0.65 - 0.05$ (clear cover) = 0.6 m.

$$\therefore \frac{M_u}{bd^2} = \frac{165.9433 \times 10^6}{1000 \times 600^2} = 0.461$$

From SP16 table – 3, $P_t = 0.75$.

$$V_u = 1.5 \times 0.6\,pl = 1.5 \times 0.6 \times 122.921 \times 3\,\text{kN} = 331.886 \times 10^3\ \text{N}.$$

So, $\tau_c = 0.56$ N/mm² for $P_t = 0.75$.

$$\tau_v = \frac{V_u}{bd} = \frac{331.886 \times 10^3}{1000 \times 600} = 0.55 < \tau_c.$$

$$A_{st} = \frac{0.75}{100} \times 1000 \times 600 = 4500\,\text{mm}^2.$$

Let us provide 25Ø bars

$$\text{Spacing required } s = \frac{1000}{4500} \times 491 = 109.11\,\text{mm c/c}$$

So, provided spacing = 100 mm c/c

$$A_{st}\text{ (provided)} = \frac{491}{100} \times 1000 = 4910\,\text{mm}^2.$$

$$\text{Maximum (+ve) B.M.} = \frac{pl^2}{16} = \frac{122.921 \times 3^2}{16} = 69.143\,\text{kN-m} = 69.143 \times 10^6\ \text{N-mm}$$

$$\text{Factored moment} = \frac{pl^2}{16} = 103.714 \times 10^6\ \text{N-mm}$$

$$\therefore \frac{M_u}{bd^2} = \frac{103.714 \times 10^6}{1000 \times 600^2} = 0.288$$

Adopting $P_t = 0.5$

$$A_{st} = \frac{0.5}{100} \times 1000 \times 600 = 3000\,\text{mm}^2.$$

Let us provide 20Ø bars

$$\text{Spacing required } s = \frac{1000}{3000} \times 314 = 104.67\,\text{mm c/c}$$

So, provided spacing = 100 mm c/c

$$A_{st}\text{(provided)} = \frac{1000}{100} \times 314 = 3140\,\text{mm}^2.$$

Design of Stem (Vertical Slab)

The stem acts as a continuous slab. Consider 1 m strip at B.

The intensity of earth pressure is given by

$$P_h = K_a \gamma H_1 = \frac{1}{3} \times 18 \times 7.35 = 44.1\ \text{kN/m}^2.$$

(where revised value of $H_1 = 8 - 0.65 = 7.35\,\text{m}$.

∴ Negative B.M. in slab near counterfort is

$$M_1 = \frac{P_h \cdot l^2}{10} = \frac{44.1 \times 3^2}{10} = 39.69\,\text{kN-m} = 39.69 \times 10^6\ \text{N-mm}$$

Factored moment $M_u = \gamma_f \cdot M_1 = 1.5M_1 = 1.5 \times 39.69$ kN-m = 59.535 kN-m.

$$\therefore\ d = \sqrt{\frac{M_u}{Q_u b}} = \sqrt{\frac{59.535 \times 10^6}{2.76 \times 1000}} = 146.87\,\text{mm}$$

Providing effective cover = 50 mm, total depth = 146.87 + 50 = 196.87 mm.

However, provide total depth = 450 mm, so that available d = 450 − 50 = 400 mm.

This increased thickness will keep the shear stress within limits so that additional shear reinforcement is not required.

$$\therefore \frac{M_u}{bd^2} = \frac{59.535 \times 10^6}{1000 \times 400^2} = 0.372$$

Adopting $P_t = 0.3$

Shear force, $V_u = 1.5 \times 0.6\, p\, l = 1.5 \times 0.6 \times 44.1 \times 3$ kN = 119.07×10^3 N.

$$\tau_v = \frac{V_u}{bd} = \frac{119.07 \times 10^3}{1000 \times 400} = 0.298 < \tau_c.$$

So, $\tau_c = 0.38$ N/mm$^2 > \tau_v$ for $P_t = 0.3$.

$$A_{st} = \frac{0.3}{100} \times 1000 \times 400 = 1200\,\text{mm}^2.$$

Let us provide 16Ø bars

$$\text{Spacing required } s = \frac{1000}{1200} \times 201 = 167.5\,\text{mm c/c}$$

So, provided spacing = 150 mm c/c

$$A_{st}(\text{ provided }) = \frac{1000}{150} \times 201 = 1340\,\text{mm}^2.$$

Design of Main Counterfort

Let us assume thickness of counterfort = 500 mm. The counterforts will thus be spaced @ 300 cm c/c. They will thus receive earth pressure from a width of 3 m and downward reaction from the heel slab for a width of 3 m (Figure 12.19).

At any section at depth h below the top A, the earth pressure acting on each counterfort will be

$$= \frac{1}{3} \times 18 \times h \times 3 = 18h\,\text{kN / m} \quad (12.4)$$

Similarly, net downward pressure on heel at C is

$$= 134.1 + 0.65 \times 25 - 24.929 = 125.421\ \text{kN/m}^2.$$

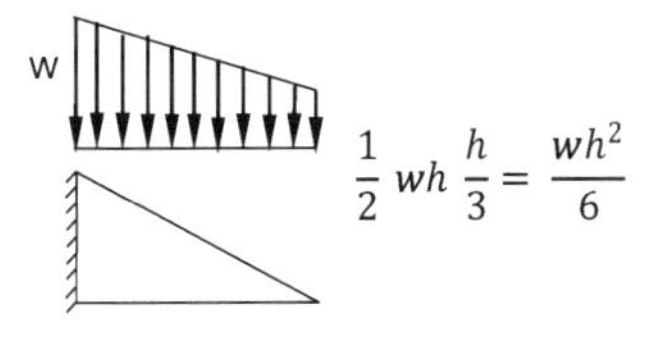

FIGURE 12.19 Net pressure on counterfort.

And thet at B is = 134.1 + 0.65 × 25–117.465 = 32.885 kN/m².

Hence reaction transferred to each counterfort will be

At C, 125.421 × 3 = 376.263 kN/m
At B, 32.885 × 3 = 98.655 kN/m.

The variation of horizontal and vertical forces on the counterfort is shown in figure 12.19.

The critical section for the counterfort will be at F, since below this, enormous depth will be available to resist bending.

$$\text{Pressure intensity at } b = 6.5\,\text{m is } = 18 \times 6.5 = 117\,\text{kN/m}$$

$$\text{Shear force at } F = \frac{1}{2} \times 117 \times 6.5 = 380.25\,\text{kN}$$

$$\text{B.M.} \quad M = 380.25 \times \frac{6.5}{3} = 823.875\,\text{kN-m} = 823.875 \times 10^6\ \text{N-mm.}$$

$$\text{Factored moment} = 1.5 \times 823.875\ \text{kN-m} = 1235.8125\ \text{kN-m.}$$

$$\therefore d = \sqrt{\frac{M_u}{Q_u b}} = \sqrt{\frac{1235.8125 \times 10^6}{2.76 \times 500}} = 946.32\,\text{mm}$$

$$\therefore \text{Total depth} = 946.32 + 50 = 996.32\ \text{mm.}$$

Angle θ of face AC is given by

$$\tan\theta = \frac{2.35}{7.35} = 0.3197$$

$$\therefore \theta = 17.73°.$$

$$\sin\theta = 0.3045, \qquad \cos\theta = 0.9524$$

$$\text{Depth } F_1C_1 = AF_1 \sin\theta = 6.5 \times 0.3045 = 1.979\ \text{m} = 1979\ \text{mm.}$$

$$\therefore \text{Depth } FG = 1979 + 450 = 2429\ \text{mm.} \approx 2430\ \text{mm.}$$

This is much more than required.

Assuming that steel reinforcement is provided in two layers and providing a clear cover of 50 mm and 20 mmØ bars, the effective depth may be assumed approximately to be equal to 2350 mm.

$$\therefore \quad \frac{M_u}{bd^2} = \frac{1235.8125 \times 10^6}{500 \times 2350^2} = 0.447$$

Adopting $P_t = 0.25$

$$A_{st} = \frac{0.25}{100} \times 500 \times 2350 = 2937.5\,\text{mm}^2.\ \text{Using, } 25\,\text{mm Ø bars, } AØ = 491\,\text{mm}^2.$$

$$\text{No of bars} = \frac{2937.5}{491} = 5.98 \approx 6 \text{ nos. bars}$$

Provide this in two layers.

$$\text{Effective S.F.} = Q - \frac{M}{d'}\tan\theta \text{ where } d' = \frac{d}{\cos\theta} = \frac{2350}{0.9524} = 2467.45\,\text{mm}$$

$$\therefore \quad \text{Effective S.F.} = Q - \frac{M}{d'}\tan\theta$$

$$= 380.25 \times 10^3 - \frac{823.875 \times 10^6}{2467.45} \times 0.3197 = 273503\,\text{N}.$$

$$\therefore \quad \tau_v = \frac{273{,}503}{500 \times 2467.45} = 0.222\,\text{N/mm}^2.$$

$$100\frac{A_s}{bd} = \frac{100 \times 6 \times 491}{500 \times 2467.45} = 0.24\%$$

$$\therefore \quad \tau_c = 0.362\,\text{N/mm}^2.$$

The height h where half of the reinforcement can be curtailed will be equal to $\sqrt{H} = \sqrt{8} \approx 2.8\,\text{m}$ below A, i.e. at point H. To locate the position of point of curtailment on AC, draw HI parallel to FG. Thus half the bars can be curtailed at I. However, these should be extended by a distance of $12\,\phi = 240\,\text{mm}$. beyond I, i.e. extended up to I_1. The location of H, corresponding to I_1 can be located by drawing line I_1H_1 parallel to FG. It should be noted that I_1G should not be less than 58.3 $\phi = 1166\,\text{mm}$. Similarly, other bars can be curtailed, if desired (Figure 12.23).

Design of Horizontal Ties

The vertical stem slab has a tendency to separate from the counterfort, and hence should be tied to it by horizontal ties.

At any depth h below the top, force causing sepearation $= \frac{1}{3} \times 18 \times h \times 2.5$

$15h$ kN/m.

At $h = 6.5\,\text{m}$, force $= 15 \times 6.5 = 97.5$ kN/m.

$$\therefore \quad \text{Steel area required} = \frac{97.5 \times 1000}{0.87 f_y} = 270\,\text{mm}^2 \text{ per meter height.}$$

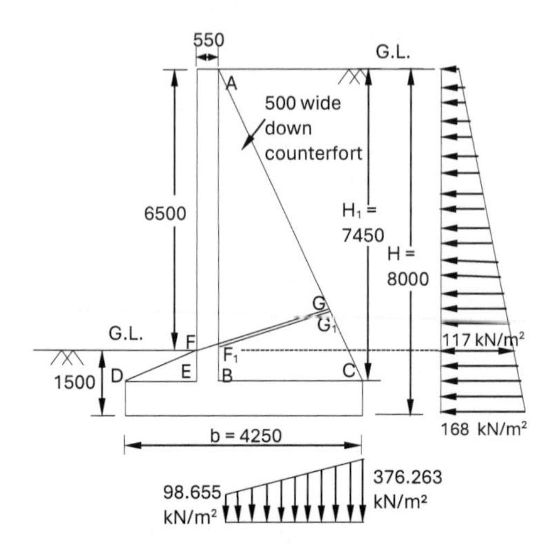

FIGURE 12.20 Earth pressure and soil reaction on counterfort.

Using 8 mm ϕ 2-legged ties, $A\text{Ø} = 2 \times \dfrac{\pi \times 8^2}{4} = 100\,\text{mm}^2$ Spacing $= \dfrac{1000 \times 100}{270} =$ 370.37mm c/c. Adopt spacing 300 mm c/c.

Design of Vertical Ties

Similar to the stem slab, heel slab has also tendency to separate from the counterfort, due to net downward force, unless tied properly by vertical ties. The downward force at C will be $376.263 \times \dfrac{2.5}{3} = 313.55$ kN/m at C and $98.655 \times \dfrac{2.5}{3} = 82.21$ kN/m at B. Near end C, the heel slab is tied to counterforts with the help of main reinforcement of counterforts.

$$\therefore \quad \text{Steel area at } C = \frac{313.55 \times 1000}{0.87 f_y} = 868.44\,\text{mm}^2$$

Using 12 mm ϕ 2 legged ties, $A\text{Ø} = 2 \times \dfrac{\pi \times 12^2}{4} = 226\,\text{mm}^2$

$$\text{Spacing of ties at } C = \frac{1000 \times 226}{868.44} = 260\,\text{mm c/c.}$$

Hence, provided spacing is 250 mm c/c.

$$\text{Steel area at } B = \frac{82.21 \times 1000}{0.87 f_y} = 227.7\,\text{mm}^2$$

$$\text{Spacing of ties at } B = \frac{1000 \times 226}{227.7} = 992.53\,\text{mm c/c.}$$

Hence, provided spacing is 500 mm c/c.

Thus the spacing of vertical ties can be increased gradually from 250 mm c/c at C to 500 mm c/c at B.

Design of Shear Key

The wall is unsafe for sliding, and hence shear key will have to be provided, as shown in Figure 12.21 below. Let the depth of key = a.

Intensity of passive pressure P_P developed in front of the shear key depends upon the soil pressure p in front of key (Figures 12.22 and 12.23).

$$p_P = K_P \cdot P = 3 \times 139.59 = 418.77\,\text{kN/m}^2.$$

$$\text{Total passive pressure } P_P = p_p \cdot a = 418.77$$

$$\text{Sliding force at level } D_1C_1 = \frac{1}{3} \times \frac{18}{2} \times (8+a)^2$$

or,

$$P_H = 3(8+a)^2$$

Weight of soil between of base and D_1C_1

$$\approx 4.25 \times a \times 18 = 76.5a\,\text{kN}$$

$$\Sigma W = 469.305 + 76.5a$$

Hence for equilibriun of wall, permitting F.S. = 1.4 against sliding,

We have

or,

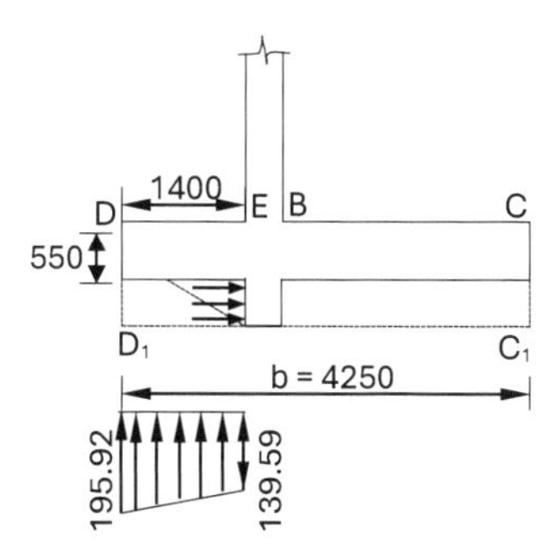

FIGURE 12.21 Soil reaction on toe slab and shearkey.

$$1.4 = \frac{0.9\,\mu\,\Sigma W + P_p}{P_H}$$

or,

$$1.4 = \frac{0.432(469.305 + 76.5a) + 418.77a}{3(8 + a)^2}$$

or,

$$a^2 - 91.578a + 15.7286 = 0$$

$$\therefore \; A = 0.172\,\text{m} = 172\,\text{mm}.$$

Hence, keep $a = 0.3\,\text{m}$ and width $= 0.3\,\text{m}$.
Hence provide 16Ø @150c/c

Reinforcement Details

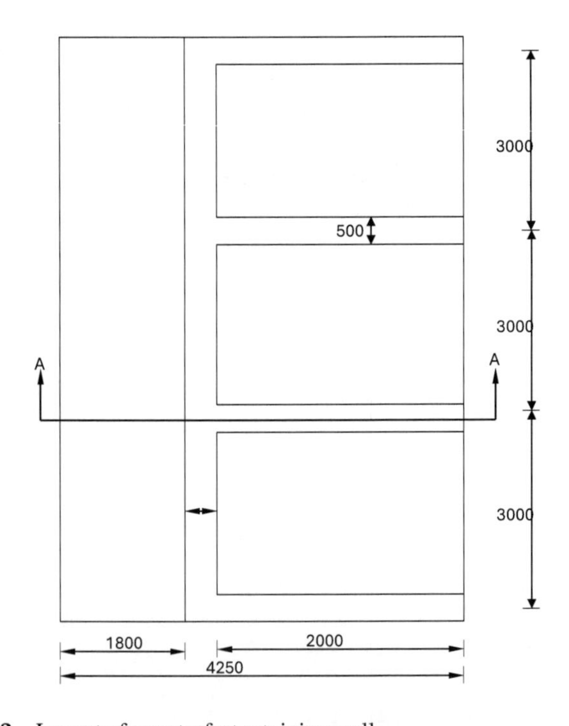

FIGURE 12.22 Layout of counterfort retaining wall.

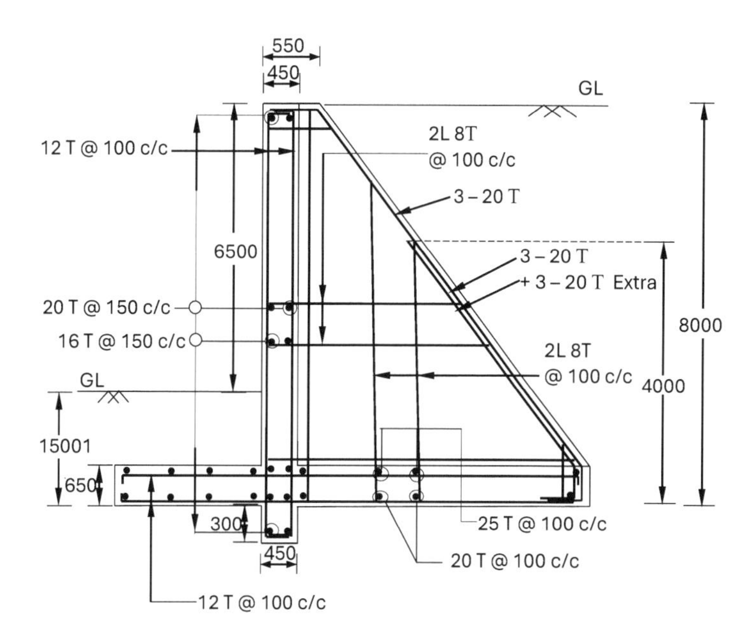

FIGURE 12.23 Details of reinforcement of counterfort retaining wall (sec A-A).

13 Design of Overhead Water Tanks

PREAMBLE

Water tanks are important public utility and industrial structure. The design and construction methods used in reinforced concrete are influenced by the prevailing construction practices, the physical property of the material, and the climatic conditions. Before taking up the design, the most suitable type of staging of tanks and correct estimation of loads including statically equilibrium of structure particularly in regards to overturning of overhanging members are made. The design is made considering the worst possible combination of loads, moments and shears arising from vertical loads and horizontal loads acting in any direction when the tank is full as well as empty. In this chapter, the analysis of Intze tank and static and dynamic water pressure is provided. Hydrodynamic effect (sloshing) has been taken into consideration to assess seismic effect as per IS1893. The parameters required to design an elevated water tank are location, capacity, and shape; analysis of different structural elements of circular water tank and staging & foundation against gravity, wind load, and seismic load is provided. Generally, tension crack width is kept within the limit. A crack width of 0.1 mm has been accepted as permissible value in liquid retaining structures. While designing liquid retaining structures, IS3370 (Parts I–IV) is considered.

Elevated water tanks are supported on staging, which may consist of R.C.C tower or R.C.C column braced together. The staging has to carry the entire load of entire tanks with water and is subjected to wind loads. The purpose for which the water tanks will be used, human consumption or industrial determines the concerns for materials that do not have side effects for human. Wind and earthquake design considerations allow the design of water tank parameters to survive seismic and high wind events.

INTZE type is a very special type of elevated tank used for very large capacities with structural advantages. Circular tanks for very large capacities prove to be uneconomical when flat bottom slab is provided. Intze type tank consists of top dome supported on a ring beam that rests on a cylindrical wall. The walls are supported on ring beam and conical slab. Bottom dome will also be provided which also supported by ring beam. The conical and bottom dome is made in such a manner that the horizontal thrust from the conical base is balanced by that from the bottom dome. The conical and bottom domes are supported on circular beam which is in turn, supported on a number of columns. For large capacities, the tank is divided in two compartments by means of partition walls supported on a circular beam (Figures 13.1–13.4).

DOI: 10.1201/9781003208204-13

FIGURE 13.1 Column-supported intze type tank.

DIFFERENT TYPES OF OVERHEAD TANKS

FLUID – STRUCTURE INTERACTION

Liquid storage overhead is extensively used in water supply in most of the towns and cities. Few fundamental problems are associated with these kinds of tanks involve. Hydrodynamic forces on the wall of a vibrating water tank is an extremely important issue from dynamic behavior point of view. Sloshing means any motion of the liquid inside the reservoir. Estimation of hydrodynamic pressure distribution and natural frequencies are the real challenge for specially thin-walled tanks. Parameters have a direct effect on the dynamic stability of the tank. In regard to the hydrodynamic pressure of water, there are two important aspects. One aspect is directly proportional to the acceleration of the tank. The aspect is caused by the part of the water moving with the same tank velocity, known as **convective**, and represents the free-surface-water motion. Mass-Spring-Dashpot is generally applied to represent the sloshing part. This phenomenon has been studied experimentally as well as theoretically for a long time, particularly the linear and nonlinear effects of sloshing for both in viscid and viscous liquids, mostly

FIGURE 13.2 Column-supported circular tank.

on cylindrical and rectangular geometry. However, predictions of large-amplitude sloshing are still not fully developed. At higher water depths, large standing waves are generally formed near resonance frequency. Partially filled tanks are prone to sloshing, create high localized impact pressure on tank walls. It supplies energy to sustain the sloshing thus causing high-intensity random motion in the free surface during this motion. It may be simple planer, non-planer, rotational, irregular beating, symmetric, asymmetric, quasi-periodic, or chaotic. In general, the amplitude of sloshing depends upon the nature, amplitude, and frequency of the tank structure.

Seismic analysis of liquid storage tanks accounts for the hydrodynamic forces exerted by the fluid on the tank wall. Knowledge of these hydrodynamic forces is essential in the seismic design of tanks. Evaluation of hydrodynamic forces

FIGURE 13.3 Shaft-supported funnel tank.

requires suitable modeling and dynamic analysis of tank-liquid system. These mechanical models convert the tank-liquid system into an equivalent spring-mass system. Design codes use these mechanical models to evaluate the seismic response of tanks. While using such an approach, various other parameters also get associated with the analysis. Some of these parameters are: Pressure distribution on tank wall due to lateral and vertical base excitation, time period of tank in lateral and vertical mode, effect of soil-structure interaction, and maximum sloshing wave height. Design codes have provisions with varying degree of details to suitably evaluate these parameters. In this study, the following codes are considered: ACI 350.3, NZSEE guidelines and IITK-GSDMA Guidelines for seismic design of liquid storage tanks. These codes use the mechanical model developed by G W Housner, which is discussed in theory part. Eurocode 8 mentions the mechanical model of Veletsos and Yang (1977) as an acceptable procedure for rigid circular tanks. An important point while using a mechanical model pertains to combination rule used for adding the impulsive and convective forces. Except Eurocode 8, all the codes suggest square root of sum of square (SRSS) rule to combine impulsive and convective forces. Eurocode 8 suggests the use of absolute summation rule. For evaluating the impulsive force, mass of tank wall and roof is also considered along with impulsive fluid mass. ACI 350.3 and Eurocode 8 suggest a reduction factor to suitably reduce the mass of tank wall. Such a reduction

FIGURE 13.4 Shaft-supported circular conical tank.

factor was suggested by Veletsos (1984) to compensate the conservativeness in the evaluation of impulsive force (Figure 13.5).

DESIGN EXAMPLE OF OVERHEAD WATER TANKS

Static Design of INTZ Type Tank

Capacity of tank = 1,000,000 L = 1000 m^{3}, Grade of concrete = M25, Grade of steel = Fe 500

Let base diameter of top dome $D = 12.5$ m

Base diameter of bottom dome $D_0 = 8$ m (Approximately $0.6D$)

Rise of top dome $h_1 = 2.5$ m (Approximately $D/8$)

Rise of bottom dome $h_2 = 2$ m (Approximate $D_0/6$)

Height of conical portion $h_0 = (12.5 - 8)/2 = 2.25$ m

Radius of curvature of bottom dome $(R_2) = \left(D_0^2 + h_2^2\right) / 2h_2 = 5\text{m}$

We know,

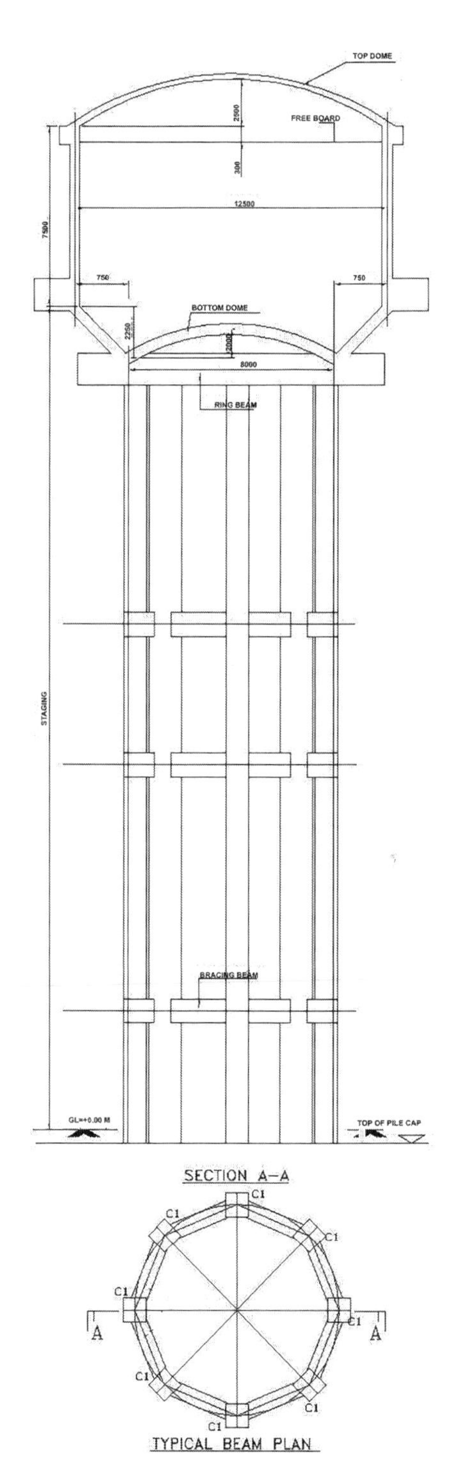

FIGURE 13.5 Plan and section of intz type tank.

$$V = (3.14D^2H/4) + 3.14\ h_0\left(D^2 + D_0^2 + Dh_0\right)/12 + 3.14\ h_2^2(3R_2 - h_2)/3$$

Substituting known values, we get, $H = 7.5$ m

Design Top Dome

Thickness of dome slab = 100 mm (minimum thickness to be considered)

Live load of dome = 1.5 kN/m² (considering accessible roof as per IS875-Part III)

$$\text{Self weight of dome } = 0.1 \times 25 = 2.5\ \text{KN}/\text{m}^2$$

$$\text{Total load } (w) = (1.5 + 2.5) = 4\ \text{kN/m}^2$$

$$\text{Radius of curvature of top dome } (R_1) = \left(D_0^2 + h_1^2\right)/2h_1 = 9.0625\ \text{m}$$

Therefore, $\sin\phi = D/2R_{1=}0.69$ m, $\cos\phi = 0.724$

$$\text{Meridional force } = T_M = wR/(1+\cos\phi) = 21.02\ \text{kN}/\text{m}$$

$$\text{Meridional stress } = (21.02 \times 1000)/1000 \times 100) = 0.2102\ \text{N}/\text{mm}^2(\text{compressive})$$

(since thickness of dome = 100 mm)

Considering M25 grade concrete

Referring to Table 2 of IS3370- Part-2–2009

$$\text{Permissible concrete stress in direct compression } = 6\ \text{N}/\text{mm}^2 >$$

$$0.2102\ \text{N}/\text{mm}^2\text{OK}$$

$$\text{Hoop force } = T_H = wR_1(1/(1+\cos\phi) - \cos\phi) = -5.22\ \text{kN/m}$$

(negative sign indicates Tensile)

$$\text{Hoop stress } = (-5.22 \times 1000)/(1000 \times 100) = -0.0522\ \text{N}/\text{mm}^2\ (\text{Tensile})$$

Referring to Table 1 of IS3370- Part-2–2009

Permissible stress in direct tension for M25 grade concrete = 1.3 N/mm² > 0.0522 N/mm² OK

Minimum reinforcement = 0.3% of gross area, has to be provided as stresses are within permissible limits.

Referring to Table 4 of IS3370- Part-2–2009

$$\text{Permissible stress for steel reinforcement } = 130 \text{ N / mm}^2$$

$$\text{Steel required for hoop tension } = T_H/\sigma_{st} = 5.22 \times 10^3 \ / \ 130 = 40.15 \text{ mm}^2/\text{m width}$$

$$\text{Minimum steel required } (A_{st})_{min}$$

$$= 0.3 \times 1000 \times 100 / 100 = 300\text{mm}^2 \ / \text{ m width } > \text{Steel required}$$

$$\text{Spacing of reinforcements } = 3.14 \times 8^2 \times 1000 / 4 \times 300 = 166 \text{ mm.}$$

Let us provide 8 ϕ @150 c/c both circumferentially and along the meridian

Design Top Ring Beam RB_1

$$P_1 = \text{ Horizontal component of meridional stress}$$

$$= T_M \cos\phi = 21.02 \times 0.724 = 15.22 \text{ kN/m}$$

$$\text{Hoop tension in ring beam } RB_1 = P_1 D / 2 = 15.22 \times 12.5 / 2 = 95.125 \text{ kN/m}$$

$$A_s = \text{ Steel required } = 95.125 \times 10^3 \ / \ 130 = 731.73 \text{ mm}^2$$

Providing 4–16 ϕ, Fe 500 TMT bars ($A_{st} = 804 \text{ mm}^2$)
As per IS456:2000 $\sigma_{cbc} = 8.5 \text{ N/mm}^2$ for M25 grade concrete

$$\text{Modular ratio } m = 280 / 3\sigma_{cbc} = 11$$

We have

$$\sigma_{ct} = T / (\text{BD} + (m-1)A_s$$

$$1.3 = (95.125 \times 10^3) / (\text{BD} + (11-1)804)$$

$$\text{BD} = 65{,}133.1 \text{ mm}^2$$

Let us provide width of beam = 350 mm × 400 mm
Provide 2L-8ϕ stirrups @150 mm c/c

Design of Cylindrical Tank Wall

$$P = \text{ Maximum hoop tensile stress at RB}_2 \text{ beam level}$$

$$= p_w \text{HD} / 2 = 9810 \times 7.5 \times 12.5 / 2$$

$$= 459{,}843.75 \text{ N/m width}$$

$$A_{\text{st(required)}} = 459{,}843.75 / 130 = 3538 \text{ mm}^2 / \text{m width}$$

$$\text{Providing steel at each the faces } = 3538 / 2 = 1769 \text{ mm}^2 \text{ per face}$$

Providing 16ϕ TMT bars

$$\text{Spacing of reinforcement required } = 201 \times 1000 / 1769 = 113.62 \text{ mm}$$

Providing steel 16 mm TMT bar @100c/c at both the faces.

$$A_{\text{st(provided)}} = 2 \times 201 \times 1000 / 100 = 4020 \text{ mm}^2 / \text{m width.}$$

We have

$$\sigma_{\text{ct}} = P / \left(Bt + (m-1)A_s\right.$$

t = Thickness of cylindrical portion and B = 1 m = 1000 mm
i.e. 1.3 = 459,843.75/(Bt + (11 − 1) × 4020)

$$Bt = 313526 \text{ mm}^2$$

$$t = 313.5 \text{ mm}$$

$$t_{\min} = \text{ Minimum thickness } = 3H + 5 = 3 \times 7.5 + 5 = 27.5 \text{ cm} = 275 \text{ mm}$$

Let us provide t = 350 mm uniform and 16 ϕ TMT bar @100 mm c/c at both faces at RB_2 level.

Provide thickness = 200 mm at RB_1 level.

$$\text{Minimum distribution of steel } = 0.3\%$$

$$A_{\text{st}} = 0.3 \times 1000 \times 350 / 100 = 1050 \text{ mm}^2 / \text{m}$$

$$A_{\text{st}} \text{ per face } = 1050 / 2 = 525 \text{ mm}^2 / \text{m}$$

Let us provide 10 ϕ

$$\text{Spacing} = 78 \times 1000 / 525 = 148 \text{ mm}$$

So, provide distribution steel 10ϕ@125 mm c/c at both faces.

Design of Ring Beam RB_2

$$V_1 = \text{Vertical load transferred from top dome}$$

$$= T_M \sin\phi = 21.02 \times 0.69 = 14.5 \text{ kN / m}$$

$$V_2 = \text{Self weight of rib } RB_1 = 0.6 \times 0.9 \times 25 = 13.5 \text{ kN / m}$$

$$V_3 = \text{Self weight of tank wall } = 0.35 \times 7.5 \times 25 = 65.625 \text{ kN / m}$$

$$V_4 = \text{Self weight of } RB_2 \text{ (Assuming } 600 \text{ mm} \times 900 \text{ mm)}$$

$$= (0.9 - 0.325) \times 0.6 \times 25 = 8.625 \text{ kN / m}$$

Therefore, W = Total vertical load = 14.5 + 13.5 + 65.625 + 8.625=102.25 kN/m
Inclination of conical portion $\beta = 60^\circ$ with horizontal.
So, inclination with vertical portion = 90 – 60 = 30°

$$H_1 = W \tan 30^\circ = 102.25 \times \tan 30^\circ = 59 \text{ kN / m}$$

$$H_2 = p_w H B_R = 9.81 \times 7.5 \times 0.6 = 44 \text{ kN / m}$$

$$H_1 + H_2 = 59 + 44 = 103 \text{ kN / m}$$

$$\text{Hoop tension in } RB_2 = HD / 2 = 103 \times 10^3 \times 12.5 / 2 = 643.75 \times 10^3 \text{ N / m}$$

$$A_s = 643.75 \times 10^3 / 130 = 4952 \text{ mm}^2$$

$$\sigma_{ct} = T / \left(BD + (m-1)A_s\right.$$

$$= 643.75 \times 10^3 / (1000 \times 600 + 10 \times 4910)$$

$$= 0.99/\text{mm}^2 < 1.3 \text{ N/mm}^2 \text{OK}$$

Provide 4L -8mmϕ stirrups @150 c/c

Design of Conical Portion

Weight of top dome, cylindrical, RB_1, RB_2 and live load= 73,100 N/m

Total weight of water on the wall conical portion

$$= 3.14\left(D^2 - D_0^2\right)hp_w / 4 + 3.14h_0 p_w \left(D^2 + D_0^2 + DD_0\right)/12$$

$$= 3.14\left(12.5^2 - 8^2\right)\times 7.5\times 9800/4 + 3.14\times 2.25\times 9800\left(12.5^2 + 8^2 + 12.5\times 8\right)/12$$

$$= 7{,}173{,}994 \text{ N}$$

$$\text{Weight of conical portion } = 3.14(D + D_0)t\ t_0 p_c / 2$$

$$= 3.14(12.5 + 8)\times 2.25\times 0.8\times 2500/2\times \cos 60° = 2{,}896{,}650 \text{ N}$$

$$W_2 = (3.14DW + W_W + W_S)/3.14D_0 \text{ per m width}$$

$$= (3.14\times 12.5\times 73{,}100 + 7{,}173{,}994 + 2{,}896{,}650)/(3.14\times 8)$$

$$= 515{,}120 \text{ N per metre width}$$

$$\text{Meridional thrust } = T_0 = W_2/\cos\phi = 515{,}120/\cos 60°$$

$$= 1{,}030{,}240 \text{ N per metre width}$$

$$\text{Meridional stress } = 1{,}030{,}240/(800\times 1000) = 1.29 \text{ N/mm}^2 < 1.3\text{N/mm}^2\text{OK}$$

Let us consider a level at a height h' from RB_3 level

$$D = D_0 + (D - D_0)h/h_0 = 8 + 2h'$$

$$P = \text{ Intensity of water pressure } = (h + h_0 - h)p_w = (9.75 - h')9800 \text{ N/m}$$

$$\text{Self-weight of wall } = 0.8\times 25\times 10^3 = 20\times 10^3 \text{ N/m}^2$$

We know hoop tension $P_0 = \{P/\cos\phi + q\tan\phi\}\ D'/2$

$$P_0 = \left\{(9.75 - h)9800/\cos 60° + 20\times 10^3 \tan 60°\right\}(8 + 2h')/2$$

$$= (225{,}741 - 19{,}600h)/(4 + h) = 902{,}964 + 14{,}7341h - 19{,}600h'^2$$

To get maximum hoop tension $dP_0/dh = 0$

$$147{,}341 - 2 \times 19{,}600\ h = 0$$

$$h' = 3.76 \text{ m} \left(\text{Above RB}_3 \text{ level}\right)$$

Hence, maximum hoop tension at $h = h_0 = 2.25$ m

$$P_0 = 902{,}964 + 147{,}341 \times 2.25 - 19{,}600 \times 2.25^2$$
$$= 1{,}135{,}256.25 \text{ N / m}$$

$$A_{st} = 1{,}135{,}256.25 / 130 = 8733 \text{ mm}^2 \text{ per m width}$$

$$A_{\text{st in each face}} = 8733 / 2 = 4366 \text{ mm}^2 \text{ per m width}$$

Providing 25ϕ reinforcement

$$\text{Spacing of reinforcement } = 490.625 \times 1000 / 4366 = 112.37 \text{ mm}$$

Let us provide 25ϕ @100 mm c/c both faces

$$\text{Therefore, } A_{\text{st in each face}} = 490.625 \times 1000 / 100 = 4906 \text{ mm}^2$$

We have $\sigma_{ct} = T/(BD + (m - 1)A_s = 1{,}135{,}256.25/(1000 \times 800 + 10 \times 4906 \times 2) =$ 1.264 MPa < 1.3 MPa

As per IS3370-2009, minimum reinforcement (Clause A-1.2)

$$f_{\min} = f_{ce}/f_y = 1.15 / 500 = 0.0023 \; \left(\text{since, for M25 } f_{ce} = 1.15\right)$$

$$A_{S\min} = 0.0023 \times 800 \times 1000 = 1840 \text{ mm}^2 \text{ / m width}$$

Providing16ϕ, Spacing of reinforcement = $201 \times 1000/1840 = 109$ mm

Providing 16ϕ @100 mm c/c in along the inclined face

Design of Bottom Dome

$$\text{Radius of curvature of bottom dome } (R_2) = (D_0^2 + h_2^2) / 2h_2 = 5 \text{ m}$$

$$H_0 = h + h_0 = 7.5 + 2.25 = 9.75 \text{ m}$$

Weight of water on bottom dome W_0

$$= \left\{\left(3.14 D_0^2 H_0 / 4\right) + 3.14 h_2^2 \left(3R_2 - h_2\right) / 3\right\} p_w$$

$$= \{(3.14 \times 82 \times 9.75 / 4) + 3.1422(3 \times 5 - 2) / 3\}9800$$

$$= 4,269,215\text{N}$$

$\sin\phi = D_0/2R_2 = 4/5 = 0.8$, $\cos\phi = 0.6$

$$\text{Total surface area of bottom dome } = 2 \times 3.14 \times R_2 \times h_2$$

$$\text{Weight of the bottom dome } = (2 \times 3.14 \times R_2 \times h_2)t_2 p_c$$

Let us consider thickness of the bottom dome = 250 mm

$$W_T = W_0 + (2 \times 3.14 \times R_2 \times h_2)t_2 p_c = 4,269,215$$

$$+(2 \times 3.14 \times 5 \times 2) \times 0.25 \times 25,000$$

$$= \ 4,661,914 \text{ N}$$

$$P_2 = \text{ Intensity of load } = W_T / 2 \times 3.14 \times R_2 \times h_2 = 4661914 / 2 \times 3.14 \times 5 \times 2$$

$$= 74197 \text{ N/m}^2$$

$$\text{Hoop stress } = P_2 R_2 / 2t_2 = (74197 \times 5) / (2 \times 0.25) = 0.742 \text{ N / mm}^2\text{OK}$$

$$\text{Meridional stress } = T_2 = W_T / 3.14 \times D_2 \times \sin\phi$$
$$= 4,661,914 / (3.14 \times 8 \times 0.8)$$

$$= 231,865 \text{ N per m width}$$

$$\text{Meridional stress } = 231,865 / 250 \times 1000 = 0.93 \text{ N / mm}^2 \text{ ok}$$

$$f_{\min} = f_{ce} / f_y = 1.15 / 500 = 0.0023 \quad (\text{since , for M25} f_{ce} = 1.15)$$

$$A_{s\min} = 0.0023 \times 250 \times 1000 = 575 \text{ mm}^2 / \text{m width}$$

Let us provide 10ϕ TMT bar

$$\text{Spacing } = 78.5 \times 1000 / 575 = 136 \text{ mm}$$

Provide 10 mm TMT @125 c/c along the meridian and circumference (Table 13.1)

Design Ring Beam RB_3

$$P = \text{Horizontal force } = T_0 \sin\beta - T_2 \cos\phi,$$

$$\cos\phi = 0.6, \;\; \sin\phi = 0.8$$

$$\sin\ \beta = \sin 60°\left(\text{inclination of conical portion } \beta = 60^{\beta} \text{ with horizontal}\right)$$

$$\text{Inward thrust from conical portion } P_i = T_0 \sin\beta = 1,030,240 \sin 60^{\circ}$$

$$= 892,214 \text{ N / m}$$

$$\text{Outward thrust from bottom dome } P_o = T_2 \cos\phi = 231,865 \times 0.6$$

$$= 139,119 \text{ N / m}$$

$$\text{Hoop compression } = (P_i - P_o) D_0 / 2 = (892,214 - 139,119) \times 8 / 2$$

$$= 3,012,380 \text{ N / m}$$

Let us provide a beam 600 mm × 1200 mm

$$\text{Meridional stress } = 3,012,380 / 1200 \times 600 = 4.2 \text{ Mpa} < 6 \text{ MPa ok}$$

Vertical Load on RB3

$$q_v = \text{ Net vertical force } = T_0 \cos\beta + T_2 \sin\phi°$$

$$= 1,030,240 \times \cos 60° + 231,865 \times 0.8 = 700,612 \text{ N / m}$$

$$W = \text{ vertical load on } RB_3 (\text{Including self-weight })$$

$$= 700,612 + 0.6 \times 1.2 \times 25 \times 10^3 = 718,612 \text{ N / m}$$

Ring beam RB_3 supported by eight columns

$$2\alpha = 45°\alpha = 22.5°$$

C_1, C_2, $C_{3,}$ coefficient from Reynold's handbook

$$C_1 = 0.066, C_2 = 0.03, C_3 = 0.005$$

$$W\ r^2 2\alpha = 718{,}612 \times 4^2 \times 3.14 / 4 = 9{,}025{,}767 \text{ Nm } (\text{since } D_0 = 8\text{m})$$

$$\text{Maximum negative Bending moment at support } = M_0 = C_1(Wr^2 2\alpha)$$
$$= 595{,}701 \text{ Nm}$$

$$\text{Maximum positive Bending moment at mid span } = M_e = C_2(Wr^2 2\alpha)$$
$$= 270{,}773 \text{ Nm}$$

$$\text{Maximum Torsional moment } = M_T = C_3(Wr^2 2\alpha) = 45.129 \text{ Nm}$$

$$\text{Maximum shear force at support } V = wr\alpha = 718{,}612 \times 4 \times 3.14 / 8 = 118{,}221 \text{ N}$$

$$\text{Shear force at any point } = V_\varphi = wr(\alpha - \phi)$$

At 9.5° shear force $V_m = 718{,}612 \times 4 \times (22.5 - 9.5)3.14/180 = 651{,}861$ N

Bending moment at the point of maximum torsional moment $\phi = 9.5^\circ$

$$M_m = wr^2(\alpha \sin\phi + \alpha \cot \alpha \cos\phi - 1) = 718{,}612 \times 4^2\{(3.14/8)\sin 9.5$$
$$+ (3.14/8)\cot 22.5 \cos 9.5\} - 1$$
$$= -1478 \text{ Nm } (\text{Hogging})$$

$$\text{Equivalent Moment as per IS 456} = M_e = M + T(1 + D/b)/1.7$$
$$= \{1478 + 45{,}129(1 + 1200/600)/1.7\}$$
$$= -81{,}117 \text{ Nm}$$

$$\text{Equivalent shear as per IS 456} = \text{V} + 1.6\text{T}/\text{b} = 651{,}861 + 1.6 \times 45{,}129/0.6$$
$$= 772{,}205 \text{ N}$$

Considering beam dimension 600 mm × 1200 mm
Applying working design methodology as per IS456

TABLE 13.1
Design Moments and Shear Force of Ring Beam Supported on Columns

Location	Bending Moment (Nm)	Shear Force (N)	Torsional Moment (Nm)
At support	59,50	1,128,221	0
At midpoint	20,3	0	0
At 9.5°	–148	651,861	45,129

$\sigma_{cbc} = 8.5\ \text{N/mm}^2$, Modular ratio $m = 280/3\sigma_{cbc} = 11$

$$K = (m\sigma_{cbs})/(\sigma_{st} + ma_{cbs}) = (11 \times 8.5)/(130 + 11 \times 8.5)$$

$$= 0.418\ \text{J} = (1 - k/3) = 0.861$$

$$Q = 0.5 \times \sigma_{cbc} \times J \times k = 0.5 \times 8.5 \times 0.861 \times 0.418 = 1.529\ \text{N/mm}^2$$

$$d = 1200 - 50 - 10 - 10 = 1130\ \text{mm}$$

$$M_r = Qbd^2 = 1.529 \times 0.6 \times 1130^2 = 1{,}171{,}428\ \text{Nm} > \text{Design moment OK}$$

$$\text{Support moment} = 595.7 \times 10^6\ \text{N-mm}$$

We know $\sigma_{st} A_{st}\, d\,(1 - k/3) = 595.7 \times 10^6$ or, $130 \times A_{st} \times 1130 \times 0.861 = 595.7 \times 10^6$, $A_{st} = 4710\ \text{mm}^2$

Provided 6 nos 32 ϕ TMT bar, ie. A_{st} provided = 4824 mm^2

For mid-span moment $M = 270.8 \times 10^6$ and we have, $\sigma_{st} A_{st}\, d\,(1 - k/3) = 270.8 \times 10^6$ or, $130 \times A_{st} \times 1130 \times 0.861 = 270.8 \times 10^6$, $A_{st} = 2141\ \text{mm}^2$

Provided 4 nos. 32 ϕ TMT bar along the bottom of the beam, A_{st} provided = 2940 mm^2

$$\text{Maximum shear at support} = 1{,}128{,}221\ \text{KN} = 1128.2\ \text{KN}$$

$$P_t = (A_{st}/bd) \times 100 = (4824 \times 100)/(600 \times 1130) = 0.71\%$$

Permissible shear stress $\tau_c = 0.35\ \text{N/mm}^2$

$$\text{Actual shear stress}\ \ 7_v = \left(1128.2 \times 10^3\right)/(600 \times 1130) = 1.664\ \text{N/mm}^2\ \tau$$

$\tau_v > \tau_c$, Therefore, shear reinforcements are to be designed

$$V_e = (7_v - 7_c)bd = (1.664 - 0.35) \times 600 \times 1130 = 891 \times 10^3 \text{ N}$$

Please replace 7 by τ
Providing, 6*L* -12mm stirrups

$$A_{st} = 6 \times 3.14 \times 12 \times 12 / 4 = 678 \text{ mm}^2,$$

$$S = \text{ Spacing required } = \sigma_{st} A_{sv} d / V_e = (1130 \times 130 \times 678) / (819 \times 10^3) = 121 \text{ mm}$$

Providing 6*L* -12 mm TMT bar @ 100 mm c/c.

Seismic Analysis of Overhead Water Tank

Applying IS1893 (Part-II) -2016

Thickness of top dome = 100 mm thick, Dimension of top ring beam is 200 mm × 350 mm

Thickness of cylindrical wall = 350 mm, Thickness of conical portion = 250 mm

RB_2 – 600 mm × 1000 mm, RB_3–600 mm × 1200 mm, M25 grade concrete, Seismic zone –IV, Z = Zone factor = 0.24, I = Importance factor = 1.5, R = Response reduction factor = 2.5 considering specially moment resisting frame (SMRF), soft soil, h = 8.6 m, Height of column = 16 m, total mass of water = 10^6 kg

For this tank, $h/D = 8.6/12.5 = 0.685$

m_i = Impulsive mass, m_e = convective mass, K_e = Spring stiffness
From fig 2(a), IS 1893-Part II -2016, For $h/D = 8.6/12.5 = 0.685$

$m_i/m = 0.69$ Therefore, $m_i = 0.69 \times 10^6$ kg

$m_C/m = 0.3$ Therefore, $m_c = 0.3 \times 10^6$ kg

$hc/h = 0.625$ Therefore, $h_c = 8.6 \times 0.625 = 5.35$ m

$h_c^*/h = 0.5$ Therefore, $h_c^* = 0.5 \times 8.6 = 6.45$ m

$h_i/h = 0.35$ Therefore, $h_i = 0.35 \times 8.6 = 3.225$ m

$h_i^*/h = 0.65$ Therefore, $h_i^* = 0.65 \times 8.6 = 5.805$ m

Mass of the Empty Container

$$\text{Mass of Top dome } = 2 \times 3.14 \times R^2(1 - \cos\phi)t\rho_w$$

$$= 2 \times 3.14 \times 9.0625^2(1 - 0.724) \times 0.1 \times 2500 = 35{,}588 \text{ kg}$$

$$\text{Mass of Ring beam RB1} = 0.2(0.35-0.2)\times 2500\times 2\times 3.14\times 12.5/2$$

$$= 2943.75 \text{ kg}$$

$$\text{Mass of Cylinder wall} = 3.14Dt_{\text{average}}\,H_{pc} = 3.14\times 12.5\times 0.2625\times 7.5\times 2500$$

$$= 193{,}183 \text{ kg}$$

$$\text{Mass of Ring beam RB2} = 3.14D\times 0.6\times (1-0.325)\times 2500 = 39{,}741 \text{ kg}$$

$$\text{Mass of conical portion} = 3.14(D+D_0)tt_0\,p_{\text{conc}}\;/2$$

$$= 3.14(12.5+8)\times 2.25\times 0.8\times 2500/2\times \cos 60$$

$$= 289{,}665 \text{ kg}$$

$$\text{Mass of Bottom dome} = (2\times 3.14\times 5\times 2)\times 0.25\times 2500 = 39{,}250 \text{ kg}$$

$$\text{Mass of Ring beam RB3} = 3.14D_0B\;d\;\,p_c = 3.14\times 8\times 0.6\times 1.2\times 2500$$

$$= 45{,}239 \text{ kg}$$

$$\text{Mass of empty tank} = 35{,}588+2943.75+193{,}183+39{,}741+289{,}665$$

$$+39{,}250+45{,}239$$

$$= 645{,}609.75 \text{ kg} = 646 \text{ Tonne}$$

Mass of water = 1000 Ton, Total weight = 1646 Tonne

Providing 8 columns, therefore, each column carries load = 1646/8 = 205.5 Tonnes

Considering, 10% extra load on account of a column lo = 20 Tonnes

$$\text{Design vertical load} = (205.75+20) = 225.75 \text{ T}$$

Increasing 25% load considering for effect of moments (in the absence of analyzed value)

$$= 225.75\times 1.25 = 282.18 \text{ T}$$

$$\text{Factored vertical load} = 1.5\times 282.18 = 423.28 \sim 424 \text{ Tonnes} = 4240 \text{ kN}$$

Considering 600 × 600 square column with 1% reinforcement

We know from IS 456, according to Limit State design Methodology,

$$P_u = 0.4 \times 25 \times 600 \times 600 + 0.67 \times 500 \times 3600$$
$$= 4806 \times 10^3 \text{ N} = 4806 \text{ kN} > 4240 \text{ kN, OK}$$

$$I_{xx} = I_{yy} = 8 \times (600^4 / 12) + 600 \times 600$$
$$\left[0^2 \times 2 + 4\{(8000/2)\cos 45\}^2 + 2 \times (8000/2)^2\right] \text{mm}^4$$
$$= 2.31 \times 10^{13} \text{mm}^4 = 23.1 \text{ m}^4$$

Due to heavy mass at the top of the water tank, slope at the top of this cantilever type structure is very very low (Table 13.2).

Therefore, considering stiffness of staging $K_s = 12\ EI/l^3$

For M25 grade concrete, $E = 25{,}000$ N/mm^2

$$K_s = 12EI/1^3 = (12 \times 25{,}000 \times 2.31 \times 10^{13}) / (20 \times 10^3)^3 = 866{,}250 \text{ N/mm}$$
$$= 8662.5 \times 10^5 \text{ N/m}$$

Considering 600 × 600 square column and height of column $l = 16$ m

$$\text{Weight of all the column} = 144{,}000 \text{ kg}$$

RCC bracing beams 300 mm × 600 mm are provided @ 4m c/c

$$\text{Therefore, weight of bracing } (300 \text{ mm} \times 600 \text{ mm})$$
$$= 4 \times 8 \times (3.14 \times D_0 - 0.6 \times 8) \times 0.3 \times 0.6 \times 2500$$
$$= 292{,}608 \text{ kg}$$

$$\text{Total weight of staging} = 144{,}000 + 292{,}608 = 436{,}608 \text{ kg}$$

As per IS 1893-Part II -2016

$$m_s = \text{Mass of empty water} + \text{ one third mass of staging}$$
$$= 645{,}609.75 + 145{,}536 = 791{,}145.75 \text{ kg}$$

$$T_1 = 2 \times 3.14\left\{(m_i + m_s)/k_s\right\}^{0.5}$$
$$= 2 \times 3.14\left\{\left(6.9 \times 10^5 + 7.9 \times 10^5\right) / \left(8662.5 \times 10^5\right)\right\}^{0.5} = 0.259 \text{ sec}$$

$$\text{Time period of convective mode } T_c = c_c (D / g)^{0.5}$$

For $h/D = 8.6/12.5 = 0.685$, $C_c = 3.2$ (as per IS 1893-Part II – 2016)

$$T_c = 3.2(12.5 / 9.81)^{0.5} = 3.61 \text{ sec}$$

Design Horizontal Seismic Coefficient

As per clause 4.5 for impulsive mode

$$A_{hi} = (Z / 2)(I / R)(S_a / g)$$

$$T_i = 0.259 \text{ sec}$$

For soft soil, $(S_a/g) = 2.5$ as per IS 1893(Part - I), clause 6.4.2

$$\text{Therefore, } A_{h\,I} = (Z / 2)(I / R)(S_a / g)$$
$$= (0.24 / 2)(1.5 / 2.5) \times 2.5 = 0.18$$

Design seismic coefficient for convective mode

$$A_{hc} = (Z / 2)(I / R)(S_a / g)$$

$$T_c = 3.2(12.5 / 9.81)^{0.5} = 3.61 \text{ sec}$$

For soft soil $(S_a/g) = 0.462\,5$ as per IS 1893(Part - I), clause 6.4.2
Therefore, $A_{hc} = (Z/2)\,(I/R)\,(S_a/g) = (0.24/2)\,(1.5/2.5) \times 0.462 = 0.033$

Base Shear

Base shear at the bottom of the staging in impulsive mode

$$V_i = A_{\text{hi}} (m_i + m_s) g \quad \text{as per IS 1893(Part - II) - Clause 4.6.2}$$

$$= 0.18 \times (6.9 \times 10^5 + 7.9 \times 10^5) \times 9.8 = 2{,}610{,}720 \text{ N} = 2611 \text{ kN}$$

Similarly, base shear in convective mode

$$V_c = A_{\text{hd}} (m_c) g \quad \text{as per IS 1893(Part – II) – Clause 4.6.2}$$
$$= 0.033 \times 3.7 \times 10^5 \times 9.81 = 119{,}780 \text{ N} = 119.78 \text{ kN}$$

$$V = (V_i^2 + V_c^2)^{0.5} = 2614 \text{ kN as per IS 1893(Part - II) - Clause 4.6.2}$$

Base Moment

Overturning moment at the base of staging in an impulsive mode

$$M_i^* = A_{hi}\left[m_i\left(h_i^* h_s\right) + m_s h_{cg}\right]g \quad \text{as per IS 1893(Part -II) - Clause 4.7.2}$$

$$m_i = 0.69 \times 10^6 \text{ kg}$$

$$h_I^* = 0.675 \times 8.6 = 5.805 \text{ m}$$

$$\begin{aligned} m_s &= \text{Mass of empty water +one third mass of staging} \\ &= 645{,}609.75 + 145{,}536 = 791{,}145.75 \text{ kg} \end{aligned}$$

$$h_s = 20 \text{ m}$$

$$\begin{aligned} h_c = (&35{,}588 \times 11.78 + 2943.75 \times 9.57 + 193{,}183 \times 6.325 + 39{,}741 \times 2.75 \\ &+ 289{,}665 \times 1.5 + 39{,}250 \times 0.15 - 45{,}239 \times 0.6) / 645{,}609.75 = 3.38 \text{ m} \end{aligned}$$

$$h_{cg} = h_s + h_c = 20 + 3.38 = 23.38 \text{ m}$$

$$\begin{aligned} M_i^* &= A_{hi}\left[m_i\left(h_i^* + h_s\right) + m_s h_{cg}\right]g \\ &= 0.18\left[0.69 \times 10^6 (5.805 + 20) + 791{,}145.75 \times 23.38\right] \times 9.81 = 64{,}103 \text{ kN-m} \end{aligned}$$

Similarly, overturning moment for the convective mode

$$M_c^* = A_{hc}\left[m_c\left(h_c^* + h_s\right)\right]g$$

$$m_c = 0.37 \times 10^6 \text{ kg}, \; h_c^* = 0.75 \times 8.6 = 6.45 \text{ m}$$

$$\begin{aligned} \text{Therefore, } M_c^* &= A_{hc}\left[m_c\left(h_c^* + h_s\right)\right]g \\ &= 0.033 \times \left[0.37 \times 10^6 (6.45 + 20)\right] \times 9.81 = 3169 \text{ kN-m} \end{aligned}$$

$$\text{Total overturning moment } = \left(M_i^{*2} + M_c^{*2}\right)^{0.5} = 64{,}181 \text{ kN-m}$$

Sloshing Wave Height

As per IS 1893(Part-II):2016- clause 4.11

Maximum sloshing wave height

$$d_{\max} = (A_h)_c \text{ RD} / 2 = 0.033 \times 2.5 \times 12.5 / 2 = 0.515 \text{ m}$$

Wind Analysis of Overhead Water Tank

Applying IS 185 (Part-III) -2015

Supporting the tank with eight columns equally spaced along the circumference (8 m diameter)

Each column carry load = 225.5 T, Column dimension = 600 mm × 600 mm

Design Wind Pressure

As per IS 85 – Part – III – 2015, Basic wind speed $V_b = 4$ m/sec

Design wind velocity at a height z, $V_z = V_b\, k_1\, k_2\, k_3\, k_4$,

Considering for tank, $k_a = 0.8$, for staging = $k_a = 0.8$

Wind Force on Top Dome and Cylindrical Portion

Height/Diameter = 10/12.5=0.8 < 2 (considering rough surface) $V_z\, b$= 53.81 × 12.5 > 6

Force coefficient for circular shape = 0. as per IS 85 (Part – III) -2015

Wind force on top dome and cylinder wall

$$= (7.5 + 2.5 / 2) \times 12.5 \times 1.38 \times 0.7 = 105.65 \text{ kN}$$

Wind Force on Conical Portion

Height/Diameter = 2.25/{(12.5 +8)/2} = 0.2 < 2 (considering, rough surface)

$$V_z b = 51.29 \times \{(12.5 + 8) / 2\} > 6$$

Force coefficient = 0. as per IS 85 (Part – III) – 2015

$$\text{Wind force on conical portion } = \{(12.5 + 8) / 2\} \times 2.25 \times 1.26 \times 0.7 = 20.34 \text{ kN}$$

Wind Force on Bottom Ring Beam

Height/Diameter = 1.2/8=0.15 < 2 (considering, rough surface)

$$V_z b = (51.29 \times 8) > 6$$

Force coefficient = 0. as per IS 85 (Part – III) – 2015

$$\text{Wind force on bottom ring beam } = 8 \times 1.2 \times 1.26 \times 0.7 = 8.46 \text{ kN}$$

Wind Force on Stagging

Wind force on column (consider only frontal area of wind)

Height/Diameter = 20/8 = 2.5 < 5 (considering rough surface)

$$V_z b = (50.79 \times 8) > \dot{6}$$

Force coefficient = 0.8 as per IS 85 (Part – III) – 2015

Wind force on column = 5 × 0.6 × 20 × 1.23 × 0.8 = 59.04 kN (considering only five frontal column)

Wind force on bracing beams (300 mm × 600 mm)

Height/breadth = 0.6/3.14 = 0.19 {since (3.14 × 8/8) = 3.14}

$$V_z b = (50.79 \times 3.14)$$

Force coefficient = 1, Octagonal shape as per as per IS 85 (Part – III) – 2015

Wind force on bracings = 4 × 0.3 × 3.14 × 1.23 × 1 = 4.89 kN (4nos/column @4m C/c)

Total Horizontal Wind force = 198.38 KN

Assuming contra-flexure at mid-height of the columns and fixity at the base due to raft foundations, the moment at base of the column is

$$M_1 = (198.38 \times 4) / 2(\text{since } @4\text{c/c})$$
$$= 396.76 \text{ kN-m}$$

If M_l = Moment at the base of the column due to wind loads

$$= 105.65 \times 25 + 20.34 \times 21.5 + 8.46 \times 20.06 + 1.22 \times 16 + 1.22 \times 12$$
$$+1.22 \times 8 + 1.22 \times 4$$
$$= 3297.06 \text{ kN-m}$$

If V = Reaction at the base of exterior columns

$$M_1 = \Sigma M + (V / r_1) \times \left[\Sigma r^2\right]$$

$$\text{or } 3297.06 = 396.76 + (V / 4)\left[\ 2 \times 4^2 + 4\left(4 / 2^{0.5}\right)^2\ \right]$$

TABLE 13.2
Wind Forces on Over Head Water Tank

Height(z)	V_b	K_1	K_2	K_3	K_4	V_z	P_z (N/m²)	K_d	K_a	K_c	K_d (kN/m²)
33.34	4	1.0	1.0	1	1	53.81	13.329	1	0.8	1	1.38
23.34	4	1.0	1.02	1	1	51.29	158.5.26	1	0.8	1	1.26
20	4	1.0	1.01	1	1	50.9	154.95	1	0.8	1	1.23

$$V = 181.26 \text{ kN}$$

$$\text{Total load on column in base } = 4785.76 + 181.26 = 4967.02 \text{ kN}$$

$$\text{Moment in each column at base } = 396.76 / 8 = 49.59 \text{ kN-m}$$

Stress in Column

$$\text{Axial load } P = 4967.02 \text{ kN}$$

$$\text{Bending moment } M = 49.59 \text{ kN-m}$$

$$\text{Eccentricity } = e = M / P = 49.59 \times 10^6 / (4967.02 \times 10^3) = 9.98 \text{ mm}$$

Since eccentricity is very small, direct stresses are predominant.
Try to use 1% steel so $A_{sc} = (600 \times 600)\ (1/100) = 3600 \text{ mm}^2$
Providing, 16 bars with lateral ties 12 mm @150mmc/c

$$A_{sc} = 16 \times 1256 = 20{,}096 \text{ mm}^2$$

$$A_e = \text{BD} + 1.5\, mA_{sc} \quad [\text{since, } m = 9.33 \text{ for M30}]$$
$$= 600 \times 600 + 1.5 \times 9.33 \times 20{,}096 = 641344 \text{ mm}^2$$

$$I_e = BD^3 / 12 + 1.5m\ A_{sc}h^2$$
$$= 600^4 / 12 + 1.5 \times 9.33\left[\left(2 \times 1256 \times 250^2\right) + 4 \times 1256\left(250 / 2^{0.5}\right)\right]$$
$$[\text{ since } h = (600 / 2) - 50]$$
$$= 1.3 \times 10^{10} \text{mm}^4 \text{ [50 mm clear cover]}$$

$$\text{Direct compressive stress } = \sigma_{cs} = (4{,}967.02 \times 10^3) / 641{,}344 = 7.74 \text{ N / mm}^2$$

$$\text{Bending stress } = \sigma_{cb} = (49.59 \times 10^6 \times 300) / (1.3 \times 10^{10}) = 1.14 \text{ N / mm}^2$$

For M30 grade concrete, Permissible $\sigma_{cc} = 8$ N/mm², Permissible $\sigma_{cb} = 10$ N/mm²

Permissible stress in concrete may be increased by 33.33% while considering the wind effect

$$\left(\sigma'_{cc}/\sigma_{cc}\right) + \left(\sigma'_{cb}/\sigma_{cb}\right) = 7.74 / 1.33 \times 8 + 1.14 / 1.33 \times 10 = 0.813 < 1, \text{OK}$$

Design of Bracings

$$\text{Moment in bracing } = 2 \times \text{ moment in column } \times 2^{0.5} = 2 \times 49.59 \times 2^{0.5} =$$

$$140.26 \text{ kN-m}$$

Dimension of bracing beams 300 mm × 600 mm

$B = 300$ mm, $d = 600 - 50 = 550$ mm

Moment of resistance of section is $M_1 = 1.529 \times 300 \times 550^2 = 138{,}56{,}50$ N-mm = 138.5 kN-m

$$\text{Balance moment } M_2 = 140.26 - 138.75 = 1.51 \text{ kN-m}$$

$$\text{Modular ratio } m = 280 / 3\sigma_{cbc} = 9.33 \text{ for M30}$$

$$K = \left(m\sigma_{cbc}\right) / \left(\sigma_{st} + m\sigma_{cbs}\right) = (9.33 \times 10) / (130 + 9.33 \times 10) = 0.417$$

$$J = (1 - k / 3) = 0.86, A_{st\,1} = M_1 / \sigma_{stj} d = \left(138.75 \times 10^6\right) /$$

$$(130 \times 0.86 \times 550) = 2253.46 \text{ mm}^2$$

$$A_{st2} = (1.51 \times 10^6) / (130 \times 0.86 \times 550) = 24.55 \text{ mm}^2$$

$$A_{st} = A_{st1} + A_{st2} = 2253.46 + 24.55 = 2278.01 \text{ mm}^2$$

Provided 4–32 ϕ Fe 500 TMT bars at top and bottom, since wind effect is reversible.

$$\text{Length of bracing } (c / c) = 3.14 \times 8 / 8 = 3.14 \text{ m}$$

$$\text{Maximum shear force in bracing } = \text{ Moment in bracing } / (0.5 \times \text{length of bracing })$$

$$= 140.26 / (0.5 \times 3.14) = 89.33 \text{ kN}$$

$$7 = (89.33 \times 10^3) / 300 \times 550 = 0.54 \text{ N / mm}^2$$

$$(100A_{stt}) / bd = (100 \times 3215.36) / (300 \times 550) = 1.94 \text{ N / mm}^2$$

c = 3.5 N/mm² for M30 grade concrete and percentage of tension steel from IS456 < c, Therefore, minimum shear reinforcement 2L-10 ϕ @150 mm c/c is provided.

Seismic Analysis of Overhead Water Tank

Base diameter of the top dome (D) = 14 m, Height of the cylindrical portion (H) = 5 m

Diameter of bottom dome (D_o) = 8 m, Rise of top dome (H_1) = 1.6 m

Rise of bottom dome (H_2) = 1.4 m, Height of conical portion (H_0) = 3 m

Radius of curvature of top dome (R_1) = 16.1 m

Radius of curvature of bottom dome (R_2) = 6.414 m, top dome thickness = 100 mm

Dimension of the top ring beam = 300 mm × 525 mm

Thickness of the cylindrical wall = 325 mm, size of the bottom ring beam = 600 mm × 1000 mm

Thickness of bottom dome = 200 mm, Thickness of conical portion = 250 mm

Diameter of eight nos. of circular columns = 500 mm, Height of the column staging = 20 m

Dimension of the bracing beams = 300 mm × 500 mm, Total mass of the tank m= 922,000 kg

Height of the Cylindrical portion h = 6 m, Diameter of the Cylindrical portion D = 14 m

$$h / D = 6 / 14 = 0.428$$

From fig.2 (a) of IS 1893:2016- Part II,

$$m_i/m \text{ ratio is} = 0.42 \text{ and } m_c/m = 0.4932 \text{ for } h/D = 0.428$$

Therefore, m_i = 440,049 kg and m_c = 454,26 kg

From Fig. 2(b) of IS 1893:2014-Part II, for the value of h/D = 0.428

$$h_c/h = 0.5828 \text{ ie. } h_c = 3.4928 \text{ m}$$

$$h_c^* / h = 0.5828 \; h_c^* = 3.4928 \text{ m}$$

$$h_i/h = 0.35 \text{ ie. } h_i = 2.241 \text{ m}$$

$$h_i^* / h = 0.9226 \text{ ie. } h_i^* = 5.5286 \text{ m}$$

Mass of Empty Container

*Substituting the values of parameter available, the following calculations are made

$$\text{Mass of top dome } = 2\pi R_1 h_1 t\rho = 40{,}474.4 \text{ kg}$$

$$\text{Mass of top ring beam } = \pi(D+b)bd\rho = 17{,}680.14 \text{ kg}$$

$$\text{Mass of cylindrical wall } = \pi DHt\rho = 178{,}587.5 \text{ kg}$$

$$\text{Mass of bottom ring beam } = \pi(D+b)bd\rho = 68{,}766 \text{ kg}$$

$$\text{Mass of conical portion } = 4762.4 \text{ kg}$$

$$\text{Mass of bottom dome } = 2\pi R_2 h_2 t\rho = 28{,}197.2 \text{ kg}$$

$$\text{Mass of circular ring beam } = \pi D_0 bd\rho = 45{,}216 \text{ kg}$$

Hence, total mass of empty container $=$

$$40{,}474.4 + 17{,}680.14 + 178{,}587.5 + 68{,}766 + 64{,}762.4 + 28{,}197.2 + 45{,}216$$

$$= 443{,}683.96 \text{ kg}$$

Mass of Column and Bracing

$$\text{Mass of } 8 \text{ numbers of columns } = 8 \times 3.14 \times 0.5 \times 0.5 \times 20 \times 2500 / 4 = 78{,}500 \text{ kg}$$

$$\text{Mass of bracing } = 4 \times (3.14 \times 8 - (8 \times 0.5)) \times 0.3 \times 0.5 \times 2500 = 31{,}680 \text{ kg}$$

Stiffness of Staging

As, $Ks = \frac{12EI}{L^3}$ where, I = Moment of inertia of the combined column system and E = modulus of elasticity of concrete $= 5000\sqrt{f_{ck}}\ \text{N}/\text{mm}^2, f_{ck} = 30\ \text{N/mm}^2$

I (for entire staging)

$$= \frac{\pi \times 500^4}{64} + \left(\frac{\pi \times 500^2}{4}\right)\left[0^2 \times 2 + 4 \times \left(\frac{8000}{2}\cos 45\right)^2 + 2 \times \left(\frac{8000}{2}\right)^2\right]$$

$$= 1.258 \times 10^{13}\,\text{mm}^4$$

or $I = 12.58\ \text{m}^4$

$$\text{Hence, } K_s = \frac{12 \times 27{,}386 \times 1.258}{20^3} \times 10^6 = 517{,}040{,}305.8\ \text{N}/\text{m}$$

Time Period

As per clause 4.3.1.3, time period in impulsive mode is

$$T_i = 2\pi\sqrt{\frac{m_i + m_s}{K_s}}$$

In this case, m_i = impulsive liquid mass = 440,049 kg and m_s = mass of empty container + 1/3 mass of staging = 440,049 + 1/3(8,500+31,680) = 480,410.63 kg

$$K_s = \text{ stiffness of staging } = 517{,}040{,}305.8\ \text{N}/\text{m}$$

Substituting the values

$$\text{Time period, } T_i = 0.2649\ \text{sec}$$

Similarly, time period in convective mode is,

$$T_c = C_c\sqrt{\frac{D}{g}}$$

From fig.5 of IS 1893-Part II-2016, for $h/D = 0.428$, $C_c = 3.41$

D = Inside diameter of tank

Substituting the values

$$\text{Time period in convective mode } = T_c = 4.082\ \text{sec}$$

Design Horizontal Seismic Coefficient

As per clause 4.5 of 1893:2016- Part II,

$$\text{Design horizontal seismic coefficient in impulsive mode} = \ (A_h)i = \frac{Z \times I \times S_a}{2 \times R \times g}$$

where Z = zone factor = 0.24 for zone IV (Table 2 of IS 1893: 2016: part I)
I = Importance factor = 1.5 (Table 1 of IS 1893: 2016- Part II)
R = Response reduction factor = 2.5 (Table 2 of IS 1893- Part II)
S_a/g = Design horizontal seismic coefficient = 2.5, obtained based on the time period
T_i = 0.2649 sec and medium type of soil (clause 6.4.2 of IS 1893- Part I)

Substituting different values already obtained, $(A_h)_i = 0.18$
Similarly, as per clause 4.5

$$\text{Design horizontal seismic coefficient in convective mode} = (A_h)c = \frac{Z \times I \times S_a}{2 \times R \times g}$$

where Z = zone factor = 0.24 for zone IV (Table 2 of IS 1893: 2016: part I)
I = Importance factor = 1.5 (Table 1 of IS 1893: 2014: part II)
R = Response reduction factor = 2.5 (Table 2 of IS 1893: part II)
S_a/g = Design horizontal seismic coefficient = 0.26, obtained based on the time period
T_c = 4.082 sec and medium type of soil (clause 6.4.2 of IS 1893- Part I)

Finally after calculations, $(A_h)_c = 0.018$

Base Shear Calculation

As per clause 4.6.2 of IS 1893: part II, Base shear in impulsive mode,

$$V_i = (A_h)i(m_i + m_s)g$$

So, $V_i = 0.18 \times (440{,}049 + 480{,}410.63) \times 9.81 = 1625.34$ kN

Similarly, Base shear in convective mode $= V_c = (A_h)c\ m_c g$

$$V_c = 0.018 \times 454{,}726 \times 9.81 = 80.29 \text{ kN}$$

Total base shear by SRSS method (IS 1893- Part I)

$$V = \sqrt{V_i^2 + V_c^2}$$

Hence, $V = 162.33$ KN

Base Moment Calculation

Overturning moment at the base of staging in impulsive mode (clause 4.2)

$$M_i^* = (A_h)i\left[m_i\left(h_i^* + h_s\right) + m_s h_{cg}\right]$$

Substituting all the values already obtained, finally, M_i^* = 36,802.98 KN-m

Overturning moment in convective mode,

$$M_c^* = (A_h)c\ m_c\left(h_c^* + h_s\right)g$$

or,

$$M_c^* = 1886.37 \text{ kN-m}$$

Substituting different values already obtained,

Finally, the total moment, $M^* = \sqrt{M_i^{*2} + M_c^{*2}}$ = 36,851.29 kN-m

Sloshing Wave Height

As per clause 4.11, the maximum wave height due to sloshing is,

$$d_{\max} = (A_h)c\ RD/2$$

Substituting different values already obtained, $d_{\max}$ = 0.315 m

Hydrodynamic Pressure

Impulsive Hydrodynamic Pressure

Impulsive hydrodynamic pressure as per clause 4.9.1 of IS 1893- Part II

$$P_{iw}(y) = Q_{iw}(y)(A_h)i.\rho gh\cos\varnothing$$

$$Q_{iw}(y) = 0.866\left[1 - \left(\frac{y}{h}\right)\right]\tanh\left(0.866\frac{D}{h}\right)$$

where

Maximum pressure will occur at $\varnothing = 0$

At the base of wall, $y = 0$, $Q_{iw(y=0)} = 0.8362$

Substituting different values already obtained,

Impulsive force at the base of wall is $P_{iw(y=0)} = 8.848$ kN/m^2

Now, the impulsive hydrodynamic pressure at the base of slab is

Substituting different value already obtained,

$$P_{ib} = 0.866(A_h)i\rho gh \frac{\sinh\left(1.732\frac{x}{h}\right)}{\cosh\left(0.866\frac{l}{h}\right)}$$

Impulsive hydrodynamic pressure at the base of slab = P_{ib} = 8.848 kN/m^2

Convective Hydrodynamic Pressure

As per clause 4.9.2, Convective hydrodynamic pressure on wall,

$$P_{cw} = Q_{cw}(y)(A_h)c.\rho gD\left[1-\frac{1}{3}\cos^2\varnothing\right]\cos\varnothing$$

where

$$Q_{cw}(y) = 0.5625\frac{\cosh\left(3.674\frac{y}{D}\right)}{\cosh\left(3.674\frac{h}{D}\right)}$$

At the base of the wall, $y = 0$; Therefore, $Q_{cw} = 0.2238$

Maximum pressure occurs at, $\varnothing = 0$

Substituting different values already obtained,

Convective pressure at the base of the wall, $P_{cw} = 0.30$ kN/m^2

At, $y = h$; $Q_{cw(y=h)} = 0.5625$

Substituting different values already obtained,

Convective pressure at $y = h$, $P_{cw} = 1.3905$ kN/m^2

Similarly, Convective hydrodynamic pressure at the base of slab, $y = 0$ given as per clause, 4.9.2

$$P_{cb} = Q_{cb}(x)(A_h)c.\rho gD$$

where

$$Q_{cb}(x) = 1.125\left[\frac{x}{D} - \frac{4}{3}\left(\frac{x}{D}\right)^3\right]\sec h\left(0.3674\frac{h}{D}\right)$$

Substituting different values already obtained, $Q_{cb} = 0.3$

Hence, the convective hydrodynamic pressure at the base of slab is = 0.914 kN/m^2

Equivalent Linear Pressure Distribution

As per clause 4.9.4 of IS 1893- Part II, for stress analysis on the tank wall instead of more exact analysis, equivalent linear pressure can be considered (Figures 13.6–13.9).

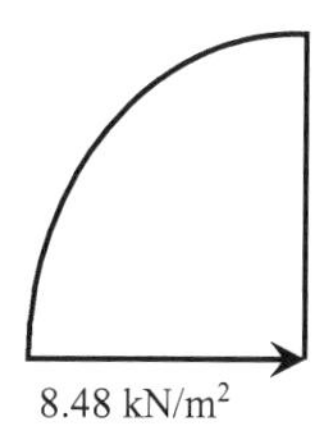

FIGURE 13.6 Actual impulsive pressure distribution on wall.

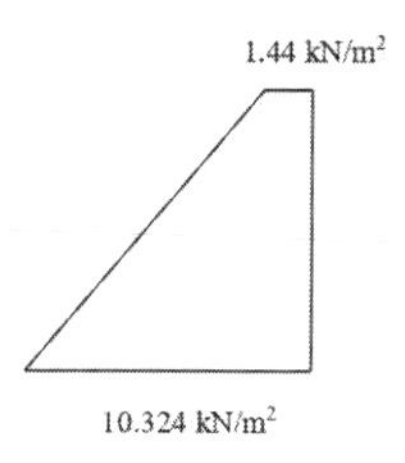

FIGURE 13.7 Equivalent impulsive pressure distribution on wall.

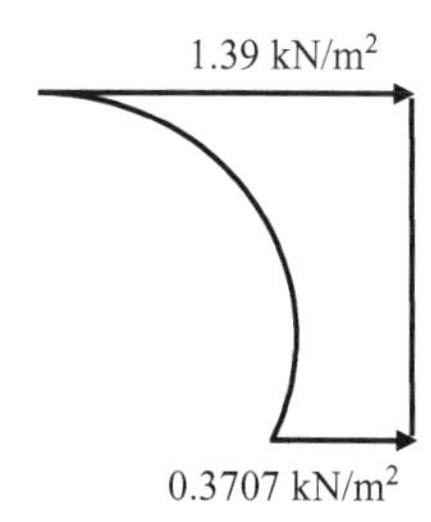

FIGURE 13.8 Actual convective pressure distribution on wall.

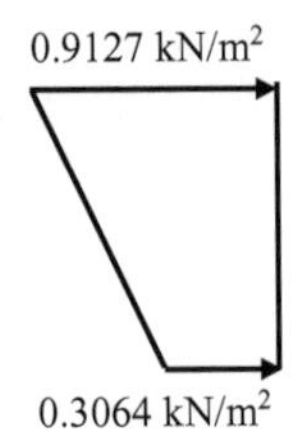

FIGURE 13.9 Equivalent convective pressure distribution on wall.

Base shear due to impulsive liquid mass per unit circumferential length is given by

$$q_i = \frac{(A_h)i.m_i g}{\pi D/2}$$

Substituting different values already obtained, $q_i = 35.35$ kN/m²

Pressure at bottom and top is given by $a_i = \frac{q_i}{h^2}(4h - 6h_i)\, b_i = \frac{q_i}{h^2}(6h_i - 2h)$

Substituting different values already obtained, $a_i = 10.324$ kN/m² and $b_i = 1.44$ kN/m²

Similarly, equivalent pressure distribution in convective mode as per clause 4.9.4 of IS1893: part II,

$$q_c = \frac{(A_h)c.m_c g}{\pi D/2}$$

Substituting different values already obtained, $q_c = 3.653$ kN/m

Pressure at top and bottom is given by $a_c = \frac{q_c}{h^2}(4h - 6h_c)\, b_c = \frac{q_c}{h^2}(6h_c - 2h)$

Now, $a_c = 0.3064$ kN/m² and $b_c = 0.912$ kN/m²

Pressure Due to Wall Inertia

Pressure on wall due to its inertia as per clause 4.9.5, is given by

where, ρ = is the mass density of the tank

$$P_{ww} = (A_h)i.t.\rho_m.g$$

Substituting different values already obtained, $P_{ww} = 1.4625$ kN/m²

This pressure is uniformly distributed along the wall height.

Pressure Due to Vertical Excitation

Hydrodynamic pressure on the tank wall due to vertical ground acceleration, as per clause 4.10.1 is

$$\text{where,}\quad P_v = A_v\left[\rho gh\left(1-\frac{y}{h}\right)\right] \quad A_v = \frac{2}{3}\left[\frac{Z}{2}.\frac{I}{R}.\frac{S_a}{g}\right]$$

Here, for the zone IV, $Z = 0.24$

$$I = \text{ Importance factor } = 1.5$$

$$R = \text{ Response reduction factor } = 2.5$$

As per clause 4.10.1 of code, in the absence of more refined analysis, time period for vertical mode of vibration for all types of tank can be considered = 0.3 sec

Hence $S_a/g = 2.5$ for hard type of soil and $T = 0.3$ sec

Therefore, $A_v = 0.16884$

At the base of the wall, $y = 0$; $P_v = 9.925$ kN/m^2

Maximum Hydrodynamic Pressure

As per clause 4.10.2, max hydrodynamic pressure by SRSS method,

$$P = \sqrt{(P_{\text{iw}} + P_{\text{ww}})^2 + P_{\text{cw}}^2 + P_v^2}$$

Substituting the values of parameter already available,

Pressure at the base of the wall, $P = 14.316$ kN/m^2

$$\text{Max Hydrostatic pressure } = \rho gh = 10 \times 9.81 \times 6 = 58.86 \text{ kN / m}^2$$

Hence, max hydrodynamic pressure is about **24.3541%** of hydrostatic pressure.

Need for Future Study

Most of the investigations aim at study of hydrodynamic behavior of elevated water tanks during multiple degree earthquake excitations experimentally. The values for studied parameters i.e. sloshing frequency, hydrodynamic pressure, base shear, tank acceleration, and sloshing height are calculated for the same tank analytically following the standard codes and compared with the work of G W Housner (1954) and findings of other researchers.

It is the combination of fluid–structure interaction and soil–structure interaction. Fluid–structure interaction is the interaction of some movable or deformable structure with an internal or surrounding fluid flow, which in our case is an elevated steel water tank. The deformations of a structure during earthquake shaking are affected by interactions between three linked systems: the structure, the foundation, and the geologic media underlying and surrounding the foundation. A seismic soil–structure interaction analysis evaluates the collective response of these systems to a specified free-field ground motion.

All the codes discussed in this chapter suggest higher design seismic force for tanks by specifying lower values of the response modification factor or its equivalent factor in comparison to the building system. There are substantial differences, however, in the manner and extent to which design seismic forces are increased in various codes. American codes and standards provide a detailed classification of tanks and are assigned a different value of the response modification factor. In contrast, Euro code 8 and NZSEE do not have such detailed classification, although NZSEE has given classification for ground-supported steel tanks. Provisions on soil-structure interaction are provided in NZSEE and Euro code 8 only (Figures 13.10 and 13.11).

Reinforcement Details of Overhead Water Tank

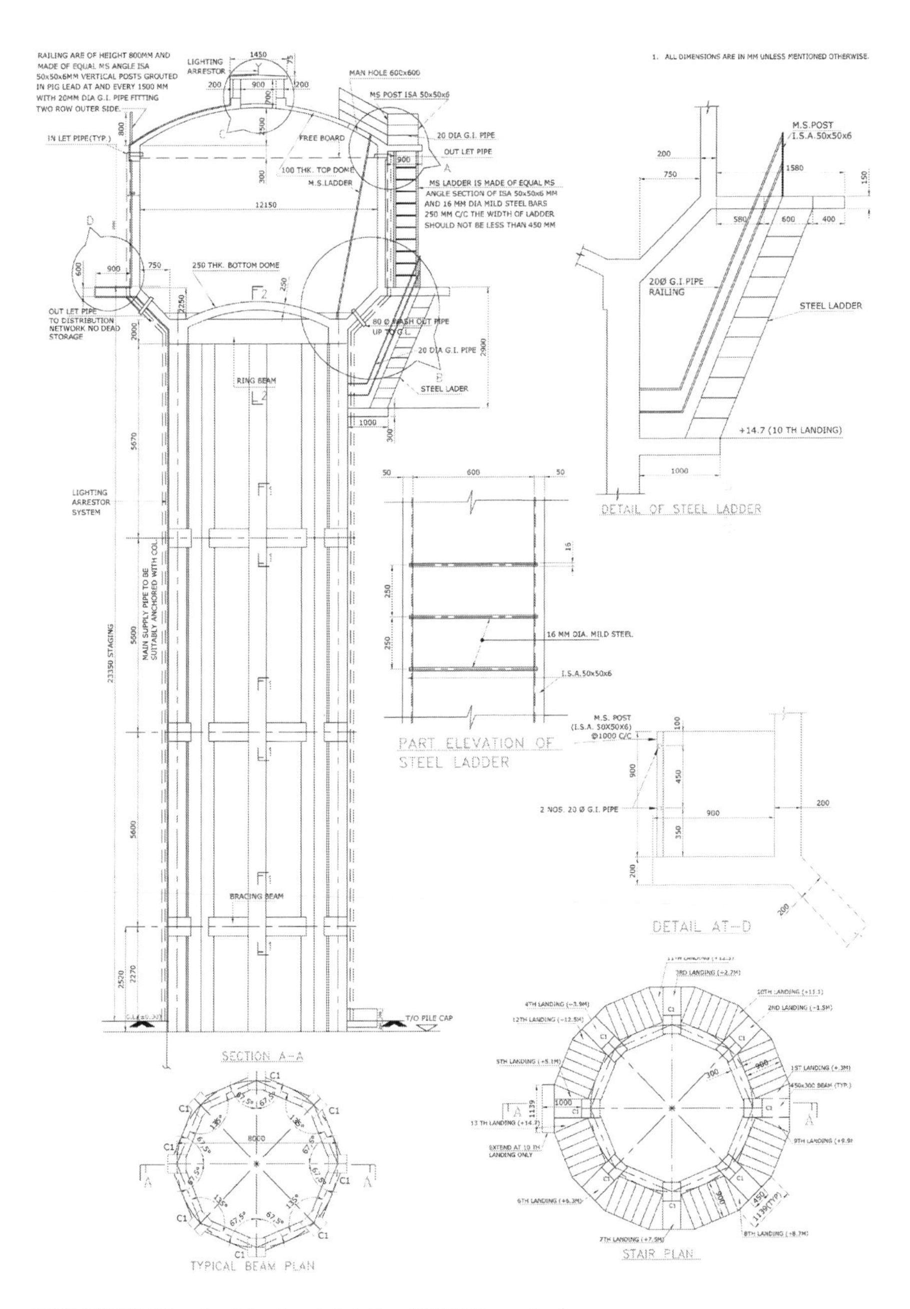

FIGURE 13.10 Architectural details of INTZ type tank.

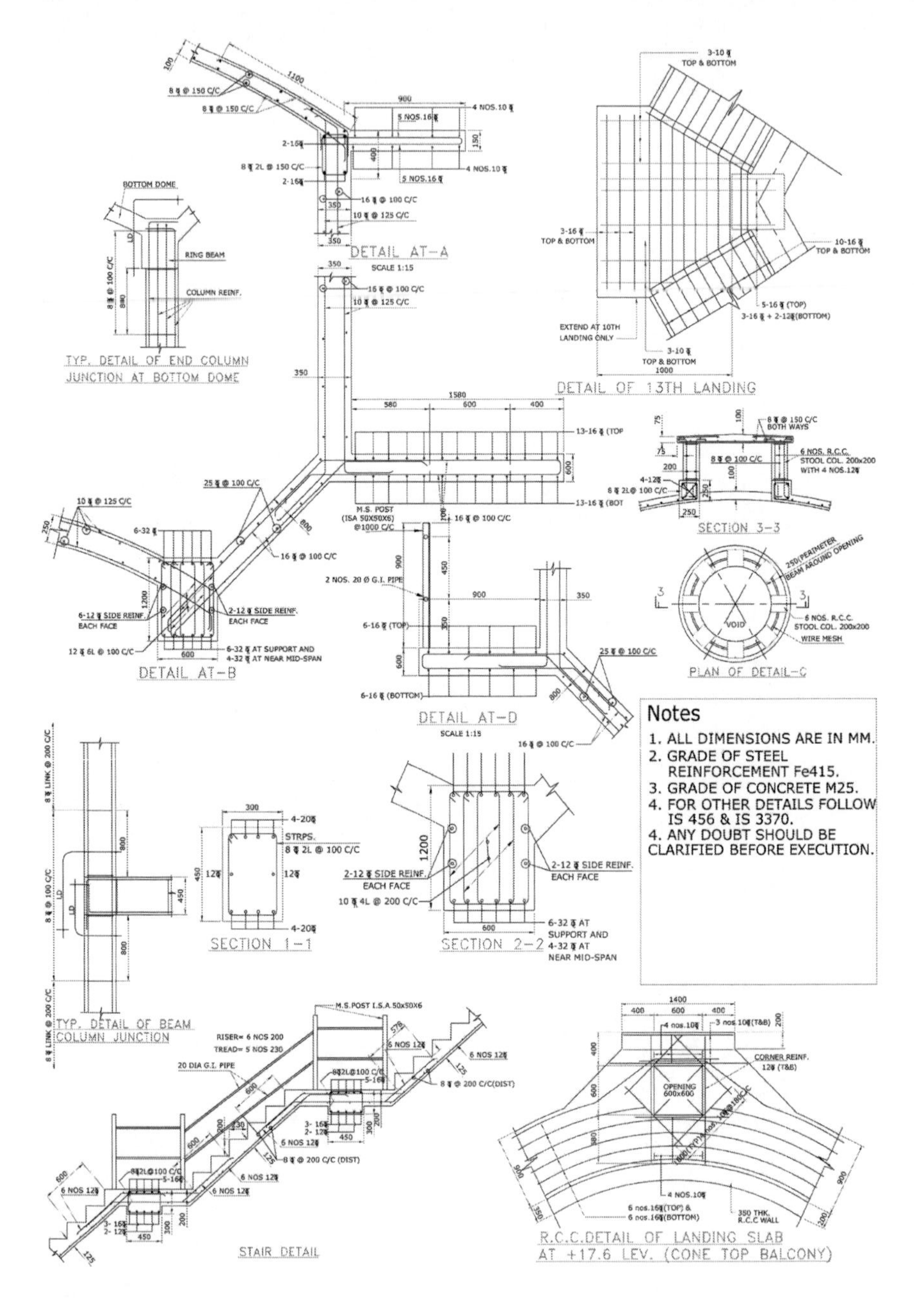

FIGURE 13.11 Structural details of INTZ type tank.

14 Comparison of Design Methodology and Output as per IS, BS, EURO, and ACI Codes

Acknowledging and after detailed studies, a number of tables and graphs are provided, and different important parameters of design of RCC structures are compared, which has got tremendous effect on structural design (Figure 14.1 and Tables 14.1–14.9)

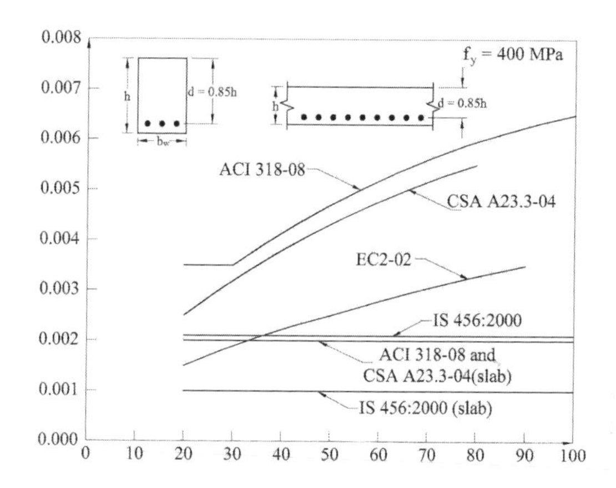

FIGURE 14.1 Comparison of minimum flexural reinforcement provision of different codes.

TABLE 14.1
Minimum Size Requirement for Seismic Beam and Column

Sl. No.	Code	Beams		Column	
		d_0 (min.), mm	B/D (min)	d_0 (min.), mm	B/D (min)
1.	ACI 318:11	250	0.30	300	0.4
2.	EC 8:1998	200	0.25	250	0.4
3.	IS 13920:2016	200[a]	0.3	300 or 15d_b	0.4

B, D, Breadth and depth of the member respectively; d_0, Largest longitudinal reinforcement bar diameter of beam.

[a] 300 mm for beam when span > 5m and column clear height > 4m.

DOI: 10.1201/9781003208204-14

TABLE 14.2
Minimum Steel Requirement for Beams as per Various Codes

Requirement	Code Provision as per				
	IS 456, Clause 26.5	**ACI 318[b]**	**CSA A23.3[b]**	**Eurocode 2**	**NZS 3101[b]**
Minimum tensile steel for flexure[a]. $\frac{A_s}{b_s\ d} \geq$	$\frac{0.85}{f_y}$ For T-sections, use b_w alone	$\frac{0.224\sqrt{f_{ck}}}{f_y} \geq \frac{1.4}{f_y}$ For T-sections, use $2b_w$ or b_f whichever is smaller alone	$\frac{0.18\sqrt{f_{ck}}}{f_y}$ For T-beams b_w is replaced by a value in the range $1.5b_w$ to $2.5b_w$	$\frac{0.26\ f_{ctm}}{f_y}$ ≥ 0.0013 For T-beams, b_w is taken as the mean breadth.	$\frac{0.224\sqrt{f_{ck}}}{f_y} \geq \frac{1.4}{f_y}$ For T-sections, b_w is taken as smaller of b_w or width of the flange.
Maximum tensile steel for flexure ≤	$0.04bD$	Net tensile strain in extreme tensile steel ≥ 0.005. This will result in approximately $p_t = 15.5\frac{f_{ck}}{f_y} \leq 2.5$	Tension reinforcement limited to satisfy $\frac{X_u}{d} \leq \frac{700}{700 + f_y}$	$0.04bD$	$\frac{0.9\ .\ f_{ck} + 10}{6\ .\ f_y} \leq 0.025$
Minimum shear reinforcement	$\frac{0.4}{0.87\ f_y}$ When $\tau v > 0.5\ \tau c$	$\frac{0.056\sqrt{f_{ck}}}{f_y} \geq \frac{0.35}{f_y}$ When applied shear is greater than 0.5 × concrete strength.	$\frac{0.054\sqrt{f_{ck}}}{f_y}$ When applied shear is greater than concrete strength.	$\frac{0.08\sqrt{f_{ck}}}{f_y}$ When applied shear is less than shear strength of concrete.	$\frac{0.09\sqrt{f_{ck}}}{f_y}$ When applied shear is greater than 0.5 × concrete strength.
Spacing of minimum stirrup ≤	$0.75d \leq 300$ mm	$0.5d \leq 600$ mm $0.25d \leq 300$ mm, When $V_s > 0.3\sqrt{f_{ck}}\ b_w.d$	$0.63d \leq 600$ mm $0.325d \leq 300$ mm, When $V_u > \varphi$ $(0.1 f_{ck}\ b_w\ d)$	$0.75d \leq 600$ mm	$0.5d \leq 600$ mm $0.25d \leq 300$ mm, When $V_s > 0.3\sqrt{f_{ck}}\ b_w.d$

A_s, Minimum tensile steel in flexure; A_{sv}, Minimum shear reinforcement; b, Breadth of beam; b_f, Breadth of flange; b_w, Breadth of web; D, Depth of beam; d, Effective depth of beam; f_{ck}, Characteristic compressive strength of concrete; f_{ctm}, Mean axial tensile strength = $0.30(f_{ck})^{0.666}$; f_y, Characteristic yield strength of reinforcement; p_t, Percentage of tensile steel; S_v, Spacing of vertical stirrup; T_c, Design shear strength of concrete; V_s, Nominal shear carried by vertical shear reinforcement; V_u, Factored shear force; x_u, Depth of neutral axis; τ_v, Nominal shear stress; φ, Resistance factor for concrete in shear = 0.65.

[a] Alternatively, it may be at least one-third greater than that required by the analysis, as per ACI code clause 10.5.3.

[b] The cylindrical strength is assumed to be 0.8 times the cube strength.

TABLE 14.3
Minimum Steel Requirement for Column as per Various Codes

Sl. No.	Code	Longitudinal Steel Minimum (%)	Longitudinal Steel Maximum (%)	Minimum Transverse Steel (Spiral),
1.	ACI 318:11	$> 1\, A_g$	$< 8\, A_g$	$> 0.09\, s\, D_k \left(\frac{A_g}{A_k} - 1\right)\frac{f_{ck}}{f_{yt}}$ $> 0.096\, s\, D_k\, \frac{f_{ck}}{f_{yt}}$
2.	IS 456:2000	$> 0.8\, A_g$	$< 6\, A_g$ *	$> 0.09\, s\, D_k \left(\frac{A_g}{A_k} - 1\right)\frac{f_{ck}}{f_{yt}}$
3.	IS 13920	$> 0.8\, A_g$	< 6 Ag *	$> 0.09\, s\, D_k \left(\frac{A_g}{A_k} - 1\right)\frac{f_{ck}}{f_{yt}}$ $> 0.096\, s\, D_k\, \frac{f_{ck}}{f_{yt}}$

TABLE 14.4
Effective Flange Width, *bf*, as per Different Codes

Type of Beam	IS 456:2000	ACI 318:08	NZS 3101
T-Beam	Least of (a) $b_w + 6\, D_f + L_0/6$ (b) $b_w + b_0$	Least of (a) $b_w + 16\, D_f$ (b) $L_0/4$ (c) $b_w + b_0$	Least of (a) $b_w + 16D_f$ (b) $b_w + L_0/4$ (c) $b_w + 2D_1$ (d) $b_w + b_0\left(\frac{D_1}{D_1 + D_2}\right)$
L-Beam	Least of (a) $b_w + 3\, D_f + L_0/12$ (b) $b_w + b_0$	Least of (a) $b_w + 6\, D_f$ (b) $b_w + L_0/12$ (b) $b_w + b_0$	Least of (a) $b_w + 8D_f$ (b) $b_w + L_0/8$ (c) $b_w + D_1$ (d) $b_w + \left(\frac{D_1}{D_1 + D_2}\right)$

TABLE 14.5
Effective Stiffness for Beam and Columns a per Different Codes

Type of Member	IS 456:2000	ACI 318:2011
Beam	I_g, I_r, or I_{cr}	$0.35\, I_g$ For T-beam take as two times the I_g of web, i.e., $2\left(b_w h_c^3 / 12\right)$
Column	I_g, I_r, or I_{cr}	$0.70\, I_g$

TABLE 14.6
Comparison of Shear Design Provision of Different Codes

Requirement	Code Provision as per				
	IS 456	ACI 318	CSA A23.3[a]	Eurocode 2	NZS 3101[a]
Minimum shear reinforcement $\frac{A_s}{b_w S_v} \geq$	$\frac{0.4}{0.87 f_y}$ when $\tau_v > 0.5\,\tau_c$	$\frac{0.9\sqrt{f_{ck}}}{16 f_y} \geq \frac{0.33}{f_y}$ when applied shear is greater than 0.5 × concrete strength	$\frac{0.054\sqrt{f_{ck}}}{f_y}$ when applied shear is greater than concrete strength	$\frac{0.08\sqrt{f_{ck}}}{f_y}$ when applied shear is greater than strength of concrete	$\frac{0.9\sqrt{f_{ck}}}{16 f_y}$ when applied shear is greater than 0.5 × concrete strength
Spacing of minimum stirrup ≤	$0.75d \leq 300$ mm	$0.5d \leq 600$ mm and $0.25d \leq 300$ mm, when $V_s > \sqrt{f'c}\, b_w d/3$	$0.63d \leq 600$ mm and $0.32d \leq 300$ mm when $V_u > \varphi_c f'c\, b_w d/11.4$	$0.75d \leq 600$ mm	$0.5d \leq 600$ mm and $0.25d \leq 300$ mm, when $V_s > \sqrt{f'c}\, b_w d/3$

[a] The cylinder strength is assumed to be equal to 0.8 times the cube strength.

TABLE 14.7
Development Length in Terms of L_d/d_p for HYSD bars (f_y = 415 MPa)

Name of Code/Formula	Grade of Concrete				
	M20	M25	M30	M35	M40
IS 456:2000 limit state $L_d = \frac{d_b\, f_s}{1.177\,(f_{ck})^{2/3}}$	47	41	38	33	29.7
IS 456:2000 working stress	45	40	36	33	30
BS 8110–1:97[a] Type 1 deformed ($\beta_1 = 0.4$)[b]	50	45	41	38	35.7
$L_d = \frac{d_b\, f_s}{4\,\beta_1 \sqrt{f_{ck}}}$ Type 2 deformed ($\beta_1 = 0.5$)	40	26	33	30.5	28.5
DIN 1045-1: 2001, Bond class I	39.2	33.4	30	26.5	24.4
Bond class II	56	47.7	43	38	35
AS 3600:2001+ $L_d = \frac{k_1\, k_2\, f_s\, A_b}{0.89\,(2a + d_b)\sqrt{f_{ck}}} \geq 25\, k_1\, d_b$	52	47	43	39	37

+ Assuming that clear cover, a = bar size, d_b; $k_1 = 1.0$ and $k_2 = 2.2$.

[a] Lap length to be increased by 40%–100% depending on the position of the bar.

[b] β_1 is a coefficient defined in BS 8110 and depends on the type of bar.

TABLE 14.8
Comparison of L_d/d_b for Bars in Tension (for Fe 415grade Steel)

Name of Code	Bar Diameter (mm)	Grade of Concrete M20	M25	M30	M35	M40
IS 456: 2000	All bars	47	40	38	33	29.7
ACI 318: 08	<19	49.5	44.2	40.4	37.4	35
$C_c = 1.5d_b$	>22	61.9	55.3	50.5	46.8	43.7
ACI 318: 08	<19	74.4	66.4	60.6	56	52.3
$C_c = 1.0d_b$	>22	93	83	75.8	70	65.6
Eq. (7.15) $C_c = 1.0d_b$; $\omega = 1$	All bars	71.2	66.4	62.5	59.2	56.5

TABLE 14.9
Tolerable Crack Widths According to ACI 224R-01, CEB-FIP Model Code-1990, and IS456

Sl. No.	Exposure Condition	Tolerable Crack Widths, mm ACI 224 R-01	CEB-FIP-90[a]	IS456-2000
1.	Low humidity, dry air, or protective environment	0.40	0.4–0.6	0.30
2.	High humidity, moist air, or soil	0.30	0.2–0.3	0.20
3.	De-icing chemicals	0.175	0.10–0.15	0.10
4.	Sea water and sea water spray	0.15	0.10–0.15	0.10
5.	Water retaining structures	0.10	–	–

Note: Lower crack width limit is for cases with minimum cover; upper limit = 1.5 × minimum cover.

List of Relevant Codes, Guidelines, and Handbooks

IS 269: 1989 — Specification for 33 Grade ordinary Portland cement
IS 8112: 1989 — Specification for 43 Grade ordinary Portland cement
IS 12269: 1987 — Specification for 53 Grade ordinary Portland cement
IS 8041: 1990 — Specification for rapid hardening Portland cement
IS 455: 1989 — Specification for Portland slag cement
IS 1489: 1991 — Specification for Portland pozzolana cement

-**Part I**: Flyash based, **Part II:** Calcined clay-based

IS 8043: 1991 — Specification for hydrophobic Portland cement
IS 12600: 1989 — Specification for low heat Portland cement
IS 12330: 1988 — Specification for sulfate resisting Portland cement
IS 8042: 1978 — Specification for Portland white cement
IS 8043: 1991 — Specification for hydrophobic Portland white cement
IS 6452: 1989 — Specification for high alumina cement for structural use
IS 6909: 1990 — Specification for super-sulfated cement
IS 4031: 1988 — Methods of physical tests for hydraulic cement
IS 383: 1970 — Specification for coarse and fine aggregates from natural sources for concrete
IS 9142: 1979 — Specification for artificial lightweight aggregates for concrete masonry units
IS 2386 (Parts 1–8) — Methods of tests for aggregate for concrete
IS 3025 (Parts 17–32) — Methods of sampling and test (physical and chemical) for water and wastewater
IS 9103: 1999 — Specification for admixtures for concrete
IS 3812: 1981 — Specification for flyash for use as pozzolana and admixture
IS 1344: 1981 — Specification for calcined clay pozzolana
IS 10262: 1982 — Recommended guidelines for concrete mix design
IS 7861 (Part 1): 1975 — Code of Practice for extreme weather concreting: Part 1 — Recommended practice for hot weather concreting
IS 4926: 1976 — Ready-mixed concrete (first revision)
IS 1199: 1959 — Methods of sampling and analysis of concrete
IS 516: 1959 — Methods of tests for strength of concrete
IS 5816: 1999 — Method of test for splitting tensile strength of concrete cylinders
IS 383: 1970 — Specification for coarse and fine aggregates from natural sources for concrete

IS 9142: 1979 — Specification for artificial lightweight aggregates for concrete masonry units
IS 2386 (Parts 1–8) — Methods of tests for aggregate for concrete
IS 3025 (Parts 17–32) — Methods of sampling and test (physical and chemical) for water and wastewater
IS 9103: 1999 — Specification for admixtures for concrete
IS 3812: 1981 — Specification for flyash for use as pozzolana and admixture
IS 1344: 1981 — Specification for calcined clay pozzolana
IS 432 (Part 1): 1982 — Specification for mild steel and medium tensile steel bars for concrete reinforcement
IS 1786: 1985 — Specification for high-strength deformed steel bars for concrete reinforcement
IS 1566: 1982 — Specification for hard-drawn steel wire fabric for concrete reinforcement
IS 2062: 1999 — Steel for general structural purposes- Specification
IS 1608: 1995 — Mechanical testing of metals – Tensile testing
IS 10262: 1982 — Recommended guidelines for concrete mix design
SP 23: 1982 — Design of Concrete Mixes
IS 875 — Code of practice for design loads (other than earthquake) for buildings and structures
Part 1: Dead loads
Part 2: Imposed (live) loads
SP 16: 1980 — Design Aids (for Reinforced Concrete) to IS 456: 1978
SP 24: 1983 — Explanatory Handbook on IS 456: 1978
SP 34: 1987 — Handbook on Concrete Reinforcement and Detailing
IS 13920: 1993 — Ductile detailing of reinforced concrete structures subjected to seismic forces
IS 3370 (Part 1): 1965 — Code of Practice for the storage of liquids: Part 1-IV — 2020
IS 7861 (Part 1): 1975 — Code of Practice for extreme weather concreting: Part 1 — Recommended practice for hot weather concreting
IS 4926: 1976 — Ready-mixed concrete (first revision)
IS 1199: 1959 — Methods of sampling and analysis of concrete
IS 516: 1959 — Methods of tests for strength of concrete
IS 5816: 1999 — Method of test for splitting tensile strength of concrete cylinders (first revision)
IS 3370 (Part 1): 1965 — Code of Practice for the storage of liquids: Part 1 — General
IS 1343: 1980 — Code of Practice for Prestressed Concrete (first revision)
IS 432 (Part 1): 1982 — Specification for mild steel and medium tensile steel bars for concrete reinforcement (third revision)
IS 1786: 1985 — Specification for high-strength deformed steel bars for concrete reinforcement (third revision)

IS 1566: 1982 — Specification for hard-drawn steel wire fabric for concrete reinforcement (second revision)
IS 2062: 1999 — Steel for general structural purposes- Specification (Fifth revision)
IS 1608: 1995 — Mechanical testing of metals – Tensile testing (second revision).

Notations

A_ϕ	Area of round reinforcing steel bar $= \pi d_b^2/4$
A_c	Area of concrete, concrete area of the assumed critical section of flat slab $= b_0 d$.
A_t	Transformed concrete area of the cracked section $= A_c + (m-1)A_{st}$
A_g	Gross cross-sectional area of concrete
A_h	Design horizontal acceleration spectrum value
A_s	Area of steel, mm^2
A_{sc}	Area of compression steel
A_{st}	Area of tension steel
A_{st+}	Reinforcement for positive moments
A_{st-}	Reinforcement for negative moments
$A_{st,min}$	Minimum tensile steel, mm^2
A_{sv}	Area of stirrup, also the projected concrete failure area of a single anchor or a group of anchors, for the calculation of strength in shear
A_g	Gross area, equal to the total area ignoring any reinforcement
a_{cr}	Distance from the point at which crack width is determined to the surface of the nearest reinforcing bar
a_{ce}	Deflection due to creep
a_{cs}	Deflection due to shrinkage
b	Breadth of beam or shorter dimension of a rectangular column
b_f	Breadth of flange
b_o	Perimeter of the critical section
b	Breadth of web
C_c	compressive force in the compression steel in a doubly reinforced concrete beam
C_s	compressive force in the concrete of a doubly reinforced concrete beam
C_c	Compressive force due to concrete stress block.
C_r	Strength reduction factor for a slender column
C_{sc}	Compressive force in steel reinforcement
c	Cohesion of soil
c_c	Clear cover
c'	Effective cover
b'	Effective side cover
D	Overall depth of beam/slab or diameter of a column; dimension of a rectangular column in the direction under consideration
DL	Dead load
D_b	Overall depth of the beam
D_c	Diameter of the circular column
D_f	Thickness of flange/depth of foundation slab,
D_p	Diameter of the pile

d	Effective depth of a beam or slab
d'	Cover for compression/tension steel
E	Young's modulus
E_c	Modulus of elasticity of concrete
E_s	Modulus of elasticity of steel reinforcement
f	Permissible bending stress
f_1, f_2	Principal stresses
f_{ck}	Characteristic cube compressive strength of concrete
f_{cr}	Modulus of rupture of concrete (flexural tensile strength)
f_d	Design strength
f_m	Mean value of the normal distribution
f_s	Stress in steel
f_{sc}	Stress in compression steel
f_y	Characteristic yield strength of steel rebar or wire
g	Acceleration due to gravity
H	Height of the building
I	Second moment of area (moment of inertia)
I_{eff}	Effective moment of inertia of the cracked section considering equivalent area of tension and compression reinforcement
I_g	Gross moment of inertia of cross section
I_r	Moment of inertia of the cracked section
I_s	Moment of inertia of the slab
J	Calculated property of the assumed critical section analogous to the polar moment of inertia = $(I_x + I_y)$
jd	Leaver arm distance
K	Stiffness of member
K_a	Coefficient of the active earth pressure
K_b	Flexural stiffness of the beam = EI_b/L_b
K_c	Flexural stiffness of the column = EI_c/L_{cs}
K_p	Coefficient of the passive earth pressure
K_s	Flexural stiffness of the slab = EI_s/L_s
k_c	Modification factor for compression reinforcement to be applied on basic L/d ratio
k_s	Modulus of sub-grade reaction, N/mm^3
L	Length of a column or beam between adequate lateral restraints or the unsupported length of a column, effective span of beam, span length of slab, mm
LL	Live load or imposed load
l_0	Distance between the points of contra flexure (zero moments)
L_c	Longer clear span
L_d	Development length of bar in tension
L_x	Length of the shorter side of a slab
L	Unsupported length of column
L_y	Length of the longer side of a slab
lex	Effective length of a column, bending about xx-axis
ley	Effective length of a column, bending about yy-axis

M	Maximum moment under service loads
Mcr	Cracking moment
Mu	Design moment for limit state, design factored moment
Mux	Design moment about x-axis
$Mux1$	Maximum uniaxial moment capacity of the section with axial load, bending about xx-axis
$Muy1$	Maximum uniaxial moment capacity of the section with axial load, bending about yy-axis
$Me1$	Equivalent bending moment
Mu,lim	Limiting moment of resistance
m	Modular ratio
M_{cr}	Cracking moment
M_{ax}, M_{ay}	Additional moment in slender columns about x-axis and y-axis, respectively
M_d	Design moment
M_m	Mid-span moment
M_o	Total static moment in flat slabs/flat plates
$M_{u,lim}$	Limiting moment of resistance of a singly reinforced section
M_{ux}, M_{uy}	Factored moment about x-axis and y-axis, respectively
$M_{w,lim}$	Limiting moment of resistance of the web
M_x, M_y	Moment about x-axis and y-axis, respectively
M'_x, M'_y	Bending moments considering torsional effects in slabs
m	Modular ratio [$=280/(3\sigma_{cbc})$ as per working stress method] $= E_s/E_c$
P	Axial load
Pu	Design axial load for limit state, design (factored load)
p_c	Percentage of compression reinforcement = $100\ A_{sc}/bd$
p_t	Percentage of tension reinforcement = = $100\ A_{st}/bd$
p_b	Balanced steel ratio
p_c	Percentage of compression steel = $A_{sc}/bd \times 100$
$p_{c,lim}$	Limiting compression reinforcement for the balanced section
p_f	Probability that the load exceeds the characteristic load Q
p_h	Perimeter of the closed stirrup, mm
p_t	Percentage of tension steel = $A_{st}/bd \times 100$
$p_{t,lim}$	Limiting percentage tensile steel
$p_{v,max}$	Maximum amount of shear reinforcement for ductile failure
q_a	Safe or allowable bearing capacity of soil
SBC	Safe bearing capacity of soil
T	Torsional/twisting moment
T_u	Factored torque
V	Shear force
V_{az}	Shear transferred across the crack by interlock of aggregate particles
V_c	Nominal shear resistance provided by concrete
V_{cr}	Shear force at which the diagonal tension crack occurs
V_{cz}	Shear in the compression zone of concrete
V_d	Shear resisted by dowel action of the longitudinal reinforcement, also the design shear force

V_e	Equivalent shear force including torsion
V_s	Nominal shear carried by vertical shear reinforcement
V_u	Factored shear force, kN
V_{us}	Shear to be resisted by shear reinforcements, kN
W_{cr}	Crack width, mm
W_f	Weight of the foundation and structure
W_{max}	Maximum crack width
W_s	Weight of soil or backfill
w	Uniformly distributed load per unit area
w_d	Distributed dead load per unit area
w_l	Distributed imposed load per unit area
w/c	Water/cement ratio
w_x	Share of the load w in the short direction
w_y	Share of the load w in the long direction
x	Depth of neutral axis
x_u	Depth of neutral axis at the ultimate failure of under-reinforced beam, mm
$x_{u,lim}$	Limiting depth of neutral axis, mm
y	Distance of extreme fiber from the neutral axis, mm
Z	Modulus of section
z	Lever arm distance
γ_s	Unit weight of earth or soil
γ_f	Partial safety factors for load, also the factor used to determine the unbalanced moment transferred by flexure at slab-column connections,
γ_m	Partial safety factor for material
γ_v	Factor used to determine the unbalanced moment transferred by eccentricity of shear at slab-column connections = $(1 - \gamma_f)$
γ_w	Weight of wall material, kg/m^3 also the unit weight of water = 10 kN/m^3
Δ	Deflection of beam or column, mm
δ	Factor to increase/decrease design shear strength of concrete for considering the effect of axial compressive/tensile force, deflection/displacement, mm, and angle of wall friction between pile and soil
ε_{cu}	Ultimate compressive strain in concrete
ε_s	Strain in steel
ε_{sc}	Strain in compression steel
ε_{sh}	Design shrinkage strain in concrete
μ	Coefficient of friction
μ_Δ	Displacement ductility factor = Δ_u/Δ_y
μ_ϕ	Curvature ductility factor = φ_u/φ_y
ν	Poisson's ratio
ρ	A_{st}/bd
ρ'	A_{sc}/bd
ρ_c	Unit weight of concrete

ϕ_{sh}	Shrinkage curvature
ϕ_y	Yield curvature
σ or s	Standard deviation
σ^2	Variance
σ_R	Standard deviation of resistance
σ_Q	Standard deviation of the loads
σ_{cbc}	Stress in concrete in bending compression
σ_{cc}	Stress in concrete in direct compression
σ_{c}	Stress in steel in compression
σ_{st}	Stress in steel in tension
σ_{sv}	Sensile stress in shear reinforcement
σ_{sc}	Compressive stress in concrete at the level of centroid of compression reinforcement
τ	Shear stress
τ_b	Average bond stress
τ_{ba}	Anchorage bond stress
τ_{bd}	Design bond stress
τ_c	Design shear strength (stress) of concrete
τ_{ce}	Enhanced design shear strength of concrete
τ_{cp}	Punching shear strength of concrete
τ_{cr}	Average critical shear stress at which the diagonal tension crack appears, characteristic bond stress of adhesive anchor in cracked concrete
$\tau_{c,\text{max}}$	Maximum shear stress in concrete with shear reinforcement
τ_c	Design shear strength of concrete in walls
$\tau_{s,\text{max}}$	Maximum stress in shear reinforcement
τ_t	Torsional shear stress
τ_v	Nominal shear stress = V_u/bd
τ_{vw}	Nominal shear stress in walls
τbd	Design bond stress
τmax	Maximum shear stress in concrete with shear reinforcement
ϕ	Creep coefficient
xu	depth of the effective compression block in a concrete beam
d	effective depth from the top of a reinforced concrete beam to the centroid of the tensile steel
d'	effective depth from the top of a reinforced concrete beam to the centroid of the compression steel
fy	Proof/yield stress or strength
γc	unit weight of concrete
ϕ	Diameter of steel reinforcements

Note: All other symbols are explained appropriately in the text.

Fundamental Questions

1. What is the difference between structural design and structural analysis?
2. What are the basic principles of structural design?
3. What is a 'durable concrete'?
4. Is there any relationship between permeability and durability of concrete?
5. What is the role of concrete in regard to corrosion of reinforcing steel?
6. What is 'creep of concrete'?
7. How do you define grade of reinforcing steel?
8. How do you define grade of concrete?
9. Discuss the applications of working stress method and Limit State method.
10. What is the difference between deterministic design and probabilistic design?
11. What is the importance of probabilistic design of a structure?
12. What is meant by limit state? Discuss.
13. Why partial safety factor for concrete is greater than that for reinforcing steel in limit state design methodology?
14. What is the fundamental assumptions normally made in elastic flexural theory? Is it valid in limit state design methodology?
15. Explain the concept of 'transformed section'.
16. Why deflection of beam occurs due to shrinkage of concrete?
17. Why modular ratio for compression reinforcement, as compared to tension reinforcement, is more?
18. Is it true that concrete resists no tensile stresses in reinforced concrete beams under bending?
19. Why moment-curvature relationship for reinforced concrete beams, is extremely important.
20. How do you define 'Balanced section' in working stress method (WSM) as well as in limit state method (LSM)?
21. Why over-reinforced is avoided in LSM?
22. Explain the concept of "effective flange width".
25. What are factors that influence the effective flange width in a T-beam?
26. Why code imposed minimum and maximum limits of percentage of flexural reinforcement?
27. Why code imposed minimum and maximum limits of spacing of flexural reinforcement?
28. Why code imposed minimum and maximum limits of spacing of shear reinforcements?
29. What are the advantages and disadvantages of providing more clear cover to reinforcement?
30. Is it possible to control deflection by limiting span/effective depth ratios as described in codes? Is this clause need revision? Discuss.
31. When doubly reinforced design is preferred?

32. When under-reinforced design is preferred?
33. Discuss flexural cracks, diagonal tension cracks, flexural-shear cracks, and splitting cracks?
34. How shear is carried by a reinforced concrete section?
35. What is 'shear span' and how does it influence the mode of shear failure?
36. How nominal shear stress for beams with variable depth is calculated? Discuss.
37. What is 'Dowel action of longitudinal reinforcements'?
38. How does the presence of an axial force (tension or compression) influence the shear strength of concrete?
39. Stirrups may be open or closed. Discuss.
40. Stirrups may be 'vertical' or inclined. Discuss.
41. The shear resistance of bent-up bars is considered if stirrups are also provided. Why?
42. Why maximum shear stress $\tau_{c,\max}$ is imposed?.
43. What is 'Truss analogy model' in regard to design against shear?
49. What should be policy regarding curtailment of tension reinforcement?
50. What is the difference between equilibrium torsion and compatibility torsion?
51. Inclined stirrups and bent-up bars are not considered suitable for shear reinforcement.
52. Discuss the different modes of failure under combined flexure and torsion.
53. Discuss torsion–shear interaction of reinforced concrete beams.
54. Explain the difference between flexural bond and development bond.
55. What is the significance of 'development length'?
56. What is 'Anchorage bond'?
57. What is the criterion for radius of bend reinforcing bar? Discuss.
58. Why is a welded splice? Discuss.
59. How do you decide appropriate location of a splice in the tension reinforcement in a beam/slab?
60. What is *redistribution* of bending moments?
61. Discuss 'Substitute frame method'.
62. Explain 'Redistribution of moment'.
63. Discuss "Serviceability limit states" in the structural design
64. Why deflections is to be limited?
65. What is the difference between short-term and long-term deflection?
66. Why shrinkage of concrete is responsible for deflections to occur in reinforced concrete beams and slabs?
67. What is the role of reinforcement in compression regarding deflections due to shrinkage/creep?
68. How temperature affects deflections in reinforced concrete beams/slabs?
69. Discuss factors influencing crack widths in beams/slabs?
70. What is the conceptual difference between direct design method and equivalent frame method, in regard to design of flat slabs?

71. Why the drop panel and the column capital, are provided, in flat slab construction?
72. What is the conceptual difference between design of slender compression member and short compression member?
73. What is the role of lateral ties in reinforced concrete columns?
74. What is the difference between unsupported length and effective length of a compression member?
75. What is the difference between braced and unbraced columns?
76. What is the implication of minimum eccentricity of load, to be considered in column design as per code?
77. What is the implication of limits to the minimum and maximum reinforcement in columns?
78. Discuss the load-carrying aspects of tied columns with spiral columns, under axial load.
79. Discuss the design output of 'rectangular section with reinforcement distributed equally on four sides' and. 'Rectangular section with reinforcement provided on the opposite sides only'
80. What are the advantages of providing pedestals to columns?
81. When combined footings are preferred over isolated footings?
82. What is the purpose of a retaining wall?
83. When Counterfort retaining walls are suitable over Cantilever type retaining walls?
84. What is the difference between active pressure and passive pressure of earth on walls?
85. What is a surcharged load?
86. Why shear key is provided?
87. What is the importance of 'Ductility' in reinforced concrete structure?
88. How do we assess the ductility of a reinforced concrete structure?
89. What are the special detailing provisions in IS 13920?
90. Why special confining reinforcement in columns and beams are suggested to make ductile frames?
91. What are the detailing requirements of beam-column joints as per IS13920?
92. Why the code is limiting neutral axis depth in LSM?
93. Why is ductility considered important in beam design? How can we achieve the required ductility by the design methods?
94. What are the advantages of using design charts presented in SP 16? Can they be used for the design of non-rectangular sections?
95. Why is it necessary to limit x_u/d in the design of singly reinforced beams? Can the condition be relaxed in doubly reinforced beams? State the reasons.
96. Under what circumstances are doubly reinforced beams used? What are the advantages of doubly reinforced beams over single beams?
97. How will you decide whether a doubly reinforced section is under- or over-reinforced?

98. What is the minimum percentage of steel to be provided as compression steel to consider the beam as doubly reinforced?
99. What are the minimum and maximum percentages of tension and compression reinforcement in doubly reinforced beams?
100. How is it determined whether a beam of given dimensions is to be designed as doubly reinforced?
101. What is the difference between an L- and a T-beam?
102. How is the effective flange width calculated for a T-beam using the IS code?
103. What is meant by shear lag in T-beams?
104. When is a T-beam designed as a rectangular beam?
105. What are the possible positions of neutral axis in the design of T-beams?
106. Describe the method of locating the position of neutral axis in T-beams.
107. T what location of neutral axes will the flange of a T-beam be subjected to non-linear stress distribution?
108. What is the role of transverse reinforcement in the slab portion of T-beams?
109. Is the minimum percentage of tension steel in a T-beam different from a rectangular beam? Is it determined based on web width or flange width?
110. What is the maximum percentage of steel that is allowed in T-beams?
111. Give the approximate formula that is used to determine the area of steel for T-beams subjected to factored moment.
112. What is the flange width on either side of a web, the reinforcement of which can be considered to act as tension reinforcement when a negative bending moment is acting on the beam?
113. What is the distance over which the reinforcement should be distributed to control flexural cracking in T-beam flanges?
114. What are deep beams? When a beam is considered a deep beam according to IS 456?
115. List the design procedure of deep beams according to the IS code.
116. Describe the detailing to be adopted in simply supported deep beams according to IS 456:2000.
117. What is the percentage and spacing of steel to be provided as vertical and horizontal reinforcement in deep beams as per the IS and ACI codes?
118. How are bearing stresses checked in deep beams?
119. What are hidden beams? How are they designed?
120. What is the difference in structural action between a normal beam and a lintel?
121. Distinguish between grade and plinth beams.

References

1318M- *Building Code Requirements for Structural Concrete and Commentary*, American Concrete Institute, Farmington Hills, Michigan, 2008

ACI Committee 116, *Cement and Concrete Terminology*, Special Publication SP-19, American Concrete Institute, Detroit, Michigan, 1967.

ACI Committee 201, Guide to Durable Concrete, *Journal ACI*, Vol. 74, 1977, pp. 573–609.

ACI Committee 209, *Prediction of Creep, Shrinkage and Temperature Effects in Concrete Structures*, SP–27, American Concrete Institute, Detroit, Michigan, 1971.

ACI Committee 224, Control of Cracking in Concrete Structures, *Journal ACI*, Vol. 69, No. 12, December 1972, pp. 717–752.

ACI Committee 224, Cracking of Concrete Members in Direct Tension, *Journal ACI*, Vol. 83, No. 1, Jan–Feb. 1986, pp 3–13.

ACI Committee 318, *Commentary on Building Code Requirements for Structural Concrete* (ACI 318R-95), American Concrete Institute, Detroit, Michigan, 1995.

ACI Committee 336, Suggested Design Procedures for Combined Footings and Mats, *Journal ACI*, Vol. 63, No. 10, Oct. 1966, pp. 1041–1057.

ACI Committee 435, (Subcommittee 1), Allowable Deflections, *Journal ACI*, Vol. 65, No. 6, June 1968, pp 433–444.

ACI Standard 318–89, *Building Code Requirements for Reinforced Concrete*, American Concrete Institute, Detroit, Michigan, 1989.

ACI–ASCE Committee 326, Report, Shear and Diagonal Tension, *Journal ACI*, Vol. 59, Jan., Feb., and Mar., 1962, pp. 1–30, 277–334 and 352–396.

ACI-ASCE Committee 426, *Suggested Revisions to Shear Provisions for Building Codes*, American Concrete Institute, Detroit, 1978.

ACT-ASCE Committee 352R-02, *Recommendations for Design of Beam-Column Connections in Monolithic Reinforced Concrete Structures*, American Concrete Institute, Farmington Hills, 2002.

ASCE-ACI Committee 426, The Shear Strength of Reinforced Concrete Members, *Journal of the Structural Division*, Vol. 99, June 1973, pp. 1091–1187.

Beeby, A.W., *Modified Proposals for Controlling the Deflection by Means of Ratios of Span to Effective Depth*, Publication No. 42, Cement and Concrete Association, London, 1971.

Bresler, B., Design Criteria for Reinforced Concrete Columns under Axial Load and Biaxial Bending, *Journal ACI*, Vol. 57, November 1960, pp. 481–490.

Chen, B.F. and Chiang, H.W., Complete 2D and Fully Nonlinear Analysis of Ideal Fluid in Tanks, *Journal of Engineering Mechanics ASCE*, Vol. 125, 1999, pp. 70–78.

CIRIA Guide 2, *The Design of Deep Beams in Reinforced Concrete*, Construction Industry Research and Information Association and Ove Arup and Partners, London, 1977.

CP 110-Part 1, *Code of Practice for Structural Use of Concrete*, British Standards Institution, London, 1972.

CSA Standard A23.3–94, *Design of Concrete Structures*, Canadian Standards Association, Rexdale, Ontario, Canada, 1994.

Desai, R.K., Design of Shear Reinforcement for Reinforced Cement Concrete Beams as per IS 13920 Design Aids, 2003.

Design Aids (for Reinforced Concrete) to IS 456:1978, Special Publication SP:16, Bureau of Indian Standards, New Delhi, 1980.

Ellingwood, B., *Reliability Basis for Load and Resistance Factors for R C Design*, NBS Building Science Series 110, National Bureau of Standards, Washington DC, 1978.

Explanatory Handbook on Indian Standard Code of Practice for Plain and Reinforced Concrete (IS 456:1978), Special Publication SP:24, Bureau of Indian Standards, New Delhi, 1983.

Fergusson, P.M., *Reinforced Concrete Fundamentals*, John Wiley and Sons, New York.

Gouthaman, A. and Menon, D., Increased Cover Specifications in IS 456 (2000) – Crack-width Implications in RC Slabs, *Indian Concrete Journal*, Vol. 75, Sept. 2001, pp. 581–586.

Handbook on Concrete Reinforcement and Detailing, Special Publication SP 34, Bureau of Indian Standards, New Delhi, 1987.

Hognestad, E., Hanson, N.W. and McHenry, D., Concrete Stress Distribution in Ultimate Strength Design, *Journal ACI*, Vol. 52, Dec. 1955, pp. 455–479.

Hsu, T.T.C., *Unified Theory of Reinforced Concrete*, CRC Press, Boca Raton, 1993.

IITK- GSDMA, *Guidelines for Seismic Design of Liquid Storage Tank*, Indian Institute of Technology, Kanpur, 2005.

IS 13920, *Indian Standard Code of Practice for Ductile Detailing of Reinforced Concrete Structures Subjected to Seismic Forces*, Bureau of Indian Standards. New Delhi.

IS 1893 (Part I): 2016, *Criteria for Earthquake Design of Structures – Part I: General Provisions and Buildings* (Fifth revision), Bureau of Indian Standards, New Delhi, 2002.

IS 4326: 1993, *Code of Practice for Earthquake Resistant Design and Construction of Buildings*, Bureau of Indian Standards, New Delhi, 1993 (reaffirmed 1998).

IS:1080, *Code of Practice for Design and Construction of Simple Spread Foundations* (First revision), Bureau of Indian Standards, New Delhi, 1980.

IS:1904, *Code of Practice for Structural Safety of Buildings: Shallow Foundations* (Third revision), Bureau of Indian Standards, New Delhi, 1986.

IS:3370 (Part 4) 904, *Code of Practice for Concrete Structures for the Storage of Liquids, Part 4: Design Tables* (Third revision), Bureau of Indian Standards, New Delhi, 1967.

Iyengar, K.T.S. and Viswanatha, C.S., *Tor-Steel Design Handbook for Reinforced Concrete Members with Limit State Design*, Torsteel Research Foundation in India, Mumbai, 1990.

Mehta, P.K. and Monteiro, P.J.M., *Concrete: Microstructure, Properties and Materials*, Indian edition, Indian Concrete Institute, Chennai, 1997.

Mehta, P.K., Concrete Durability: Critical Issues for the Future, *Concrete International*, Vol. 19, No. 7, 1997, pp. 27–33.

Murata, K. and Miyajima, M., Influence of Receiving Water Tank Sloshing on Water Distribution System, *Journal of Japan Earthquake Engineering*, Vol. 7, No. 1, 2007, pp. 27–42.

Neville, A.M. and Brooks, J.J., *Concrete Technology*, ELBS edition, Longman, London, 1990.

Neville, A.M., *Properties of Concrete*, Second edition, Pitman Publishing Co., London, 1973.

NZS 3101:2006, *Part 1: The Design of Concrete Structures, Par Commentary*, Standards New Zealand, Wellington.

Park, R. and Paulay, T., *Reinforced Concrete Structures*, Wiley and Sons, New York, 1975.

Pillai, S.U. and Menon, D., *Reinforced Concrete Design Edition*, Tata McGraw Hill Publishing Company Ltd, New Delhi, 2003.

Placas, A. and Regan, P.E., Shear Failures of Reinforced Concrete Beams, *Proceedings, American Concrete Institute*, Vol. 68, No. III, 1971, pp. 763–73.

Prakash Rao, D.S., *Design Principles and Detailing of Concrete Structures*, Tata McGraw-Hill Publishing Company Ltd, New Delhi, 1995.

Price, W.H., Factors Influencing Concrete Strength, *Journal ACI*, Vol. 47, Feb. 1951, pp. 417–432.

Purushothaman, P., *Reinforced Concrete Structural Elements — Behaviour, Analysis and Design*, Tata McGraw Hill Publication Co. Ltd., New Delhi, 1984.

Ranganathan, R., *Reliability Analysis and Design of Structures*, Tata McGraw-Hill Publication Co. Ltd., New Delhi, 1990.

Sturman, G. M., Shah, S. P. and Winter, G., Effect of Flexural Strain Gradients on Micro-Cracking and Stress-Strain Behaviour of Concrete. *Journal ACI*, Vol. 62, No.7, July 1965, pp. H05–H22.

Taranath, B.S., *Structural Analysis and Design of Tall Buildings*, McGraw-Hill International Edition, 1988.

Wang, C.K. and Salmon, C.G., *Reinforced Concrete Design*, 6th edition, John Wiley and Sons Inc., Hoboken, 2002.

Wood, R.H., *Plastic and Elastic Design of Slabs and Plates*, The Ronald Press Company, New York, 1961.

Index

For Product Safety Concerns and Information please contact our EU representative GPSR@taylorandfrancis.com Taylor & Francis Verlag GmbH, Kaufingerstraße 24, 80331 München, Germany

Printed and bound by CPI Group (UK) Ltd, Croydon, CR0 4YY
16/01/2025
01821410-0004